Thorsten Philipp, Tobias Schmohl (eds.)
Handbook Transdisciplinary Learning

Higher Education: University Teaching & Research | Volume 6

Editorial

The German term *Hochschulbildung* refers to teaching and learning at a broad range of research and art institutions in tertiary education. Within this sector, *Hochschulbildung* builds bridges between professional and social efforts on the one hand and scientific endeavours on the other. Its manifestations are shaped by social, cultural, ecological, economic, and political dynamics (e.g. internationalisation and globalization). However, it is itself also a driving force of changing living conditions (e.g., digitality, mediality, and networking).

A key feature of *Hochschulbildung* is its interconnection with research. The curriculum and teaching practices at the university should align with the most recent research findings and the scientific discourse. As part of the education system, *Hochschulbildung* contributes significantly to shaping individual and collective relationships to the world.

Teaching and studying should prepare for professional practice and promote real-work capabilities as well as participation in public life. *Hochschulbildung* ultimately aims at proficiency, employability, and citizenship. To this end, it becomes crucial to enhance subject-specific teaching by interdisciplinary and transdisciplinary forms of cooperation.

This series offers a forum for the transformation of higher education in the context of institutional and organisational development. We invite our authors and readers to exchange views about influencing factors, diagnoses and designs of teaching & learning in the 21st century.

The series is edited by Tobias Schmohl and Johannes Wildt.

Thorsten Philipp is a political scientist and lecturer for sustainability and political communication at various universities. As a member of the presidential staff at Technische Universität Berlin, his task is to promote transdisciplinary learning at the interface between science and society. His research focuses on the intersection of sustainably theories and pop music.

Tobias Schmohl is a full professor of higher education research at Technische Hochschule Ostwestfalen-Lippe in Germany. He advocates for university teaching that bridges professional/social efforts and scientific pursuits, fostering transdisciplinary skills, real-world abilities, and civic engagement.

Thorsten Philipp, Tobias Schmohl (eds.)

Handbook Transdisciplinary Learning

[transcript]

All Internet links in this publication were last verified on 2 May 2023.

Double-blind peer review: In order to ensure their quality, all book chapters were subjected to a review process with double-blind peer reviews. The reviewers are listed on p. 423.

Bibliographic information published by the German National Library

The German National Library lists this publication in the German National Bibliography; detailed bibliographic data are available in the Internet at http://dnb.d-nb.de/en

First published in 2023 by transcript Verlag, Bielefeld

Cover layout: Maria Arndt, Bielefeld
Inside cover: Trampolinhuset – Københavns flygtningemedborgerhus
Proofread: Joan Dale Lace
Typeset: Jan Gerbach, Bielefeld
Printed by: Majuskel Medienproduktion GmbH, Wetzlar
https://doi.org/10.14361/9783839463475
Print-ISBN: 978-3-8376-6347-1
PDF-ISBN: 978-3-8394-6347-5
EPUB-ISBN: 978-3-7328-6347-1
ISSN of series: 2749-7623
eISSN of series: 2749-7631

Printed on permanent acid-free text paper.

Content

* Authors contributed equally

Boundaries do not sit still.
Karen Barad

Preface

The fluidity of borders that feminist theorist Karen Barad discusses is both an opportunity and a risk. It contains liberation, renegotiation, and redistribution, as well as disorientation, confusion, and conflict. The dynamism applies to research and education as well. As we continue to navigate the ever-evolving learning sciences landscape, the need to create a community of professionals and students devoted to transdisciplinary learning grows. Hence, the *Handbook of Transdisciplinary Learning* aims to engage and inspire students, researchers, educators, and practitioners who seek a deeper understanding of the intricate connections that tie multiple disciplines in higher education.

Our publication retains the structural foundation of its 2021 German-language predecessor *Handbuch Transdisziplinäre Didaktik,* while incorporating new concepts, international expertise, postcolonial criticism, and in general a much broader perspective of the discussion's global dimension. Comprised of 37 entries, the Handbook unpacks key concepts to describe the broad panorama of transdisciplinary learning in the context of academic education. It examines the etymological origins, historical trajectories, disciplinary influences, inherent challenges, the criticisms it has provoked, and its consequences for academic education. By examining these terms through the lens of a historical-etymological "sense horizon" (Gadamer), our idea was to trace the evolution of ideas while simultaneously fostering critical dialogue and debate.

Recognizing that the pursuit of transdisciplinary learning is inherently dynamic and multifaceted, our compendium presents each entry as a point of convergence and interaction between diverse threads of thought. The Handbook seeks to provide a comprehensive understanding of transdisciplinary practices and their impact on participative learning, as well as innovative methods of information dissemination, while departing from conventional modes of scientific communication and bibliography. Although all titles of the chapters are formulated in the singular, they are in fact concealing an infinite number of divergent practices, educational attempts, and ways of thinking.

Despite all efforts to provide as comprehensive a perspective as possible on the complex dispute, we acknowledge our own limitations as editors with a predominantly Western perspective on knowledge and knowledge production in developing this publication. Grappling with issues of global applicability, colonial and neocolonial thought structures, and the cross-cultural applicability of Western concepts – such as citizen science and science shop – our Handbook may nonetheless serve as a first step toward engaging in a larger, global conversation about learning and overcoming barriers that have long impeded the expansion and exchange of ideas.

As a collaborative effort involving 113 authors and 39 scientific reviewers, the creation of this volume has been an exercise in critical thinking, concentrated analysis, and vigorous debate. We hope that this work will serve as a springboard for additional research, discussion, and development in the field of transdisciplinary learning, and that it will inspire others to join us in our pursuit of a more inclusive, interconnected approach to higher education.

As editors, we owe sincere gratitude to all contributors to this volume, who, despite the unusual concept, the tight schedule, and the unique requirements of the double-blind review procedure, took on the challenge of delving into the discussion of transdisciplinary learning based on a key concept. Numerous other experts provided the project with their professional and collaborative guidance. We particularly appreciate the reviewers' contributions to quality control during the review process. We also express our gratitude to the entire crew at transcript Publishing for their dedicated work on this Handbook and their willingness to consider our suggestions, which helped the book take its own particular, unconventional turns. Our special appreciation goes to Joan Dale Lace for her careful copyediting.

Our book received support within the framework of the Berlin University Alliance and was co-financed by funds from the Excellence Strategy of the German Federal and State Governments. Our thanks are extended to Anika Rehder and her team at TU Berlin for their valuable support. Further funds were provided from the Alliance of European Universities of Technology ENHANCE. We owe gratitude to Paul Forberger, Sibylle Groth, and Ulrike Hillemann-Delaney at TU Berlin for their dedication. Additional support was provided by Stifterverband, Randstad Foundation and the Foundation of the OWL University of Applied Sciences and Arts, Germany. We express our thanks to Hanna Daum, Andreas Bolder, Stephanie Wulfert, Sven Hinrichsen, and Jürgen Krahl for their commitment.

Finally, our particular gratitude goes to our collaborators, Johanna Falkenhagen, Nicole Hahn, and Nina Schmulius, who managed the administration of this project with the utmost accuracy, combining perseverance, enthusiasm, and meticulous care. Due to their dedicated efforts in editing and background research, it was possible to conclude this project within only 16 months. We were privileged to draw from so many resources that we would never have dared to dream of.

Berlin and Lemgo, March 2023
Thorsten Philipp and Tobias Schmohl

Embracing the Rhizome: Transdisciplinary Learning for Innovative Problem Solving

Thorsten Philipp and Tobias Schmohl

Trampoline House in Copenhagen, a hub that assists migrants and asylum seekers by providing shelter within the Danish asylum system, recently adopted the metaphor of a castle to describe its work and conditions: *Kassel Castle*, physically marked by a simple chalk circle at the international art exhibition *documenta fifteen*, was an impenetrable prison, a chain of invisible shackles and intangible politics. Inside, however, there was a hive of activity: in a theater workshop, rejected asylum seekers and young refugees penned skits about their personal experiences with the migration regime. The contestants also served as models for a fashion show by designer Dady de Maximo Mwicira-Mitali. A massage workshop led by and involving displaced people serviced the seemingly utopian practice of "massaging" the asylum system – to make it softer, more tranquil, and less stressful. The castle acted as a jail and a bulwark against an oppressive government. The guiding principles of a creative production of knowledge and practice of mutual learning became justice and freedom.

Trampoline House's artworks are featured on the endpapers of our book, not for aesthetic reasons, but rather to honor their subversive concept: Knowledge created through the arts influences society and serves as an example of the transformative power of transdisciplinary education (Kaiza 2022, 205; documenta fifteen 2022) – and in doing so, it is beneficial for sciences and educational systems as a whole. With techniques that cross traditional disciplinary lines and incorporate information from several fields, learners are given the ability to engage in critical thought and creative problem-solving.

The artistic attempt to produce knowledge resources beyond, against, and across disciplines coincides with another development: Science journalist Max Kozlov (2023) recently revealed that the number of scientific and technological research papers published has increased significantly over the past few decades. However, the disruptiveness of these papers, their impact on the status quo, has decreased, as measured by the degree to which they deviate from previous research. Recent research is less likely to cause major upheaval compared to research conducted at the end of the 20th century. Such an observation raises questions about the current state of academic research and its capacity to generate innovative solutions and acknowledge responsibility. In pursuit of novel insights and transformative knowledge, it also highlights the need for a transdisciplinary approach that challenges conventional wisdom, integrates diverse perspectives, and pushes the boundaries of established disciplines.

The purpose of this Handbook is to address this need by advocating for and demonstrating the potential of transdisciplinary learning in higher education, thereby fostering an environment that encourages the production of disruptive and transformative research. It aims to introduce transdisciplinary learning with a focus on the fundamental values of liberty and justice. By providing guidelines and strategies for implementing this educational approach, the Handbook seeks to contribute to an education culture in which both students and researchers contribute meaningfully to addressing the pressing global challenges of the present day. Transdisciplinary learning emerges as a promising means for enhancing innovation and transformative knowledge in the context of academic research's diminishing disruptiveness. It offers the potential to revitalize academia and address complex, real-world problems requiring a multifaceted approach by transcending the limitations of traditional disciplines and integrating a variety of perspectives.

Transdisciplinary learning presupposes a systemic change in various ways, particularly in basic teaching attitudes and understanding of didactics. Teachers must be willing to (1) reduce their control if they want to allow a free (inter)play of creative forces. Transdisciplinary work requires a commitment to (2) activate participation and co-creation on the threshold. As a consequence of acknowledging the plurality of knowledge paths, it is essential to (3) embrace failures, setbacks, and detours of students, learners, and teachers. Transdisciplinary practices will also change given structures in universities and contribute to (4) dismantling hierarchies and extending collective responsibility. (5) Reflective practices must be established, respected, and defended. Finally, given the plurality of actors involved, transdisciplinary practices require sound and systematic (6) feedback literacy to ensure that lessons are learned from cooperation with the practical sphere and adequate measures are taken to meet future educational challenges.

With this predisposition, our Handbook enhances a search process that was already gaining momentum, when modern life and cognition models were increasingly criticized. Gilles Deleuze and Félix Guattari, both disturbed by all attempts to dismember and categorize the world by linear and dichotomous patterns, identified the mental figure of the tree that blocked all access to the real world. The tree is a "taproot, with its pivotal spine and surrounding leaves" (Deleuze and Guattari 2013, 3), an organizational model of the trunk and secondary branches, of supporting order and derivatives, of dichotomous, structured categories and subordinate hierarchical ramifications, was debunked as an epistemic model of increasing fragmentation, hybridization, and volatility: "We're tired of trees.

We should stop believing in trees, roots, and radicles. They've made us suffer too much. All of arborescent culture is founded on them, from biology to linguistics" (Deleuze and Guattari 2013, 15). For Deleuze and Guattari, the tree had to be replaced by the "tufted root", a system of small, branching braids whose inconspicuous nodes form non-hierarchical nexuses to each other and are not subject to any categorical or binary order: "Nothing is beautiful or loving or political aside from underground stems and aerial roots, adventitious growths, and rhizomes. Amsterdam, a city entirely without roots, a rhizome-city with its stem-canals, where utility connects with the greatest folly in relation to a commercial war machine." (Deleuze and Guattari 2013, 15)

Thinking in contexts and networks, along with the utopian ideal of the rhizomatic city of Amsterdam, whose stylized canal plan illustrates the cover of our Handbook, have far-reaching implications. Science is no longer a hierarchical, dichotomous, or tree-like order to distribute and stabilize privileges, power, and status. It is rather a cooperative-egalitarian network-based process in which a variety of knowledge resources, educational biographies, and knowledge potentials gain their form. Seen from this perspective, ▸transdisciplinarity (*Vilsmaier, Merçon, and Meyer*) is not a unidirectional integration of "non-disciplinary" or "non-scientific" knowledge into research and learning. Terms of negation, which aim at denying eligibility and qualification, are unsuitable for this discourse. Rather, it is about research alliances of diverse but equal actors and about overcoming the rifts between university and society. ▸Learning in transformation (*Jahnke and Wildt*), ▸experimentation (*West, Böttger, and Tang*), and working responsibly toward ▸global citizenship (*Grobbauer and Whalen*) are defining syntagms of change.

The distinctive *plural* nature of transdisciplinary education equips students with the framework required to differentiate between diverse knowledge-based resources, evaluate their applicability to specific challenges, and devise strategies for integrating these diverse sources into their academic pursuits and research. ▸boundary work (*Vilsmaier and Thompson Klein*), originally coined for analytical purposes to address the problem of *differences*, has been adapted for boundary-crossing and boundary-spanning research to support collaboration in heterogeneous teams. It is not only about the particular characteristics of knowledge fields, but about the participants' specific features, their ability to elaborate on different objectives, roles, and tasks in collaborative processes. ▸Critical thinking (*Barth and Pfister*) is one of the most central skills in transformational dynamics. Though its roots lie in philosophy, it has significant metacognitive features, including the use of techniques for overcoming cognitive biases and navigating a variety of knowledge sources.

Whereas ▸interdisciplinarity (*Thompson Klein and Philipp*) employs a variety of disciplines to tackle a specific issue, but still holds them constrained within disciplinary boundaries, transdisciplinary learning transcends the boundaries of traditional disciplines and promotes the integration of knowledge from multiple fields such as practical or bodily experiences. This holistic approach to education fosters students' critical and creative thinking, allowing them to apply their skills and knowledge in real-world contexts (Bammer 2015). Transdisciplinary learning acknowledges the multifaceted nature of action and problems, which frequently exist independently, without any mutual relation. The current era necessitates integrative approaches that unite disparate elements to create holistic solutions (Stokols et al. 2008). However, it remains challenging to capture the essence of transdisciplinary learning within the confines of existing disciplines in higher education, highlighting the need for continued research and development of new educational paradigms (Klein 2010).

Transdisciplinary learning promotes a "new production of knowledge" (Gibbons et al. 1994) by promoting experimental and transformative research designs. It actively seeks to integrate a vast array of knowledge resources, such as professional, everyday, and implicit knowledge from various sectors, including politics, civil society, business, and culture (Nowotny et al. 2001). By doing so, transdisciplinary learning calls for a science system that operates in ▸Mode 2 (*Langemeyer and Zimpelmann*). The distinction between two modes of knowledge production sparked an international discussion in the 1990s that focused its attention on the application of research and science in modern society. With this shift, the institutionally protected sphere of research and teaching, known as *Mode 1* and established at universities and colleges, eroded. Knowledge became usable for concrete, problem-oriented solution of social challenges – often in cooperation with new partners from the societal sphere. The new production of knowledge addresses knowledge resources that are cultivated in unexpected and conflictive spheres, as the discussions around ▸indigenous knowledge (*Le Hunte, Yunkaporta, Melvold, Potts, Ross, and Allen*), ▸embodied learning (*Allen, Pratt, Le Hunte, Melvold, Doran, Kligyte, and Ross*), and ▸performative knowledge (*van den Berg and Schmidt-Wulffen*) show. By encouraging collaboration and inclusiveness, this type of learning advances the cause of justice, as it ensures that diverse perspectives are acknowledged and valued in the pursuit of knowledge.

Attempts to achieve participative learning encompass collective practices such as ▸citizen science (*Jaeger-Erben, Becker, Prūse, Mendoza, Gutberlet, and Rodrigues*), do-it-yourself cultures, and ▸fab labs (*Brandenburger, Adzaho, Mostert-van der Sar, Voigt, and Troxler*). ▸Cooperative education (*Coones, Johannsen, and Philipp*), a particular form of transdisciplinary learning, combines academic coursework with practical work experience, enabling students to apply their knowledge in business and administration settings. In fact, most study programs in higher education today provide opportunities to integrate practical experience, often in the professional field students strive for: ▸internships (*Terhart and Weyland*). Yet they are frequently not recognized as a valuable method of transdisciplinary learning, and educational quality can only be achieved if there is a mutual connection between the learning experiences in the classroom and in the practical field. The preparation before, the support during, and the reflection after an internship ensure a qualitative integration of diverse knowledge into students' academic learning journey.

The educational approach that supports the premise of transdisciplinarity encourages learners to discover solutions unrestricted by conventional wisdom in a freed setting. This way of ▸transformative learning (*Taimur and Ross*) enables learners to reflect on their experiences, beliefs, and assumptions, resulting in profound and long-lasting shifts in their perspectives and actions. The heterogeneity of actors involved, however, requires additional skills in managing the plurality: ▸Feedback literacy (*Schluer, Rütti-Joy, and Unger*), with its high relevance for all areas of life, has long since been addressed by numerous disciplines. Nonetheless, feedback is still frequently thought of in the context of education as a one-way information flow from teacher to student. The modern, socio-constructivist paradigm, in contrast, emphasizes the shared obligations of all participants in the feedback process. Feedback is therefore viewed as a dialogic exchange that is influenced by personal, interpersonal, and environmental aspects. Learners must have the attitudes and skills necessary to seek out, comprehend, and apply feedback to their learning in order to take part in these exchanges.

The digital world and its currencies are likewise impacted by the necessary growth of competencies: ▸data literacy (*Unger, Beck, and Husfeldt*) encompasses the various knowledge components required for sensitive handling of data or decisions made on the basis of data, and it enables students to collect, process, evaluate, and apply data thoughtfully. On the other hand, ▸storytelling (*Cortes Arevalo, Adamson, Fantini, Verbrugge, and Postma*) ensures the pivotal capability of human problem-solving: describing experiences or expressing ideas through language and images supports transformative co-creative learning by tapping into personal and experiential knowledge. Transdisciplinary learning, in addition, can be implemented as a form of ▸research-based education (*Koltay and Karvalics*) that integrates research activities into the learning process. ▸Participatory action research (*Alatorre Frenk, Hensler, and Merçon*), a collaborative approach to research that incorporates stakeholders in the research process, promotes co-learning and empowerment. Instead of conforming to dichotomous, tree-like structures that distribute and stabilize privileges, authority, hierarchies, and status, transdisciplinary learning promotes a collaborative environment that supports a plurality of knowledge resources, educational biographies, and knowledge potentials (Nowotny et al. 2003).

It is not surprising that the main topic of today's debates is societal transformation, as transdisciplinary learning is fundamentally about societal change for a fair future for all. The educational goal ties transdisciplinary learning to ▸education for sustainable development (*Brennan and Sabogal-Paz*) to empower people and communities with the knowledge, abilities, and attitudes required to contribute to a livable future. This emphasis highlights an additional key issue that is sometimes overlooked: societal development cannot occur without the transformation of people's values, beliefs, worldviews, and corresponding inner traits and capacities. The objective is to promote students' knowledge of their underlying motivations and highlight their ▸personal sustainability (*Parodi, Wamsler, and Dusseldorp*). Inevitably, learning is no longer a solitary or private activity: ▸engaged learning (*Chmelka, Griffith, and Weiner*), a term that emerged from service learning, is today a general postulate to encompass pedagogical strategies and to allow students to gain knowledge through meaningful community engagement. From this angle, students are both engaged citizens and, at the same time – far from the logics of the capitalist market and its exploitation goals – ▸entrepreneurs (*Mittelstädt, Mykolenko, and Wiepcke*). Therefore, the educational objective across all subject areas, not just economics, may be to promote entrepreneurial action, spirit, or behavior.

As a distinct approach within the realm of higher education, transdisciplinary learning emphasizes the significance of successful communication between the academic community and various societal sectors, including politics, civil society, culture, and business. Through the involvement of a broad variety of stakeholders, ▸science communication (*Kiprijanov and Joubert*) as a dialogue-focused and participation-oriented activity, plays a crucial part in the exchange of knowledge and research. A wide range of techniques has been established by higher education and research institutions throughout the world as well as other training providers in order to foster the knowledge and abilities required for planning and implementing open discussion and participatory scientific communication.

In a time when knowledge production increasingly transgresses national boundaries, ▸knowledge transfer (*Alhassan and Ruser*) is a practical issue of paramount significance, although the very concept is contested within academic environments of research, teaching, and learning. Substantial learning opportunities in terms of transdisciplinarity can further be explored in ▸Living labs (*Backhaus, Böschen, John, Altepost, Cloppenburg, Fahy, Gäckle, Gries, Heckwolf, Matschoss, Meyer, Münderlein, Schmitt, Sonntag, Timpe, and Gramelsberger*) and ▸real-world labs (*Parodi, Steglich, and Bylund*): integrated research and innovation processes between university and local stakeholders in a public-private-people partnership. Research no longer takes place in closed labs: society itself is the new "laboratory" (Krohn 1994) to develop sustainable solutions. Although the research on these practices has grown in importance over recent years, the involvement of students often remains poor. Most labs are research rather than education oriented. The task is to open the lab culture as a learning arena for students.

The plentiful practical experiences and a lack of reflection on its educational dimension also characterize ▸science shops (*Legris and Becker*), institutionalized by several European universities during the 1980s. Their central goal is to provide participatory research support in response to industry or civil society concerns, particularly with regard to environmental conflicts, urban development, consumption, or sustainable innovation. However, here again, their ability to contribute to transdisciplinary education depends on their degree of student involvement. A similar reform phenomenon, ▸student-organized teaching (*Bönisch, Becker, Blömer, Raj Pandey, Prūse, and Vollbehr*) offers a major opportunity to experience transdisciplinarity individually on any chosen topics. The idea is easy: Any student can set up a project workshop with other colleagues or stakeholders. This approach permits studying without any professors or research assistants, with guidance provided only by a student tutor.

The panorama of transdisciplinary learning is even broader than these institutionalized programs and also covers uncountable creative practices. Typically, ▸scrum (*Heibges, Jungnickel, and Feufel*) can be used as an agile, playful project management framework that emphasizes iterative progress, collaboration, and adaptability in order to efficiently achieve project goals. ▸Design thinking (*Taimur, Peukert, and Pearce*) is also frequently employed as it emphasizes empathy, experimentation, and iteration to create solutions. ▸Hackathons and challenge-based learning (*Massari, Roversi, Finn, Solimeo, Jatwani, Fusco, Solimeno, Cavicchi and Vignoli*), short-term events, in which obstacles must be overcome in a collective, often tech-based strategy, are an increasingly popular way to bridge the gap between academic research and practical applications. Transdisciplinary potential can also be attributed to ▸case studies (*Meyer, Brundiers, Mader, and Weiser*): they aim at helping students to find a better understanding of identified problems, by investigating their origins, extent, and dynamics in the specific context and by deriving transferable knowledge for similar and future problems.

Transdisciplinary action, by all these practices, takes place in a "Third Space" (Soja 2007) where practitioners, educators, curriculum developers, and other stakeholders work together to create learning scenarios. It is closely related to learner-centered, research-driven teaching strategies that adhere to constructivist principles, prioritize active learning, and support the growth of higher-order thinking skills – strategies aim to produce ▸scientific knowledge (*Walter and Kremer*) by methodically observing, testing, and analyzing social phenomena. Transdisciplinary research may be open to accusations of solutionism or to being branded as tendentious commissioned research due to its emphasis on quick, usable solutions to problems. Thus, in order to conduct and report this particular form of research in accordance with ethical principles and professional scientific standards, it requires its own reflection and discourse within the framework of ▸research integrity (*Alavi and Schmohl*).

Transdisciplinary learning, as our overview shows, involves a focus on practical challenges, knowledge integration, normative components, and collaboration among many stakeholders. Although transdisciplinary learning does not create a new field, it enriches existing ones. In conclusion, the contributions to this Handbook underline the importance of higher education as an environment for the personal development of both students and instructors. The primary objective of academic education is not preparation for the labor market, but the participants' personal development. The promise of comprehensive education applies to everyone. Universities are not neutral spaces, but rather sites of contentious debate. With its ability to house the rhizomatic diversity of knowledge resources, it serves as an agonistic arena (Mouffe 2013) of the various forms of cooperation that societies create in response to complex problems. By encouraging a collaborative and equitable transdisciplinary approach to learning, this Handbook is a contribution to setting the stage for conflict, passion, and difference across, beyond, and away from disciplines.

References

Bammer, Gabriele. 2015. Enhancing research collaborations: Three key management challenges. *Research Policy* 44 (5): 875–87.

Deleuze, Gilles, and Félix Guattari. 2013. Rhizome. In *A thousand plateaus: Capitalism and schizophrenia*, 1–27. London: Bloomsbury.

documenta fifteen, ed. 2022. *Trampoline House*. Available from https://documenta-fifteen.de/en/lumbung-members-artists/trampoline-house.

Gibbons, Michael, Helga Nowotny, Simon Schwartman, Peter Scott, and Martin Trow. 1994. *The new production of knowledge: The dynamics of science and research in contemporary societies*. London: Sage.

Kaiza, Aino-Kaisa. 2022. *documenta fifteen handbook*. Berlin: Hatje Cantz.

Klein, Julie Thompson. 2010. *Creating interdisciplinary campus cultures: A model for strength and sustainability*. Hoboken, NJ: Wiley.

Kozlov, Max. 2023. *"Disruptive" science has declined — and no one knows why*. Available from https://www.nature.com/articles/d41586-022-04577-5.

Krohn, Wolfgang. 1994. Society as a laboratory: The social risks of experimental research. *Science and Public Policy* 21 (3), 173–83.

Mouffe, Chantal. 2013. *Agonistics: Thinking the world politically*. London: Verso.

Nowotny, Helga, Peter Scott, and Michael Gibbons. 2001. *Re-thinking science: Knowledge and the public in an age of uncertainty. Cambridge*: Polity Press.

Nowotny, Helga, Peter Scott, and Michael Gibbons. 2003. Introduction: "Mode 2" revisited: The new production of knowledge. *Minerva* 41 (3), 179–94.

Soja, Edward W. 2007. *Thirdspace: Journeys to Los Angeles and other real-and-imagined places*. Malden, MA: Blackwell.

Stokols, Daniel, Satveer Misra, Richard P. Moser, Kimberly L. Hall, and Brian K. Taylor. 2008. The ecology of team science: Understanding contextual influences on transdisciplinary collaboration. *American Journal of Preventive Medicine* 35 (2 Suppl.), 96–115.

Boundary Work

Ulli Vilsmaier and Julie Thompson Klein

Definition

The term *boundary work* is conventionally traced in science studies to Thomas Gieryn's (1983) demarcation of science from non-science. He described boundary work as creation, relocation, and strengthening of boundaries between science and other forms of knowledge such as religion in 19th-century Britain. He aligned it with the "attribution of selected characteristics to the institution of science (i.e. to its practitioners, methods, stock of knowledge, values and work organization) for purposes of constructing a social boundary that distinguishes some intellectual activities as non-science" (1983, 7–82). For Gieryn, boundary work explores the interrelation of boundary construction and social identity and belonging to a community. With his analysis, Gieryn highlighted how boundary work can be enacted as ideological means to expand and monopolize authority through distinction and separation. Originally applied for analytical purposes, it was subsequently adapted for research spanning and crossing disciplinary and professional boundaries. Donald Fisher proposed a more generic definition: "Boundary work is defined as those acts and structures that create, maintain and break down the boundaries between knowledge units" (1993, 13–14). It encompasses claims, activities as well as institutional structures (Klein 2021). Langley et al. define boundary work as a "purposeful individual and collective effort to influence the social, symbolic, material and temporal boundaries, demarcations and distinctions affecting groups, occupations and organizations" (2019, 704). They consider boundary work a practice that clarifies differences and enables connections. The authors further distinguish actions that aim at creating, maintaining, blurring, or transforming boundaries.

Other scholars followed, broadening the concept to include applications to inter- and transdisciplinary research. In transdisciplinary research and learning, boundary work further addresses not only characteristics of knowledge fields but also sectors of society, including positionality of all participants to elaborate on different objectives, roles, and tasks through negotiation of ethical-political challenges in collaborative processes. In classifying the concept, Peter Mollinga (2010) identified three types of boundary work: (1) development of appropriate concepts of bordering, which allow us to address the multidimensionality of research; (2) configuration of adequate boundary objects as instruments and methods, through which incomplete and insecure knowledge, non-linearity, and diverging interests can be approached; and (3) creation of boundary situations where concepts, instruments, and methods can be explored in a profitable manner.

In addition, neighboring concepts relate to different dimensions and types of boundary work. Paulo Freire (1996), for example, aligned the concept with emancipatory and liberating pedagogy. He consideres "limit acts", drawing on Vieira Pinto (1960), as practices that expand perception and understanding of an existential situation people are experiencing and people *are* because they are "in a situation, … rooted in temporal–spatial conditions which mark them and which they also mark" (1996, 90). Limit acts are provoked by being challenged and reflect upon a situation. According to Freire, working on boundaries is where transformation happens. Sahr and Wardenga (2005) also locate this idea of boundaries playing a central role in understanding and appropriating the world in the origins of Geography as subject. According to Kant (cited in Hard 1993), Geography is the pre-exercise in knowledge about the world (German: *Kenntnis der Welt*) and a precondition of an understanding of the world (German: *Welterkenntnis*). It is in this formation of worldviews that the political character of boundaries brings itself to the fore. The common drawing of geographical boundaries that is anchored in polarities and an Aristotelian logic, however, is distinct to the boundary work this chapter highlights. Here, boundary work is introduced as a praxis of differentiating that brings forth connections while working in inter- and transdisciplinary research and higher education. A generic definition of boundary work consists of multiple practices related to differentiating, mediating, and negotiating different ways of knowing, acting, and being, thereby opening up conditions for joint thinking and collaboration or closing down options for co-work due to epistemological or value-based reasons. Boundary work is the praxis of making differences visible, utterable, and tangible to confirm, reinforce, transgress, transcend, or transform boundaries.

Background

Boundaries are a universal category. They are at the bottom of any formation of identity and social order. This generalization applies to the experience of the self brought to bear in forming standpoints and positionalities in the sense of locating ourselves in the world and belonging to it. Heintel et al. (2018, 1) consider boundaries, boundary demarcation, and transgression deeply internalized abstractions and actions. Nevertheless, as fundamental as the category of the boundary is, its character and constitution remain elusive. Whether boundaries are given, or processually brought forth, is the stuff big theories are made of (Vilsmaier 2018). Post-structuralist considerations have particularly informed theoretical discussion in recent decades, and shown that the ambivalence of the core concept of boundary defies clear definition. Redepenning draws attention to this ambivalence when emphasizing "boundaries are somewhat confusing [as] they limit 'something' *and* at the same time give us the instruction to overcome the limits of that 'something'" (2005, 168, own translation, italics in the original). Boundaries therefore always imply transgression. According to Cassirer (1994), boundaries can only be thought of as networks of relationships and processes that connect aspects of perception, expression, and action. Thus, bordering and ordering can be considered complementary categories (Sahr and Wardenga 2005). Every process involves acts of positioning and relationing that demarcate and transgress boundaries. While the concept of difference focuses on the one and the other, the concept of bounding shifts attention to a third process. Compared to the concept of border demarcation, bounding does not only describe demarcation but also the emergence and reconfiguration of boundaries. With his concept of Third Space, Bhabha (2004) introduces a topography that emerges from bounding and enables mediation of differences. Thereby, difference is considered a dynamic, or more precisely a diastatic, category that only comes into being in processes of differentiation (Vilsmaier 2018).

Boundaries are also at the bottom of the landscape of modern science. Disciplines only exist against the background of other disciplines they separate from. In this act of separation, Hamberger (2004) sees a transdisciplinary momentum in any discipline and Bhabha (2006) considers boundaries between disciplines as barriers to transverse or transcend when entering interdisciplinary inquiries, and at the same time "liminal forms of definition". The ambiguity of boundaries is apparent within inter- and transdisciplinary research, teaching, and learning. Boundaries between disciplines or specialized fields of knowledge structure institutionalizing distinctions as while multiple forces drive us to transgress them. We draw our professional identities from limited fields that allow us to develop a standpoint, while at the same time seeking to transcend them. Often, the impossibility of fully grasping a phenomenon from different disciplinary perspectives drives us towards boundaries.

Yet becoming aware of boundaries prompts acting upon them (Freire 1996). What matters most here is different dimensions of reference from which we attend to boundary work. This imperative is central to boundary-spanning and boundary-crossing research, allowing us to understand underlying conceptions of boundaries. Of added significance, it is crucial to consider whether boundaries are conceived as stabilizing or narrowing entities from the perspective of differentiation or from a performative conception of boundaries. The distinction pertains whether one strives towards fixating the separating elements (A and B, such as two different disciplines) as a basis of creating connections, or whether commonalities, differences, overlaps, and intersections bring forth C that not only includes but also modifies A and B. The focus is therefore on either objects or entities (e.g. disciplines) or subjects or people (e.g. researcher). If phenomena or problems require alteration of historically developed processes of ordering, of a shift or even demolishing of boundaries, this added caveat also requires attention to institutional practices and professional identities. When critiqued, what is known, customary, established, or unquestioned can hence shatter or weaken a supposedly sound terrain, or conversely be vindicated and open up to transgressing boundaries (Vilsmaier 2018).

Debate and criticism

Despite the popularity of inter- and transdisciplinarity and neighboring boundary-spanning and boundary-crossing forms of research, boundary work still receives little to no systematic attention. Transdisciplinary forms of research and learning are often more celebrated than they are founded epistemologically and conducted methodologically. Thus, they are not only vulnerable to attack but also prone to fail to transgress boundaries, integrate knowledge, or unfold transformative potential. A solid understanding of the kind of boundary spanning or crossing in research, teaching, and learning is key for conducting boundary work that enables mutual understanding of existing boundaries and also whether and how to work productively on these.

A recent book placing boundary work at the heart of inter- and transdisciplinarity brought together prior and new recognition of its centrality while extending implications (Klein 2021). Boundaries have a dual function: they demarcate different forms of expertise but are permeable and contingent as well, leading to both difference – in images of turf and territory – and interaction – in biological images of cross-fertilization. The two underlying metaphors, though, do not constitute a dichotomy. They operate simultaneously in the composite concept of an ecology of spatializing practices, illustrated by the evolving nature of disciplines as well as enclaves of trading zones and communities of practice. Thus, boundary work entails navigating and negotiating existing divisions as well as catalyzing new enclaves, while also bridging sectors of the academy, government, industry, and communities. Updated descriptions of disciplines also acknowledge their porous nature. Openness to change, however, is uneven, and lack of familiarity with inter- and transdisciplinarity limits prospects for transformational change. Michael Foucault (1995) stipulated boundaries that prescribe social order, and dualisms of normality and deviance, as well as belonging and not-belonging. In that sense, a boundary classifies, categorizes, sorts, segments, and normalizes. It also includes and excludes, privileges and de-privileges. Yet boundary remains uncrossed. Moreover, boundaries are contested and their authority is disputed. The boundary rhetoric of both inter- and transdisciplinarity, then, is complex: it "compasses acts of spanning, crossing, and bridging; processes of interacting, integrating, and collaborating; strategies of brokering, mediating, and negotiating; operations of demarcating, constructing, and refiguring; new relations of interdependence and convergence; and outcomes of breaching, transgressing, and transforming" (Klein 2021, 22–23).

Methods of boundary work support systematic approaches to elaborate on differences while differentiating and thereby laying the ground for integration (Vilsmaier 2018). The term method encompasses different types of proceedings. Methods of boundary work serve multiple purposes, ranging from creating conditions for shared thinking and acting by creating understanding for one's own and others' standpoint and positionality, support problem framing and mutual learning, and, depending on the types of research, theoretical or methodical integration or product development. They often deploy a related concept highlighted in this chapter: boundary objects. According to Star and Griesemer, who introduced the concept in 1989 in the field of Science and Technology Studies, boundary objects "are both adaptable to different viewpoints and robust enough to maintain identity across them" (1989, 387). Boundary objects can mediate between different social worlds: "They have different meanings in different social worlds but their structure is common enough to more than one world to make them recognizable means of translation" (Star and Griesemer 1989, 393). The authors distinguish between repositories, ideal types, coincident boundaries, and standardized forms. Bergmann et al. broadened the meaning of boundary objects for practices of integration in transdisciplinary research. The concept is applied to all sorts of "interfaces where actors from different fields, such as science, politics and business, meet and communicate" (2012, 105). Deployed as integration methods, different entities play out as boundary objects, such as artifacts, products, abstract ideas, common research goals, illuminating examples, or publications (Bergmann et al. 2012).

In methods of boundary work, boundary objects first and foremost serve the elaboration of differences. Peukert (2022) experimented with design prototyping as boundary work, using prototypes as boundary objects in transdisciplinary research. The dual characteristic of process character and object status could be identified as a particular quality of prototyping. Working with highly diverse research participants in a case study in rural Romania, language of form of prototypes and common abilities in constructing and modifying these turned out to be a powerful way of navigating differences and unequal means of conceptual expressions (Peukert et al. 2020).

Conceptual work is an elementary form of boundary work in transdisciplinary research. The more heterogeneous a research team, the more likely that the same concepts carry different meanings in different social worlds. Boundary concepts are key means of making connections. They are semantic anchors for developing coherent research frameworks and meaningful results that exhibit communicative, epistemic, and ethical–political dimensions. Conceptual work aims at creating mutual understanding for different semantics and roles of concepts in a research field, as well as negotiating the use of concepts. However, the process is often not explicitly placed within research processes, leading to difficulties (Hoffmann et al. 2017). Many technical terms are used in everyday language and mutual understanding is often presumed, but without exploring their meaning (Bergmann et al. 2012). These tendencies can become obstacles showing the paradoxical nature of such concepts. Quotidian usage of terms can be made productive for communication, but semantic differences are too easily glossed over (ibid. 2012). Conceptual work, however, is by no means limited to academic clarification of meanings. It has a significant political dimension where disciplinary or, more generally, scientific hegemonies and inequalities in conceptual abilities may easily play out. Here, conceptual work as boundary work has the potential to create visibility not only for different semantics of terms but also for differences amongst collaborators. In that sense, conceptual work can be defined "as the collaborative process of clarifying the meaning and use of concepts across disciplines and epistemic cultures, developing mutual understanding and balancing power inequalities amongst participants in order to support knowledge co-creation" (Juarez-Bourke and Vilsmaier 2020, 25).

Current forms of implementation in higher education

Boundary work draws on difference. In higher education, it is particularly useful when guiding students from different study fields. Multi-, inter-, and transdisciplinary classrooms present great opportunities for boundary work. Experiencing, exploring, and systematically approaching different perspectives and ways of acting upon a given problem, boundary work equips students with abilities and techniques to elaborate on and integrate different knowledges or practices. In research on environmental science education, Fortuin (2015) further distinguishes boundary-crossing skills from inter- and transdisciplinary cognitive skills and reflexive skills. Boundary-crossing skills should equip students to (1) be aware of different disciplinary, cultural, theoretical, and practical perspectives; (2) acknowledge the values of using these perspectives in addressing complex problems, and (3) use various disciplinary perspectives and connect them, to collaborate, negotiate, make decisions in intercultural settings, and deal with complexity and uncertainty (Fortuin 2015, 133).

To illustrate: in a student-driven transdisciplinary research module, conducted over several years in a Master's program of sustainability science, different forms of boundary work were applied to train students in boundary work on three levels (Vilsmaier and Lang 2015). (1) On the *personal level,* students explore and elaborate their professional profiles and identities related to their study fields. This kind of boundary work departs from the perception of others' professional characteristics. For instance, a sociologist provides a description of his or her imaginary of a sustainable chemist and vice versa, thereby laying bare often glossed over assumptions, supporting reflection and building self-awareness of the student's specialization. The process aims to uncover assumed positions from which a research or study subject is approached. Thereby, not only abstract systems of knowledge. Individual configurations of the same unfold, helping visualize researchers' positionality with regard to their situatedness within knowledge fields, paradigms, and personal situated accounts that inform study and research (Rose 1997). Within transdisciplinary research and learning, this procedure also takes on cultural and social situatedness while taking values and norms into consideration (Rosendahl et al. 2015). Boundary work allows for visualizing situated relations of researchers or learners with each other (Klein 2010). (2) On the level of *knowledge fields,* students explore their study fields by developing topographies of knowledge fields with regard to core topics, dominant theories, and common methods. Based on individual maps, student teams elaborate on commonalities while exploring differences. As a result, a map of the student team is developed that provides insights into the team's expertise, abilities, and perspectives. Already at this stage an interdisciplinary in-between space shapes and serves as a starting point for collaborative research. (3) On the level of *societal domains,* boundary work includes elaboration of differences with regard to roles, responsibilities, interest, and objectives in a transdisciplinary team. At this level, students enter the constitution of a transdisciplinary in-between space that emerges from difference (for more details see Vilsmaier and Lang 2015).

Didactical approaches that prepare students to deal with the complexity they will face conducting transdisciplinary research must pay particular attention to a *literacy of difference* – supporting students in reflecting on their own positionality and in developing an attitude of openness for mutual learning.

References

Bergmann, Matthias, Thomas Jahn, Tobias Knobloch, Wolfgang Krohn, Christian Pohl, and Engelbert Schramm. 2012. *Methods for transdisciplinary research: A primer for practice.* Frankfurt am Main: Campus.

Bhabha, Homi K. 2004. *The location of culture.* New York: Routledge.

Bhabha, Homi. 2006. *Boundaries. Differences. Passages.* Available from: https://de.scribd.com/document/590010987/Bhabha-Homi-Boundaries-Differences-Passages#.

Cassirer, Ernst. 1994. *Philosophie der symbolischen Formen. Erster Teil: Die Sprache.* Darmstadt: Wissenschaftliche Buchgesellschaft.

Fisher, Donald. 1993. *Fundamental development of the social sciences: Rockefeller philanthropy and the United States Social Science Research Council.* Ann Abor, MI: University of Michigan Press.

Fortuin, Karen P. J. 2015. *Heuristic principles to teach and learn boundary crossing skills in environmental science education.* Wageningen: Wageningen University.

Foucault, Michael. 1995. *Discipline and punish: The birth of the prison* (Alan Sheridan, trans.). New York: Vintage Books.

Freire, Paulo. 1996. *Pedagogy of the oppressed.* London: Penguin.

Gieryn, Thomas F. 1983. Boundary-work and the demarcation of science from non-science: Strains and interests in professional ideologies of scientists. *American Sociological Review* 48 (6): 781–95.

Hamberger, Erich. 2004. Transdisciplinarity: A scientific essential. *Annals New York Academy of Science* 1028 (1): 487–96.

Hard, Gerhard. 1993. Kants "Physische Geographie", wieder gelesen. In *Grenzüberschreitung. Wandlungen der Geisteshaltung, dargestellt an Beispielen der Geographie und Wissenschaftstheorie, Theologie, Religions- und Erziehungswissenschaften, Philosophie, Musikwissenschaften und Liturgie.* ed. Heyno Kattenstedt, 51–72. Bochum: Brockmeyer.

Heintel, Martin, Robert Musil, Markus Stupphannn, and Norbert Weixlbaumer. 2018. Grenzen – eine Einführung. In *Grenzen. Theoretische, konzeptionelle und praxisbezogene Fragestellungen zu Grenzen und deren Überschreitungen*, eds. Martin Heintel, Robert Musil and Norbert Weixlbaumer, 1–18. Wiesbaden: Springer VS.

Hoffmann, Sabine, Christian Pohl, and Janet G. Hering. 2017. Methods and procedures of transdisciplinary knowledge integration: Empirical insights from four thematic synthesis processes. *Ecology and Society* 22 (1): 27.

Juarez-Bourke, Sadhbh, and Ulli Vilsmaier. 2020. The semantics of transformation: Conceptual work for inter- and transdisciplinary research based on Paulo Freire's approach to literacy. *Journal für Entwicklungspolitik* XXXVI (3): 19–43.

Klein, Julie Thompson. 2010. *Creating interdisciplinary campus cultures: A model for strength and sustainability.* San Francisco: Jossey-Bass.

Klein, Julie Thompson. 2021. *Beyond Interdisciplinarity: Boundary work, communication and collaboration.* New York: Oxford University Press.

Kleinschmidt, Christoph. 2011. Einleitung: Formen und Funktionen von Grenzen. Anstöße zu einer interdisziplinären Grenzforschung. In *Topographien der Grenze. Verortungen einer kulturellen, politischen und* ästhetischen Kategorie, eds.

Christoph Kleinschmidt and Christine Hewel, 9–24. Würzburg: Königshausen & Neumann.

Langley, Ann, Kasia Lindberg, Bjorn Erik Mork, Davide Nicolini, Elena Raviola, and Lars Walter. 2019. Boundary work among groups, occupations and organizations: From cartography to process. *Academy of Management Annals*, 13 (2): 704-736.

Mollinga, Peter. 2010. Boundary work and the complexity of natural resources management. *Crop Science* 50(S1): 1–9.

Peukert, Daniela. 2022. *Design methods for collaborative knowledge production in inter- and transdisciplinary research*. Lüneburg: Leuphana University. Available at https://pub-data.leuphana.de/frontdoor/index/index/docId/1228.

Peukert, Daniela, David P. M. Lahm, Andra I. Horcea-Milcu, and Daniel J. Lang. 2020. Facilitating collaborative processes in transdisciplinary research using design prototyping. *Journal of Design Research* 18 (5–6): 294–326.

Redepenning, Marc. 2005. Über die Unvermeidlichkeit von Grenzziehung. *Berichte zur deutschen Landeskunde* 79 (2–3): 167–77.

Rose, Gillian. 1997. Situating knowledges: Positionality, reflexivities and other tactics. *Progress in Human Geography* 21 (3): 305–20.

Rosendahl, Judith, Matheus A. Zanella, Stephan Rist, and Jes Weigelt. 2015. Scientists' situated knowledge: Strong objectivity in transdisciplinarity. *Futures* 65: 17–27.

Sahr, Wolf-Dietrich, and Ute Wardenga. 2005. Grenzgänge – Ein Vorwort über Grenzen und ihre (Be-) Deutungen in der Geographie. *Berichte zur deutschen Landeskunde* 79 (2–3): 157–66.

Star, Susan Leigh, and James R. Griesemer. 1989. Institutional ecology, "translations" and boundary objects: Amateurs and professionals in Berkley's Museum of Vertebrate Zoology, 1907–39. *Social Studies of Science* 19 (3): 387–420.

Vieira Pinto, Alvaro. 1960. *Consciência e realidade nacional*. Vol. 2. Rio de Janeiro: Instituto Superior de Estudios Brasileiros.

Vilsmaier, Ulli. 2018. Grenzarbeit in integrativer und grenzüberschreitender Forschung. In *Grenzen. Theoretische, konzeptionelle und praxisbezogene Fragestellungen zu Grenzen und deren Überschreitungen*, eds. Martin Heintel, Robert Musil and Norbert Weixlbaumer, 113–34. Wiesbaden: Springer VS.

Vilsmaier, Ulli, and Daniel J. Lang. 2015. Making a difference by marking the difference: Constituting in-between spaces for sustainability learning. *Current Opinion in Environmental Sustainability* 16: 51–55.

Case Study

Stefanie Meyer, Katja Brundiers, Marlene Mader, and Annika Weiser

Definition

The term *case* has its etymological roots in the Latin *casus* and literally stands for "a fall", being understood as an "accident" in terms of "a chance, occasion, opportunity" (Harper 2022). Thus, a case study refers to the investigation of a specific event, for example a product, people, or organization, or processes such as policy-making, decision-making, and transformation. It illustrates a rich and multi-layered picture of the selected phenomenon and its setting, which is often characterized by complex circumstances. The aim of a case study is first to gain a better understanding of the phenomenon by investigating and revealing its mechanisms and origins in its specific context, and then to derive lessons learned, which may be transferable to other and future cases. Cases can be real phenomena, fictitious inquiries, or circumstances drawn from the literature. Considering the versatility of case studies, they are applied in many disciplines in research and teaching at higher educational institutes.

Yin (2018) provides general guidelines on planning, preparing, and executing case studies, and sharing results from case study research, which can be adjusted and applied to various disciplines. In general, designing a case study offers choices along several variables: scope (e.g. depth, breadth, and boundedness), function (e.g. theory-testing versus action-oriented problem-solving), and outcome (e.g. theoretical practical insights, lessons learned, policy options), as well as format (e.g. single versus multiple case studies). Single cases may contribute to either confirming or challenging a theory, or to represent a unique or extreme phenomenon, which can be a fictitious inquiry or circumstance drawn from the literature. Single cases require careful investigation to avoid misrepresentation. In contrast, using multiple cases in conjunction with quantitative research approaches allow for cross-case comparison and potential replication of research approaches, holding the possibility of providing findings that are more robust and generalizable or transferable to other cases (Adler et al. 2018; Stake 2013, 39–41). Both single and multiple case studies should use multiple sources of information to integrate knowledge from the case study with scientific and other pertinent societal discourses. While some disciplines rely on qualitative sources of evidence, such as interviews, observations, documentation (Yin 2018, 111–13), multidisciplinary approaches to complex problem-solving call for mixed-method data collection that integrates qualitative and quantitative resources, such as systems dynamics modeling and formative scenario analysis (Scholz and Tietje 2002).

Transdisciplinary case study research explores socially relevant problems in collaboration with scholars from different disciplines and practitioners from various sectors (Bergmann et al. 2012). To facilitate the collaborative study mode, a transdisciplinary case study typically contains three main phases: (1) joint problem-framing, (2) co-production of knowledge, and (3) integration and application of knowledge (Bergmann et al. 2012; Brundiers and Wiek 2013; Lang et al. 2012).

In sustainability studies, transdisciplinary case studies evolved from conducting research on sustainability problems, resulting in deeper analysis and description of the problem, to studies that also investigate solutions to these sustainability problems. This solutions-oriented research may focus on developing solutions, studying their actual implementation, or evaluating the effects of implemented solutions. In either process, researchers, students, and practitioners can select different degrees of collaboration and co-production of knowledge, ranging from informing and consulting each other to collaborating with potential implementers (e.g. policymakers or NGOs) and affected actors (Lang and Wiek 2022; Lang et al. 2012; Stauffacher et al. 2006). In higher education teaching, transdisciplinary case study approaches offer students opportunities to learn how to methodically approach complex action-oriented problem-solving processes in a collaborative format (Brundiers and Wiek 2013).

Background

The general case study method was first applied at the beginning of the 1900s in qualitative case study research in social and anthropology studies (Platt 1992). Since then, different strands evolved in parallel (Table 1). Various disciplines embraced the problem-centered case study approach, including law, business, and management, as well as educational sciences. Those studies were rather single-case-based, descriptive, and qualitative in nature and were considered as a special form of experimental, statistical, or comparative methods. At the same time, case studies were introduced as a teaching demonstration tool in disciplines such as business, law, and medicine (McNair and Hersum 1954). In these disciplines case studies are continuously used to derive and evaluate knowledge about individuals, groups, and organizations and related social and political phenomena (Carter and Unklesbay 1989).

In the 1950s, systems dynamics offered a new modeling approach for simulating the behavior of real-life social systems (Forrester 1971). Drawing on the advancement in systems thinking and complex real-world problem-solving through Checkland's (1972) soft systems methodology and Meadows' (1999) system-influencing leverage points, it became possible to understand how phenomena change through time and how to direct future developments. Since the 1980s, knowledge on the design of the case study approach and on its process has been solidified (Merriam 1998; Ragin and Becker 1992; Stake 1995; Yin 1981) and case studies were increasingly used in many more disciplines, such as planning sciences, and decision-making and risk research. This fostered the incorporation of scenario planning as a quantitative source of evidence into the environmental case study approach (Gomm et al. 2000). In the mid-1980s, environmental problem-solving emerged as a goal of education, research, and application; the transdisciplinary case study approach was developed at the Eidgenössische Technische Hochschule (ETH) in Zurich (Müller-Herold and Neuenschwander 1992). Since then, universities have adjusted case-based learning courses, providing real-world learning opportunities for students to identify the interdependencies and complexities of (research) problems and striving to integrate pertinent academic disciplines. In the 1990s, the case study approach became more collaborative as more methods and approaches for transdisciplinary collaboration were developed (Scholz et al. 2006), and as participatory action research had advanced in education, combining theory, research, and practice towards a problem-based methodology focusing on theories of action (Carr and Kemmis 1986).

In the 2000s, case studies evolved towards an integration of various types of methods and were used not only to describe and observe phenomena, but also to analyze how to change, fix, and improve contemporary situations. As sustainability science emerged as a research field, solutions- and action-oriented case study approaches were applied in various sustainability-related contexts such as disaster recovery, bioenergy, and precautionary purchasing, vulnerability assessment, or water resource management (Wiek et al. 2012). The concept of sustainability science influenced the learning situations at universities (Stauffacher et al. 2006) and spaces were created to shift the focus from the disciplinary perspective to the inter- and transdisciplinary collaboration of science and practice.

The complex challenges facing societies in the 21st century, such as the climate emergency, increasing urbanization, rising inequalities, and loss of biodiversity (IPCC 2022) expand the focus of case study analysis for transformational change towards advanced participatory research settings and stakeholder involvement as well as capacity-building and better generalization of single and multiple cases (Caniglia et al. 2021; Lang and Wiek 2022; Wiek et al. 2012).

Table 1. Developments in the history of case studies and educational innovations

	Case studies (~1900)	Transdisciplinary research (~1990)	Sustainability Science (~2000)
Learning that is...	- individual - problem-centered - context-related	- collaborative and participatory	- social and participatory - problem- and solution-oriented - competency-based - centered on real-world issues
An epistemology that is...	- descriptive and exploratory	- integrative	- transformational
The investigated phenomenon is...	- mostly a single case - often defined as monodisciplinary	- complex - requiring inter- and transdisciplinary approaches	- complex and difficult to define - requiring transdisciplinary approaches
Applications of case studies...	- in research or teaching	- that combine research and teaching	- that integrate research, teaching and goal of social transformation

Debate and criticism

The co-evolution of case studies as a research method and teaching tool in different disciplines produced diverse approaches to case study work. Thus, the literature contains numerous examples that are referenced as case studies in research, teaching, or practice, but not all of these approaches reflect the characteristics and guidelines for the use and design of case studies (Gerring 2004). Comparing these case studies in attempts to gain generalizable insights remains a challenge, because case studies differ in disciplinary approaches and scope (complexity, context), as well as function and outcome. The multi-methods design of the case study approach is still considered to be a weaker form of analysis compared to a quantitative methods approach, as the latter may promise repetitions, which may increase representativeness, reliability, and validity (Takahashi and Araujo 2020). In either case, interpreting observations requires robust data collection and analysis, as well as competencies including critical thinking and normative and ethical competencies (Takahashi and Araujo 2020). In writing up the results of their case study analysis, researchers need to differentiate results that describe the studied unit and results which may apply to a broader set of units (Gerring 2004). Thus, considering these challenges, common research protocols for case studies are needed to yield comparable data on influencing variables and to enhance the reliability of cross-case comparisons. Case studies in transdisciplinary sustainability science aim to develop context-sensitive sustainability solutions and an understanding of whether and how such solutions could be transferred to other contexts and what adaptations such transfer might entail (Forrest et al. 2020).

Case study methods are a suitable strategy for teaching and learning about problem-solving competencies to address complex issues. Case study teaching and learning in transdisciplinary settings is a demanding task. Nevertheless, the application of criteria for developing good case study teaching, such as establishing a framework for student discussion and debate, is not as strict (Yin 2018, 19). This allows for adjusting the case study to the level of students being taught, to be relevant to the course content, and to provide a way for students to practice the knowledge and skills they have learned so far in their study program. More emphasis needs to be put on better understanding context conditions across various cases to improve the generation of socially robust knowledge. The implementation of a common framework on self-directed learning and key competencies in sustainability could help support students dealing with the demanding case study work (Brundiers et al. 2021; Pearce et al. 2018; Wiek et al. 2015).

Teaching case studies also demands additional skills from instructors (e.g. coaching, supporting the learning processes). To ensure that case studies as teaching tools can best fulfill their potential, instructors need to be trained in teaching good case studies (Barnes et al. 1994; Brundiers and Wiek 2013; Weber and Kirk 2000). Weber and Kirk (2000) stressed that attempts to align demands and expectations of scientists and practice partners often lead to difficulties. To avoid burdening instructors with additional tasks related to stakeholder collaboration, it is recommended to hire a "Transacademic Interface Manager" – a person who is trained in facilitating transdisciplinary sustainability research and education in collaboration with stakeholders (Brundiers and Wiek 2013). A Transacademic Interface Manager can help facilitate alignment among researchers and stakeholders around shared purposes, by explaining different approaches and developing curricula and standardized protocols (Caniglia et al. 2021). Tools to support collaboration exist and are offered, for example, through platforms like the *td-net* (td-net 2023) and the tdAcademy (2023).

In summary: (1) *Benefits* for students in transdisciplinary and solutions-oriented sustainability case study learning include investigating real-world sustainability issues in collaboration with stakeholders with the goal of co-producing a potentially transformational solution as a motivational driver. This offers valuable opportunities for intellectual, professional, and personal growth as a sustainability change agent. (2) *Challenges* comprise collaborating with stakeholders and diverse scholars in a research-based educational setting requires reconciling different ways of knowing and working, as well as timelines and reward systems.

Finally, (3) to *embrace the challenges*, the interpersonal skills and personal strengths needed for such collaboration can be trained and are valuable professional skills in the workforce (Inner Development Goals, n.d.; Ives et al. 2020). Similarly, skills for inter- and transdisciplinary research can be developed through self-directed learning and seeking out training opportunities offered, e.g. through campus transdisciplinary living learning labs (Fam et al. 2018).

Current forms of implementation in higher education

As discussed above, various disciplines use (transdisciplinary) case studies in research, teaching, and practice and implement them in diverse ways specific to their topics, disciplines, and contexts. Contextual aspects shaping implementation include socio-ecological and cultural dimensions of place as well as the educational and political system of countries (Mieg et al. 2022, 441–43). Selected examples show that sustainability-related challenges and collaborative approaches are gaining momentum in case study implementations.

A case study in the Seychelles addressed the problem of solid waste and involved students and researchers from ETH Zurich and the University of Seychelles, as well as representatives from the Seychelles' Ministry of Environment, Energy and Climate Change, the private sector, and the community. By applying soft systems methodology and design thinking, students learned to better understand the problem and developed place-based and context-sensitive solutions together (Krütli et al. 2018; Pohl et al. 2018). Urban sustainability was addressed within a socio-ecological systems framework in case studies in southern Africa (Thondhlana et al. 2021) and Austria (Biberhofer and Rammel 2017). They investigate challenges, such as poor sanitation and power supply in different southern African urban areas and increasing CO_2 emissions and limited fossil resources in Austrian cities, respectively. Case studies in Mexico, Colombia, Nicaragua, and China (Transdisciplinary Student Team Research) investigated rural development accounting for local conditions (Acevedo-Osorio et al. 2020). Students were engaged in problem-based- and co-learning with peers to develop responsible actions that support sustainable rural development.

These examples highlight the spectrum of teaching and learning approaches ranging from student-centered, experience-based service learning to approaches that aim to increase broader social learning in transdisciplinary learning. The examples also illustrate how learning objectives for students combine, in varying degrees, content knowledge, professional skills, and sustainability competencies to create a meaningful teaching and learning environment in higher education. As such, they emphasize the need to provide good teacher training offerings when implementing such complex teaching innovations. Real-world laboratories support long-term learning and teaching at the science–society interface, engaging students with all phases of transdisciplinary research, including framing the project as well as developing and testing solutions, and evaluating processes and outcomes (Barth et al. 2017; Beecroft 2018; Singer-Brodowski et al. 2018). Thus, in conclusion, the examples show how transdisciplinary case studies may vary, ranging from a traditional focus on a deep understanding and co-development of the problem in its context to an emphasis on co-designing solution approaches and using gained knowledge to advance science and practice.

References

Acevedo-Osorio, Álvaro, Susanne Hofmann-Souki, and Juana Cruz Morales. 2020. Holistic competence orientation in sustainability- elated study programmes: Lessons from implementing transdisciplinary student team research in Colombia, China, Mexico and Nicaragua. *Sustainability Science* 15 (1): 233–46.

Adler, Carolina, Gertrude Hirsch Hadorn, Thomas Breu, Urs Wiesmann, and Christian Pohl. 2018. Conceptualizing the transfer of knowledge across cases in transdisciplinary research. *Sustainability Science* 13 (1): 179–90.

Barnes, Louis B., C. Roland Christensen, and Abby J. Hansen. 1994. *Teaching and the case method: Text, cases, and readings*. Brighton, MA: Harvard Business School Press.

Barth, Matthias, Daniel J. Lang, Philip Luthardt, and Ulli Vilsmaier. 2017. Mapping a sustainable future: Community learning in dialogue at the science–society interface. *International Review of Education* 63 (6): 811–28.

Beecroft, Richard. 2018. Embedding higher education into a real-world lab: A process-oriented analysis of six transdisciplinary project courses. *Sustainability* 10 (10), 3798.

Bergmann, Matthias, Thomas Jahn, Tobias Knobloch, Wolfgang Krohn, Christian Pohl, and Engelbert Schramm. 2012. *Methods for transdisciplinary research: A primer for practice*. Frankfurt: Campus Verlag.

Biberhofer, Petra, and Christian Rammel. 2017. Transdisciplinary learning and teaching as answers to urban sustainability challenges. *International Journal of Sustainability in Higher Education* 18 (1): 63–83.

Brundiers, Katja, and Arnim Wiek. 2013. Do we teach what we preach? An international comparison of problem- and project-based learning courses in sustainability. *Sustainability* 5 (4): 1725–46.

Brundiers, Katja, Matthias Barth, Gisela Cebrián, Matthew Cohen, Liliana Diaz, Sonya Doucette-Remington, Weston Dripps, Geoffrey Habron, Niki Harré, Meghann Jarchow, Kealalokahi Losch, Jessica Michel, Yoko Mochizuki, Marco Rieckmann, Roderic Parnell, Peter Walker, and Michaela Zint. 2021. Key competencies in sustainability in higher education – Toward an agreed-upon reference framework. *Sustainability Science* 16: 13–29.

Caniglia, Guido, Christoph Lüderitz, Timo von Wirth, Ioan Fazey, Berta Martín-López, Kristina Hondrila, Ariane König, Henrik von Wehrden, Niko Alexander Schäpke, Manfred D. Laubichler, and Daniel J. Lang. 2021. A pluralistic and integrated approach to action-oriented knowledge for sustainability. *Nature Sustainability* 4 (2): 93–100.

Carr, Wilfred, and Stephen Kemmis. 1986. *Becoming critical: Education, knowledge and action research.* London: Falmer.

Carter, Kathy, and Rick Unklesbay. 1989. Cases in teaching and law. *Journal of Curriculum Studies* 21 (6): 527–36.

Checkland, Peter B. 1972. Towards a systems-based methodology for real-world problem solving. *Journal of systems engineering,* 3(2), 87-116.

Fam, Dena, Abby Mellick Lopes, Alexandra Crosby, and Katie Ross. 2018. The university campus as a transdisciplinary living laboratory. Available from https://i2insights.org/2018/05/01/campus-as-transdisciplinary-living-laboratory.

Forrest, Nigel, Zoë Stein, and Arnim Wiek. 2020. Transferability and scalability of sustainable urban water solutions – A case study from the Colorado River Basin. *Resources, Conservation and Recycling* 157: 104790.

Forrester, Jay W. 1971. *World dynamics.* Cambridge, MA: Wright-Allen Press.

Gerring, John. 2004. What is a case study and what is it good for? *American Political Science Review* 98 (2): 341–54.

Gomm, Roger, Martyn Hammersley, and Peter Foster, eds. 2000. *Case study method: Key issues, key texts.* London: Sage.

Harper, Douglas. 2022. *Etymology of case.* Online etymology dictionary. Available from https://www.etymonline.com/word/case.

IPCC [Intergovernmental Panel on Climate Change], ed. 2022: *Climate change 2022: Impacts, adaptation and vulnerability. Contribution of Working Group II to the Sixth Assessment Report of the Intergovernmental Panel on Climate Change.* Cambridge: Cambridge University Press.

Inner Development Goals. n.d. Available from https://www.innerdevelopmentgoals.org.

Ives, Christopher D., Rebecca Freeth, and Joern Fischer. 2020. Inside-out sustainability: The neglect of inner worlds. *Ambio* 49: 208–17.

Krütli, Pius, Christian Pohl, and Michael Stauffacher. 2018. Sustainability learning labs in small island developing states: A case study of the Seychelles. *GAIA – Ecological Perspectives for Science and Society* 27: 46–51.

Lang, Daniel J., and Arnim Wiek. 2022. Structuring and advancing solution-oriented research for sustainability. *Ambio* 51 (1): 31–35.

Lang, Daniel J., Arnim Wiek, Matthias Bergmann, Michael Stauffacher, Pim Martens, Peter Moll, Mark Swilling, and Christopher J. Thomas. 2012. Transdisciplinary research in sustainability science: Practice, principles, and challenges. *Sustainability Science* 7: 25–43.

McNair, Malcolm Perrine, and Anita C. Hersum, eds. 1954. *The case method at the Harvard Business School: Papers by present and past members of the faculty and staff.* New York: McGraw-Hill.

Meadows, Donella. 1999. *Leverage points: Places to intervene in a system*. The Sustainability Institute. Available from http://drbalcom.pbworks.com/w/file/fetch/35173014/Leverage_Points.pdf.

Merriam, Sharan B. 1998. *Qualitative research and case study applications in education. Revised and expanded from "Case Study Research in Education"*. San Francisco: Jossey-Bass.

Mieg, Harald A., Elizabeth Ambos, Angela Brew, Dominique M. Galli, and Judith Lehmann, eds. 2022. *The Cambridge handbook of undergraduate research*. Cambridge: Cambridge University Press.

Müller-Herold, Ulrich, and Markus Neuenschwander. 1992. Vom Reden Zum Tun: Die Fallstudie in Den Umweltnaturwissenschaften. *GAIA – Ecological Perspectives for Science and Society* 1 (6): 339–49.

Pearce, BinBin, Carolina Adler, Lisette Senn, Pius Krütli, Michael Stauffacher, and Christian Pohl. 2018. Making the link between transdisciplinary learning and research. In *Transdisciplinary theory, practice and education*, eds. Dena Fam, Linda Neuhauser, and Paul Gibbs, 167–83. Cham: Springer.

Platt, Jennifer. 1992. "Case study" in American methodological thought. *Current Sociology*, 40 (1):, 17–48.

Pohl, Christian, Pius Krütli, and Michael Stauffacher. 2018. Teaching transdisciplinarity appropriately for students' education level. *GAIA – Ecological Perspectives for Science and Society* 27 (2): 250–52.

Ragin, Charles C., and Howard S. Becker, eds. 1992. *What is a case? Exploring the foundations of social inquiry*. Cambridge: Cambridge University Press.

Scholz, Roland W., and Olaf Tietje. 2002. *Embedded case study methods: Integrating quantitative and qualitative knowledge*. Thousand Oaks, CA: Sage.

Scholz, Roland W., Daniel J. Lang, Arnim Wiek, Alexander I. Walter, and Michael Stauffacher. 2006. Transdisciplinary case studies as a means of sustainability learning: Historical framework and theory. *International Journal of Sustainability in Higher Education* 7 (3): 226–51.

Singer-Brodowski, Mandy, Richard Beecroft, and Oliver Parodi. 2018. *Learning in real-world laboratories: A systematic impulse for discussion. Gaia* 27: 23–27.

Stake, Robert E. 1995. *The art of case study research*. London: Sage.

Stake, Robert E. 2013. *Multiple case study analysis*. New York: Guilford.

Stauffacher, Michael, Alexander I. Walter, Daniel J. Lang, Arnim Wiek, and Roland W. Scholz. 2006. Learning to research environmental problems from a functional socio-cultural constructivism perspective: The transdisciplinary case study approach. *International Journal of Sustainability in Higher Education* 7 (3): 252–75.

Takahashi, Adriana R. W., and Luis Araujo. 2020. Case study research: Opening up research opportunities. *RAUSP Management Journal* 55 (1): 100–11.

tdAcademy, ed. 2023. *About us.* Available from https://td-academy.org/en/tdacademy/about-us.

td-net, ed. 2023. *Network for transdisciplinary research.* Available from https://transdisciplinarity.ch/en.

Thondhlana, Gladman, Chipo Plaxedes Mubaya, Alice McClure, Akosua Baah Kwarteng Amaka-Otchere, and Sheunesu Ruwanza. 2021. Facilitating urban sustainability through transdisciplinary (TD) research: Lessons from Ghana, South Africa and Zimbabwe. *Sustainability* 13 (11): 1–18.

Weber, Mary M., and Delaney J. Kirk. 2000. Teaching teachers to teach cases: It's not what you know, it's what you ask. *Marketing Education Review* 10 (2): 59–67.

Wiek, Arnim, Barry Ness, Petra-Schweizer-Ries, Fridolin S., Brand, and Francesca Farioli. 2012. From complex systems analysis to transformational change: A comparative appraisal of sustainability science projects. *Sustainability Science* 7 (Supplement 1): 5–24.

Wiek, Arnim, Michael Bernstein, Rider W. Foley, Matthew Cohen, Nigel Forrest, Christopher Kuzdas, Braden Kay, and Lauren Withycombe Keeler. 2015. Operationalising competencies in higher education for sustainable development. In *Routledge handbook of higher education for sustainable development*, eds. Matthias Barth, Gerd Michelsen, Marco Rieckmann, and Ian Thomas, 241–60. Abingdon: Routledge.

Yin, Robert K. 1981. The case study as a serious research strategy. *Knowledge: Creation, Diffusion, Utilization* 3 (1): 97-114.

Yin, Robert K. 2018. *Case study research and applications: Design and methods.* 6th edition. Thousand Oaks, CA: Sage.

Citizen Science

Melanie Jaeger-Erben, Frank Becker, Baiba Prūse, Jimlea Nadezhda Mendoza, Jutta Gutberlet, and Eliana Rodrigues

Definition

The term *citizen science* originates from Anglo-American contexts and generally describes the procedure of involving citizens who are not institutionally anchored in academia as active participants in a scientific research process. The use of the term "citizen" (etymologically derived from the Anglo-French word *citisein* "inhabitant of a city or community", approx. 13th century), indicates a specific understanding of the persons involved, who, in the sense of the term *citoyen* coined in the French Enlightenment, actively and autonomously participate in the community and help to shape it. The tasks of citizens in this context range from collecting data to co-designing the entire research process, applying scientific quality standards, and producing scientifically usable results (Haklay et al. 2021; Pettibone et al. 2017). Citizen science as a designation for a specific form of knowledge production is mainly used in the European and North American context, where a differentiated research and funding landscape has evolved since the beginning of this century (Haklay et al. 2021). Similar approaches can be found in other parts of the world, but are framed under alternative terms such as *community science* (Conrad and Hilchey 2011) and *community-based research* (Amauchi et al. 2022). Citizen science brings together a multiplicity of approaches ranging from mass data collection events for citizens to forms of independent or self-determined research by non-academic groups or communities, calling the term itself into question (Eitzel et al. 2017).

Background

Long before the term citizen science was coined in 1989 (Kerson 1989), citizen engagement in science shaped the history of science in North America and Europe and was vital to its formalization and institutionalization (Mahr and Dickel 2019; Vetter 2011). For example, amateur experts like bishops, farmers, hunters, and so-called gentleman engaged in the collection and processing of data and information and pushed the evolution of humanities and natural science up to the 18th century (Brenna 2011; Chuine et al. 2004; Porter 1978). However, during the course of the institutionalization of science in North America and Europe in the 19th and early 20th centuries, voluntary scientific work by citizens was marginalized (Miller-Rushing et al. 2020). As disciplines differentiated, the pressure increased to legitimize and standardize scientific methods and procedures. Scientists started to claim a certain status as professionals which led to citizen science being more and more displaced as an (equal) actor from the scientific community. Nonetheless, data collection enabled by citizen engagement has retained a certain prominence in the natural sciences (Bonney et al. 2016; Resnik et al. 2015), with sometimes thousands of citizens participating in the observation of birds or insects, collecting and classifying data.

It is only since the end of the 20th century that citizen science has been increasingly associated with a programmatic call for opening up science in the European and North American context. This can be seen as an after-effect of the *Mode 2* science debate coined in the 1980s (Gibbons et al. 1994), which describes a change in the organization and epistemology of scientific knowledge production. The protagonists of the Mode 2 debate call for the participation of social groups outside science in knowledge-producing processes, in addition to a stronger reference to application. The Mode 2 debate has primarily shaped the emergence of transdisciplinary sustainability research, but is often used to justify the relevance of citizen science (Pettibone et al. 2017). Another related but different term is *Science 2.0*, which highlights new possibilities of communication and knowledge production due to the emergence and diffusion of digital tools and media (Bücheler and Sieg 2011). Science 2.0 emphasizes not only the expanded possibilities for science communication, for example through open access, but also the opportunity for increasing interactivity between research and society. This relates to initiatives in the context of *Open Knowledge* and *Open Education* which advocate for open data and open source infrastructures, not only to increase the level of transparency and reproducibility of science, but also to build the foundation for civic engagement with technologies and facilitate bottom-up technology development (Voigt 2021).

Beyond this particular development of citizen science in European and North American science systems, a separate line of participatory procedures which broadly fall into the category of citizen science has developed among others in postcolonial countries and regions. The focus here is often on the exploration and visualization of indigenous knowledge, which not only facilitates the collection of a type of data that is more closely oriented to the lifeworld of local populations (e.g. Eicken 2010; Snively and Corsiglia 2001), but also follows emancipatory objectives. These attempts can be associated with the plea for a *decolonization* of science that goes beyond (colonial) hierarchies in knowledge production and ownership (Bhawra 2022; Mistry and Berardi 2016).

The background is thus complex and citizen science is a colorful and broad landscape of phenomena rather than a clearly definable methodological practice. A variety of typologies describe different intensities of participation along the research process (see Haklay 2013; Riesch and Potter 2013; Wiggins and Crowsten 2011). Haklay (2013), for example, distinguishes between *Crowdsourcing* (mainly virtual participation and data collection), *Distributed Intelligence* (cooperation in different phases of the research process), *Participatory Science* (cooperation in all phases of research), and *Extreme Citizen Science* (research mainly led by citizens or non-academics). Wiggins and Crowston (2011) describe education as a particular form of citizen science which represents a vast variety of different types of projects, where citizen science tools are used in courses at schools, universities, museums, and other educational institutions to enhance learning and science communication. But even though most typologies try to be as exhaustive as possible, they do not cover all cases of citizen science, particularly hybrid forms, like the integration of art projects and participatory action beyond disciplinary, sectoral or national borders (see Filgueira Risso and Greco 2020).

Debate and criticism

The increased attention towards citizen science can be connected to different transitions and innovation impulses in science systems. Since the beginning of the 21st century, the debate about the social relevance of science has gained momentum and traditional hegemonies of knowledge production are questioned (Böschen 2019). In this context, citizen science is supposed to overcome looming crises of legitimacy, such as declining trust in scientific evidence (Saltelli and Funtowicz 2017) and the call for citizen-oriented or citizen-involving research becomes part of a plea for greater engagement of science in societal transformation (Schneidewind et al. 2016). Modern science can also see citizen science as an opportunity to improve science communication, where citizens are more than an audience consuming research results (Bonney et al. 2016).

While citizen science has gained some popularity in science and science policy, a main challenge to a systematic overview of the debate is the increasing confusion of formats, goals, and actors in the growing field (see Strasser et al. 2019). Authors from different disciplines have repeatedly questioned the trend to gather many different approaches under the umbrella of citizen science (Eitzel et al. 2017; Haklay et al. 2021), since it obscures not only the differences but also the problems that can be associated with citizen science. Three critical issues in the rise of citizen science can be discerned.

The first issue relates to the *quality and credibility* of citizen science practices. On the one hand, the increasing availability of funding and the inclusion of citizen science in political strategies such as the European Green Deal (European Commission 2020) are revitalizing the field and giving it necessary attention. On the other hand, these measures can also lead to citizen science degenerating more and more into a label that is used primarily because the term is currently in vogue and not because it is scientifically or socially necessary. There is a certain risk that citizen science serves as a legitimacy provider instead of an "honest" attempt at co-creating socially relevant knowledge. This is particularly delicate in the context of funded citizen science research between the so-called Global North and Global South. Since citizen science originates from a Western European–North American context, terminologies and procedures are used that reproduce a Western image of science and implicitly reproduce hierarchies and power relations between those who offer participation and those who participate. It might lead to situations where the wrong questions are posed (Vela et al. 2021) or stereotypes arise about what "indigenous" means (Eitzel et al. 2017). Particularly in projects that aim for "eye-level" research beyond hierarchies of power in knowledge production, academic scientists might still superimpose their terminologies – and with them their paradigms and values – onto the research process and reproduce the (colonial) hegemonies they intend to overcome (Vela et al. 2017).

This related to a second critical issue which considers the claim of the *promotion of democratization and participation*. Even if citizen science has the potential to break down stereotypes about "science in the ivory tower", it seems questionable whether the projects, which are mostly tied to short-term funding and whose funding is also decided by scientific and political elites, can at all overcome current hegemonies of knowledge production and injustices in the education and science system (Strasser et al. 2019). The inclusion of citizens in scientific knowledge production is not just a matter of establishing channels of communication. Citizen science requires conflict management and, for academic researchers, always means relinquishing some responsibility and control over the research process (Weng 2015). The role of expert is not reserved for academic partners. Rather, different forms of expertise need to be defined (Mistry and Berardi 2016). Engaging in citizen science requires training of representatives from society in scientific methods which transforms them into some sort of experts. The crucial question then is: Is this expertise accepted by the "professionals" or not? Furthermore, questions of intellectual property and possible conflicts of interest need to be clarified (Resnik et al. 2015). Since citizens are often expected to volunteer for research, the question of potential exploitation arises (Riesch and Potter 2013). At the same time, there is the risk of delegitimizing professional academic work and an "Uberizing" of research (Strasser et al. 2019, 67).

The third critical issue relates to the *work and roles of academic scientists* and the contexts in which they are working. Notwithstanding the mentioned opportunities, motivating citizens to engage in research collaboration is a double-edged sword. Generating and sustaining interest and motivation over an extended period of time requires target-group or even person-specific approaches which consume much time in a situation where time is rare (Aristeidou et al. 2017). Co-designing research projects requires a high degree of reflexivity and the freedom (and time or resources) to adjust methods, strategies, and forms of communication (see Figure 1). Conducting citizen science is also a risk for academic research, because it is by no means clear whether the hoped-for effects will occur at all, despite greater effort, and whether the results will meet scientific standards (Riesch and Potter 2013). Furthermore, doing citizen science is not only a question of forms and methods, but also of developing relationships. Institutionally integrated, professional scientists need to apply communication skills and to develop forms of interaction – mostly without relevant training – in order to address, activate, and continuously interact with citizens. Intense forms of collaboration and co-design require high reflexivity from each person involved, particularly when developing research questions and interpreting data (see Figure 1).

Citizen science – if designed as an action- and transformation-oriented process – is an invitation to academic scientists to leave self-referential communication contexts, make their knowledge more accessible and debatable, and increase the chance that this knowledge becomes socially relevant (Stilgoe 2009; Wildschut

Figure 1. The two-stage reflection process in a citizen science collaboration (illustration: Frank Becker)

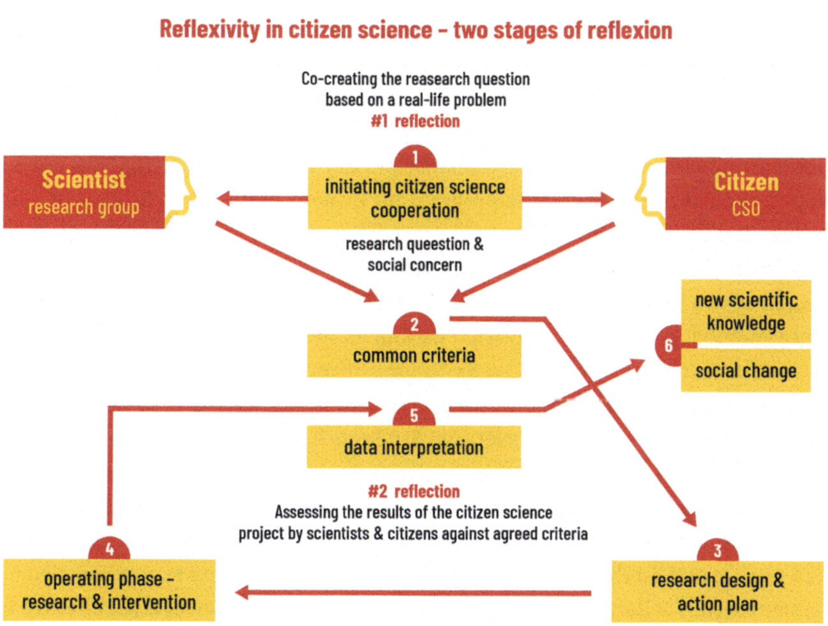

2017). This openness also entails taking the notion of the citizen (*citoyen*) in citizen science seriously, and recognizing their interest in participating in societal development and shaping their lifeworld contexts (see the Irish Citizens' Assembly as an example for framing this notion of citizen in science: The Citizens Assembly 2022).

Current forms of implementation in higher education

The (often confusingly large) variety of citizen science approaches has some advantages since it offers many opportunities of implementation in higher education. A considerable challenge for a discussion of current forms of implementation is that published literature on empirical examples is rare. A recent review of by Vance-Chalcraft et al. (2022) revealed that most published literature stems from the United States, which could only represent part of the picture. The review shows that the majority of applications of citizen science reported in the literature is on topics like ecology and environment, followed by health and medicine. At lower levels of education (introductory courses), the participation was mainly concentrated on collecting data, whereas in higher levels of education it also stretched to the analysis of data and the development and test of hypotheses. The learning objectives for using citizen science in higher education were manifold, but mostly teachers wanted to foster the students' excitement about science, bring them into contact with authentic research, and demonstrate the relevance of science to society.

Citizen science is often applied in such a way that students act as citizen scientists (e.g. Esmaeilian et al. 2018; Heigl and Zaller 2014; Oberhauser and LeBuhn 2012), but there are also examples where students applied citizen science to co-produce knowledge about their study topic (Britton and Tippins 2015). Some universities try to strengthen their profile in applied sustainability research by integrating citizen science and transdisciplinary methods like living labs explicitly in their program (e.g. MSc program on Ecology and Citizen Science at University College London). While the courses follow a disciplinary focus, they often integrate different disciplinary perspectives. The advantages of citizen science in teaching can be summarized as follows: (1) By learning and applying citizen science methods, students' motivation as well as their interest in socially relevant research can be strengthened. (2) Skills for self-reflection as a researcher and for reflecting on the role of science in society are increased. (3) Students practice communicating research to society. (4) Students learn how to engage themselves and communities affected by their research in action research. In the following we present three examples which illustrate different approaches and objectives.

An example of the role of students as citizen scientists is the paper by Esmaeilian et al. (2018). In the course described, engineering students were motivated to systematically collect examples of product designs in their own everyday lives that wear out quickly, are not very functional, or are unwieldy, and to describe them via text and images. The collections were jointly evaluated and categorized. The goal was for students to use their own observations to learn principles of sustainable design as well as sustainable production methods.

A participatory approach in applying citizen science in higher education in the Philippines was reported by Mendoza et al. (2022). The transdisciplinary endeavor, bringing together local fisherman with teachers, students, and the Bureau of Fisheries and Aquatic Resources, aimed at a better understanding of changes in the fish habitats and environmental quality at the Laguna lakes of the Philippines. Given the cooperative attitude of local collaborators, the study shows that local resource users, including teacher and students along with fishers, can be research guides in exploring further ecosystems that experience major environmental changes.

An action-oriented approach (Haklay 2013) is the UNICATA university with and for waste pickers (Gutberlet et al. 2021), which was created based on citizen science research in São Paulo, Brazil. In January 2022, a small group of academics and citizen scientists (waste pickers, NGO members) in São Paulo, Brazil resumed the idea of creating a university with and for waste pickers (UNICATA), inspired by Paulo Freire's theoretical and praxis of popular education pedagogy and peer learning (Freire 2009). The approach underlines the crucial importance for a learner-experience to influence the design of the teaching, and address a whole system of change, thinking relationally about the social, economic, and environmental aspects regarding the work of waste pickers and their livelihoods.

A form of extreme citizen science can be found in participatory ethnobotany, which can in particular be employed in teaching and education in the context of conservation. Two exemplary projects were conducted by a Brazilian team composed of members of academia and the community, acting as ethnobotanical researchers in the Atlantic Rain Forest. In such projects, members of the community are trained from both botanical and anthropological perspectives, so that they can conduct the ethnobotanical survey themselves, with technical support from the academics. One of them started in 2015 and has been carried out in two Quilombola communities (Rodrigues et al. 2020).

Nevertheless, citizen science in academic teaching encounters similar challenges as citizen science in general: contrasting conceptions of science, conflicts over different ideas of the goals of research, and the ever-present question of responsibility for the process and product of collaborative research. To facilitate experimentation with such formats at universities (or even schools), the corresponding courses should not be subject to the pressures of time, publication, and success that are common in research. Rather, they should be recognized as learning experiments that require time for negotiation and adjustment of design, and in which mistakes reveal opportunities for learning and need not be avoided at all costs through excessive standardization.

References

Amauchi, Juliana F., Maeva Gauthier, Abdolzaher Ghezeljeh, Leandro L. L. Giatti, Katlyn Keats, Dare Sholanke, Danae Zachari, and Jutta Gutberlet. 2022. The power of community-based participatory research: Ethical and effective ways of researching. *Community Development* 53 (1): 3–20.

Aristeidou, Maria, Eileen Scanlon, and Mike Sharples. 2017. Profiles of engagement in online communities of citizen science participation. *Computers in Human Behavior* 74: 246–56.

Bhawra, Jessica. 2022. Decolonizing digital citizen science: Applying the bridge framework for climate change preparedness and adaptation. *Societies* 12 (2): 71.

Bonney, Rick, Tina B. Phillips, Heidi L. Ballard, and Jody W. Enck. 2016. Can citizen science enhance public understanding of science? *Public Understanding of Science* 25 (1): 2–16.

Böschen, Stefan. 2019. Processing issues in science policy: Emerging epistemic regimes. In *Handbook on science and public policy,* eds. Dagmar Simon, Stefan Kuhlmann, Julia Stamm, and Weert Canzler, 317–35. Cheltenham: Edward Elgar.

Brenna, Brita. 2021. Clergymen abiding in the fields: The making of the naturalist observer in eighteenth-century Norwegian natural history. *Science in Context* 24 (2), 143–66.

Britton, Stacey A., and Deborah J. Tippins. 2015. Teaching with citizen science – It's more than just putting out fires! In *EcoJustice, citizen science and youth activism: Environmental discourses in science education,* eds. Michael Müller and Deborah J. Tippins, 207–22. Cham: Springer.

Bücheler, Thierry, and Jan Henrik Sieg. 2011. Understanding science 2.0: Crowdsourcing and open innovation in the scientific method. *Procedia Computer Science* 7: 327–29.

Chuine, Isabelle, Pascal Yiou, Nicolas Viovy, Bernard Seguin, Valérie Daux, and Emmanuel Le Roy Ladurie. 2004. Historical phenology: Grape ripening as a past climate indicator. *Nature* 432 (7015): 289–90.

Conrad, Cathy C., and Krista G. Hilchey. 2011. A review of citizen science and community-based environmental monitoring: Issues and opportunities. *Environmental Monitoring and Assessment* 176: 273–91.

Eicken, Hajo. 2010. Indigenous knowledge and sea ice science: What can we learn from indigenous ice users? In *SIKU: Knowing our ice,* eds. Igor Krupnik, Claudio Aporta, Shari Gearheard, Gita J. Laidler, and Lene Kielsen Holm, 357–76. Dordrecht: Springer.

Eitzel, Melissa Viola, Jessica L. Cappadonna, Chris Santos-Lang, Ruth E. Duerr, Arika Virapongse, Sarah E. West, Christopher C. Kyba, Anne Bowser, Caren B. Cooper, Andrea Sforzi, Anya N. Metcalfe, Edward S. Harris, Martin Thiel,

Mordechai Haklay, Lesandro Ponciano, Joseph Roche, Luigi Ceccaroni, Fraser M. Shilling, Daniel Dörler, Florian Heigl, Tim Kiessling, Brittany Y. Davis, and Qijun Jiang. 2017. Citizen science terminology matters: Exploring key terms. *Citizen Science: Theory and Practice* 2 (1): 1.

Esmaeilian, Behzad, Michael Rust, Praveen K. Gopalakrishnan, and Sara Behdad. 2018. Use of citizen science to improve student experience in engineering design, manufacturing and sustainability education. *Procedia Manufacturing* 26: 1361–68.

European Commission. 2020. *Enabling citizens to act on climate change, for sustainable development and environmental protection through education, citizen science, observation initiatives, and civic engagement.* Available from https://ec.europa.eu/info/funding-tenders/opportunities/portal/screen/opportunities/topic-details/lc-gd-10-3-2020.

Filgueira Risso, Ezequiel, and Ivan Greco. 2020. Cultural iceberg: The challenge of building comprehensive knowledge for social and environmental interventions in vulnerable communities in Buenos Aires. In *Co-creating actionable science: Reflections from the Global North and South*, eds. L. Gallardo Fernández, Fred Saunders, and Tatiana Sokolova, 97–120. Cambridge: Scholars Publishing.

Freire, Paolo. 2009. From pedagogy of the oppressed. *Race/Ethnicity: Multidisciplinary Global Contexts* 2 (2): 163–74.

Gibbons, Michael, Camille Limoges, Helga Nowotny, Simon Schwarzman, Peter Scott, and Martin Trow. 1994. *The new production of knowledge: The dynamics of science and research in contemporary societies.* Los Angeles: Sage.

Gutberlet, Jutta, Santiago Sorroche, Angela Martins Baeder, Patrick Zapata, and Maria José Zapata Campos. 2021. Waste pickers and their practices of insurgency and environmental stewardship. *Journal of Environment & Development* 30 (4): 369–94.

Haklay, Mordechai, Daniel Dörler, Florian Heigl, Marina Manzoni, Susanne Hecker, and Katrin Vohland. 2021. What is citizen science? The challenges of definition. In *The science of citizen science*, eds. Katrin Vohland, Anne Land-Zandstra, Luigi Ceccaroni, Rob Lemmens, Josef Perelló, Marisa Ponti, Roeland Samson, and Katherin Wagenknecht, 13–33. Cham: Springer.

Haklay, Muki. 2013. Citizen science and volunteered geographic information: Overview and typology of participation. In *Crowdsourcing geographic knowledge*, eds. Daniel Sui, Sarah Elwood, and Michael Goodchild, 105–22. Dordrecht: Springer.

Heigl, Florian, and Johann G. Zaller. 2014. Using a citizen science approach in higher education: A case study reporting roadkills in Austria. *Human Computation* 1 (2): 165–75.

Jaeger-Erben, Melanie. 2021. Citizen Science. In *Handbuch Transdisziplinäre Didaktik*, eds. Tobias Schmohl and Thorsten Philipp, 45–55. Bielefeld: transcript.

Kerson, Raymond. 1989. Lab for the environment. *Technology Review* 92 (1): 11–12.

Mahr, Dana, and Sascha Dickel. 2019. Citizen science beyond invited participation: Nineteenth century amateur naturalists, epistemic autonomy, and big data approaches avant la lettre. *History and Philosophy of the Life Sciences* 41 (4): 1–19.

Mendoza, Jimlea N., Baiba Prūse, Giulia Mattalia, Sophia Kochalski, Aimee Ciriaco, and Renata Sõukand. 2022. Fishers' perspectives: The drivers behind the decline in fish catch in Laguna Lake, Philippines. *Maritime Studies* 21 (4): 569–85.

Miller-Rushing, Abraham J., Richard B. Primack, Rick Bonney, and Emma Albee. 2020. The history of citizen science in ecology and conservation. In *Handbook of citizen science in ecology and conservation*, eds. Christopher A. Lepczyk, Owen D. Boyle, and Timothy L. V. Vargo, 17–24. Oakland, CA: University of California Press.

Mistry, Jayalaxshmi, and Andrea Berardi. 2016. Environment: Bridging indigenous and scientific knowledge. *Science* 352 (6291): 1274–75.

Oberhauser, Karen, and Gretchen LeBuhn. 2012. Insects and plants: Engaging undergraduates in authentic research through citizen science. *Frontiers in Ecology and the Environment* 10 (6): 318–20.

Porter, Roy. 1978. Gentlemen and geology: The emergence of a scientific career, 1660–1920. *Historical Journal* 21 (4): 809–36.

Pettibone, Lisa, Katrin Vohland, and David Ziegler. 2017. Understanding the (inter)disciplinary and institutional diversity of citizen science: A survey of current practice in Germany and Austria. *PLoS One* 12 (6): e0178778.

Resnik, David B., Kevin C. Elliott, and Aubrey K. Miller. 2015. A framework for addressing ethical issues in citizen science. *Environmental Science & Policy* 54: 475–81.

Riesch, Hauke, and Clive Potter. 2013. Citizen science as seen by scientists: Methodological, epistemological and ethical dimensions. *Public Understanding of Science* 23 (1): 107–20.

Rodrigues, Eliana, Fernando Cassas, Bruno E. Conde, Crenilda Da Cruz, Eduardo H. Barretto, Ginacil Dos Santos, Glyn M. Figueira, Luiz F. Passero, Maria A. Dos Santos, Maria A. Gomes, Priscila Matta, Priscila Yazbek, Ricardo J. Garcia, Silvestre Braga, Sonia Aragaki, Sumiko Honda, Thamara Sauini, Viviane S. Da Fonseca-Kruel, and Tamara Ticktin. 2020. Participatory ethnobotany and conservation: A methodological case study conducted with Quilombola communities in Brazil's Atlantic Forest. *Journal of Ethnobiology and Ethnomedicine* 16 (1): 2.

Saltelli, Andrea, and Silvio Funtowicz. 2017. What is science's crisis really about? *Futures* 91: 5–11.

Schneidewind, Uwe, Mandy Singer-Brodowski, Karoline Augenstein, and Franziska Stelzer. 2016. *Pledge for a transformative science: A conceptual framework, Wuppertal Papers* 191. Wuppertal: Wuppertal Institute.

Snively, Gloria and John Corsiglia. 2001. Discovering indigenous science: Implications for science education. *Scientific Education* 85: 6–34.

Stilgoe, Jack. 2009. *Citizen scientists: Reconnecting science with civil society.* London: Demos.

Strasser, Bruno, Jérôme Baudry, Dana Mahr, Gabriela Sanchez, and Elise Tancoigne. 2019. Citizen science? Rethinking science and public participation. *Science & Technology Studies* 32: 52–76.

The Citizens Assembly. 2022. About the Dublin Citizens' Assembly. Available from https://www.citizensassembly.ie/en/dublin-assembly/about.

Vance-Chalcraft, Heather D., Allen H Hurlbert, Jennifer Nesbitt Styrsky, Terry A. Gates, Gillian Bowser, Colleen B. Hitchcock, Michelle Anne Reyes, and Caren B. Cooper. 2022. Citizen science in postsecondary education: Current practices and knowledge gaps. *BioScience* 72 (3): 276–88.

Vela, Luis A., Enrico Bocciolesi, Giovanna Lombardi, and Robin M. Urquhart. 2017. The social function of citizen science. In *Analyzing the role of citizen science in modern research*, eds. Murray Jennex, Luigi Ceccaroni, and Jaume Piera, 89–116. Hershey, PA: Information Science Reference.

Vetter, Jerome 2011. Introduction: Lay participation in the history of scientific observation. *Science in Context* 24 (2): 127–41.

Voigt, Maximilian. 2021. Open Education and open source for sustainable economic activity. Ökologisches *Wirtschaften* 34 (1): 25–27.

Weng, Yen Chu. 2015. Contrasting visions of science in ecological restoration: Expert–lay dynamics between professional practitioners and volunteers. *Geoforum* 65: 134–45.

Wiggins, Andrea, and Kevin Crowston. 2011. *From conservation to crowdsourcing: A typology of citizen science.* Available from https://citsci.syr.edu/content/conservation-crowdsourcing-typology-citizen-science.

Wildschut, Diana. 2017. The need for citizen science in the transition to a sustainable peer-to-peer-society. *Futures* 91: 46–52.

Cooperative Education

Wendy Coones, Thies Johannsen, and Thorsten Philipp (authors contributed equally)

Definition

Cooperative education programs are a learning practice inspired by the dual education system of the German-speaking countries. They seek to combine theory and practice, academic and professional knowledge, reflection and application. A consensus about a precise definition of cooperative education is still lacking, and many countries and higher education institutions implement these programs individually (Eames and Coll 2010, 182). A study program can be defined as 'cooperative' (lat. *co-operare* 'to work together, combine, unite') if the university and a practice partner (e.g. company, business, or social institution) are structurally intertwined, i.e. content and program organization are aligned so that whenever students apply knowledge practically the same knowledge is then reflected theoretically within the university. This requires a conjoined coordination policy between practice partners and university personnel.

In most cases, practical elements comprise from one-third to one-half of a cooperative education program. Depending on the design of a program, practical elements may be integrated at various times: before, during (most commonly), or even after the theory-oriented learning phases at university. The involvement of the practice partner is not necessarily limited to the practical elements but may also include content, methodology, and assessment. Therefore, students are continuously challenged to combine academic expert knowledge with hermeneutics, situational understanding, and reflections on the contradictions between theory and practice. This is also true for study programs which are not part of the humanities, such as engineering or health professions. Cooperative programs require students to meet the academic standards of university education while simultaneously fulfilling the expectations of an economic business or social institution. Their academic training has to satisfy the demands of two addressees.

International higher education uses a variety of descriptors to identify study programs that combine theoretical coursework with relevant practical experience. All these approaches have in common that they demonstrate intersections with cooperative education. Among these are experiential learning (Kolb 2015), practice-based professional learning (Kennedy et al. 2015), work-integrated education (Jackson 2015), work-based learning (Cunningham and Dawes 2016), Youth Participatory Action Research (Anyon et al. 2018), professional development (Webster-Wright 2009), extra-occupational doctoratea (Kaiser 2020), and degrees combining studies with work (Eames and Coll 2010, 185–88). Options between these forms can provide different access points, ranging from vocational education offered at upper-level secondary schools, employer-centered placements with curricular collaboration between universities and industry, to academically comprehensive programs integrating practice throughout. Depending on the regulations and preferences of a country's educational system, cooperative education can carve out a niche to thrive. It can offer opportunities to those students best suited to benefit from a balance of learning outcomes.

While cooperative learning is a pedagogical practice that promotes learning in groups (as opposed to individualized and competitive learning), cooperative education and transdisciplinary learning refer to a specific understanding of learning that is grounded in experience and reflection on one's doing (Gillies 2016, 39; Kolb 2015, xviii). The latter, therefore, is utilized in work-related learning (Beard and Wilson 2013, 18).

Background

Cooperative education is an attempt to offer a study degree at the interface between academic and vocational training. The concept and origins of cooperative education derive in many countries from the increasing demand for a specialized workforce, the expanding academization of professional activities, and growing criticism of the low practical relevance of academic curricula. In view of the various structures of these programs, work-based learning is increasingly considered to encompass a meaning broader than cooperative education, as it refers to a spectrum of opportunities both inside and outside the workplace. However, it also becomes so broad that didactics and learning structures are difficult to define or examine. Cooperative education that integrates co-curricular elements provides a valuable feedback loop from the world of work back into higher education.

Research and literature on cooperative education consist strongly of region-related empirical analyses, case studies, and position papers (Graf et al. 2014; Jacobs and Renandya 2019; Newhook 2016; Schiller and Leisyte 2020; Tanaka and Zegwaard 2019; Zegwaard et al. 2022). Empirical research on cooperative education focuses on student populations' previous education and socio-demographic characteristics, selection and access criteria, and expectations and motives for career choices (Hemkes et al. 2019). Analytical reflections on the interplay between scientific claims, the logic of the labor markets, the postulate of the practical turn (Stern 2003), the material culture of knowledge production, and the liberal ideal of integral education are still rudimentary (e.g. Davidson 2021).

Debate and criticism

A central challenge of cooperative education programs arises from the conflicting interests of the two learning venues: While practice partners, on the one hand, are usually focused on profitability, on fast, cost-effective, and practice-oriented training, and thus on the knowledge that can be utilized in the short term, universities, on the other hand, are focused on sustainable academic education and on developing individual, holistic, and long-term knowledge resources. This leads to a multitude of further problems. On an institutional and organizational level, divergent interests and opposing positions raise questions like: How do (implicit) ideas of practice partners, professional interest groups, chambers of commerce, and ministries impact the curricula? What influence can practice partners exert on financing and quotas for programs? On an individual or personal level, dependencies, power structures, expectations, and involving supervisors from the practice partner come into focus and lead to questions such as: To what extent is grading affected if employees of practical partners are involved in assessments, e.g. a thesis or a dissertation? What measures are taken to address conflicts of interest? Comparing public and private universities mainly gives rise to a considerable need for analysis and further research. Transdisciplinarity must be a clear part of the curricula for universities and their practice partners to work together.

Four central questions determine the further development of cooperative education. (1) In addition to financial resources, time is a scarce resource that is indispensable for successful implementation. As higher education institutions coordinate a large number of different practice partners and, in addition, are responsible for quality control, workload poses a particular challenge. Additionally, in curricular terms, transdisciplinary learning depends on appropriate reflection phases to permit students to process their experiences at the interface between science and society. In most study programs, however, transdisciplinarity is more of a phenomenon of application than reflection. Time for a systematic evaluation of transdisciplinary experiences is lacking. (2) Cooperation between universities and companies requires extraordinary efforts in coordinating the plurality of involved actors, managing the necessary participation forums, and discerning the knowledge resources needed to solve a given problem. (3) From a student's perspective, it is imperative to assess whether the practical relevance, the high probability of employment, and the salary during the program outweigh the increased demand for self-organization and efficiency. Cooperative education will remain attractive to employers if employability is its primary goal, but it will deter students from developing a learning biography that reflects their particular interests. Two of the qualification goals of a study program are in question: the ability to engage in social activities and personal development. Nevertheless, cooperative education can hardly meet the objective of enabling contextualized thinking or interdisciplinary perspectives. Cooperative education requires commitment and motivation of all participants to test, reflect, and integrate transdisciplinary techniques. (4) In terms of expansion, cooperative education programs thrive in application-driven STEM disciplines, health, social, and educational professions but seem less attractive to the humanities, arts, and social sciences. Most of these disciplines are oriented more towards theory, but also show potential for integrating practical components. Examples include empirical social research, museum studies, and teacher education. Practices of transdisciplinary learning in situ could be a promising teaching innovation when expanding into application-driven fields because they tend to be solution-focused. It seems likely that cooperative education will grow to meet the needs of regional business ecosystems and large-scale political programs like the Sustainable Development Goals (Mazzucato 2018).

From the perspective of transdisciplinary learning, cooperative education is an ambivalent issue. The concept and claim of cooperative education parallel transdisciplinarity as a collaborative, society-oriented learning experience involving experts and practitioners from different disciplines. In either case, the aim is to bridge the gap between academia and a critical public, enhance participatory processes, introduce case-specific practical expertise into abstract deliberation, promote fluidity between communities, and thereby overcome boundaries between academia and society. Cooperative education can be a means to generate knowledge inclusively, make decisions in a participatory way, and foster an understanding of the differences between everyday, practical, and scientific knowledge resources. As in all transdisciplinary practices, cooperative education can thereby contribute to encounters with or avoidance of crises in science legitimacy. However, essential hindrances arise from limited opportunities for reflection, critique, and contextualization of disciplinarity in cooperative study programs; in many cases, a narrow view of disciplines in their value for the labor market aggravates contextual awareness, openness, and perception of the plurality of scientific questions within, beyond, and across disciplines (Eames and Coll 2010, 184). The potential inadequacy or absence of reflection on disciplinarity highlights the challenge for universities to integrate transdisciplinary reflection in their curricula – instead of leaving it to the students.

Cooperative education programs offer a symbiosis between vocational and academic education. At the same time, they remain a compromise between the liberal ideal of integral education and the capitalist logic of labor exploitation (see Milley 2016). In genesis and essence, they are a product of Western education concepts, and their applicability to differing cultural systems and knowledge traditions is questionable. Although there are established practices, e.g. in Latin America (Zamora-Torres and Thalheim 2020) and Russia (Lešukov et al. 2018), and initiatives to introduce cooperative education in other world regions, e.g. as part of a "new engineering education" in China (Shen et al. 2020, 890–91), research is conducted almost exclusively in English and German. This implies that local adaptations strongly rely on Western publications. It remains to be seen whether local forms of cooperative education will prevail independently of the hegemonic discourse and address local needs as well as develop innovative approaches. In general, academic institutions should be aware of their risk of being reduced to factories to produce a trained workforce. Cooperative education, then, contributes to a general tendency of realigning the education sector along with market principles and economic purpose rationality. Even though cooperative education grew out of the world of work and gave students a practical connection to it, the implicit criticism aims to give students chances to develop deeper critical and holistic thinking.

From a student's perspective, critical assessment is also essential when deciding for a cooperative education program, as it may include different forms and degrees of practice. Not all programs advertised as cooperative education or under one of its many synonyms offer genuine cooperative education, as described above. For example, a mandatory internship would not qualify as cooperative education. While there is no binding definition, we suggest two minimum criteria that must be met: (1) An academic study program that includes practical elements encompassing one-third to one-half and awards credit points for them. (2) student, practice partner, and university have all signed a contract agreeing on obligations and funding.

Undeniably, cooperative education's innovative achievement is to synthesize the dichotomy of the education system of Western industrialized countries and to increase social mobility: In addition to the binary options of higher education and vocational training, between which it was almost impossible to mediate in the past, cooperative education provides a third – hybrid and integrating – alternative. It strengthens the education system by meeting the needs of different learning biographies, including educational and vocational profiles. It thereby anticipates the plurality of knowledge resources, participation forums, involved actors, and educational biographies that form the core of transdisciplinary learning. Moving beyond criticism and encouraging expansion into new learning approaches that include learners of all backgrounds, cooperative education programs clearly hold the potential to be holistic and reflective or to address specific societal problems. By combining practical and academic elements, transdisciplinary learning is possible, and is a natural part of application-oriented education.

Current forms of didactic implementation

The number of cooperative education programs has grown steadily since their introduction. As a result of increasing state funding, programs have become more diversified. Specific transdisciplinary study programs have not yet been established in cooperative education. There is, however, potential in methodological and didactic reflections on practice.

A multifaceted and long-institutionalized culture of cooperative education has developed mainly in German-speaking countries, where the demand for employment security, a qualified workforce, and the advanced integration of plural forms of learning during the 1970s were significant incentives (Faßhauer and Severing 2016; Graf 2015). Several European countries, such as France, Spain, and the United Kingdom, offer comparable degree programs (Schmees et al. 2019, 8). In the United States, cooperative education was introduced as early as 1906 and has increased in demand since the 1960s. In Canada, work-integrated learning expanded in the 1970s and is institutionalized today in many universities (Angerilli et al. 2005; Bowen and Drysdale 2017; Milley 2016). In Australia, one-third of the university students, on average, undertake an experience in work-integrated learning (Zegwaard et al. 2022, 1).

In this context, it is possible to distinguish not only individual countries but also educational systems, which can be divided into three ideal-typical systems based on their "institutional distance", i.e. their respective institutional frameworks (Graf et al. 2014). (1) Germany is a federal system with standardized education and vocational training systems involving chambers of commerce and social partners. (2) The US is also a federal system, but hardly standardized. Education and training take place in market-oriented environments. (3) France, on the other hand, is comparatively centralized and school-based. However, vocational education is not highly valued in secondary education. Therefore, vocational orientation has assumed a more prominent role in higher education since the 1960s (Powell et al. 2012, 410). Regardless of the institutional framework, cooperative education is considered a means to link the world of work to academic training and education (see Maassen et al. 2019).

Despite these differences, the general development shows that learning outcomes that are important to the learner, the university, the employer, and other stakeholders are possible through experiential learning with academic support and critical thinking in a higher education setting. Costley (2015) makes a strong case for applying a transdisciplinary approach to work-integrated learning in the design, facilitation, and evaluation of work-based initiatives and programs. Transdisciplinary learning in a cooperative education setting can be facilitated as the field defines its body of knowledge, researches distinctive phenomena such as learning through experience, and expands the expertise of its practitioners. The enhancement of lifelong learning skills, contributing to people's working lives, and the economy can all be linked to cooperative education. Thus, it can position itself to meet students' needs for experiential learning and society's needs for transdisciplinary thinkers.

Proper planning and curricula development in cooperative education programs can integrate transdisciplinary learning with some of the following aspects: (1) Clear learning agreements between students and placements that reflect authentic and world-relevant learning outcomes. (2) Access to superiors in reflective dialogue so students may shadow higher-level thinking. (3) Project work that requires concentration on socially relevant topics rather than commercial profit. (4) Mechanisms for personal reflection on learning in group or mentor–student settings. (5) Building program teams representing a variety of disciplines and roles. As transdisciplinary research and developing practice unfolds while society's challenges transform and emerge in a new form, so should teaching and learning evolve parallel to these real-world processes.

A university can transform the student work experience into potential transdisciplinary learning when institutions and employers enable engagement with the complex and emergent real-life challenges naturally present in genuine work experience. Cooperative education will continue to move towards more transdisciplinary learning as Stern's practical turn intertwines with the growing need for industry and education to tackle significant global challenges. If binding quality standards succeed in combining academic knowledge production with practical expertise and theoretical approaches by including experts from civil society, industry, politics, public administration, the cultural sector, etc., then cooperative education can unleash its transdisciplinary potential (Carayannis and Campbell 2009; Gibbons et al. 1994). Including educators from all parts of society in higher education and guiding students in their reflections and analyses with experience from practice will contribute to bridging the gap between academia and society. Thus, cooperative education may initiate a process of universities becoming more open and socially responsible, thereby contributing to an overdue change of attitude in higher education.

References

Angerilli, Nello, Andrea Giles, Karen McCargar, Dave O'Leary, Ken Porteous, Kate Ross, Rosemarie Sampson, and Joanne Thomas. 2005. *Co-operative education manual: A guide to planning and implementing co-operative education programs in post-secondary institutions*. Toronto: Canadian Association for Co-operative Learning.

Anyon, Yolanda, Kimberly Bender, Heather Kennedy, and Jonah Dechants. 2018. A systematic review of youth participatory action research (YPAR) in the United States. *Health Education & Behavior* (45): 865–78.

Beard, Colin, and John P. Wilson. 2013. *Experiential learning: A handbook for education, training and coaching*. London: Kogan Page.

Bowen, Tracey, and Maureen T. Drysdale, eds. 2017. *Work-integrated learning in the 21st century: Global perspectives on the future*. Bingley: Emerald Publishing.

Carayannis, Elias G., and David F. Campbell. 2009. "Mode 3" and "Quadruple Helix": Toward a 21st century fractal innovation ecosystem. *International Journal of Technology Management* 46 (3–4): 201–34.

Costley, Carol 2015. Educational knowledge in professional practice: A transdisciplinary approach. In *Transdisciplinary learning and professional practice*, ed. Paul Gibbs, 121–33. London: Springer.

Cunningham, Ian, and Graham Dawes. 2016. *The handbook of work based learning*. Abingdon: Routledge.

Davidson, Neil, ed. 2021. *Pioneering perspectives in cooperative learning: Theory, research, and classroom practice for diverse approaches to cooperative learning*. Abingdon: Routledge.

Eames, Chris, and Richard K. Coll. 2010. Cooperative education: Integrating classroom and workplace learning. In *Learning through practice*, ed. Stephen Billett, 180–96. Dordrecht: Springer.

Faßhauer, Uwe, and Eckart Severing, eds. 2016. *Verzahnung beruflicher und akademischer Bildung. Duale Studiengänge in Theorie und Praxis*. Bielefeld: Bertelsmann.

Gibbons, Michael, Camille Limoges, Helga Nowotny, Simon Schwartzman, Peter Scott, and Martin Trow. 1994. *The new production of knowledge: The dynamics of science and research in contemporary societies*. London: SAGE.

Gillies, Robyn M. 2016. Cooperative learning: Review of research and practice. *Australian Journal of Teacher Education* 41 (3): 39–54.

Graf, Lukas. 2015. Hybridisierung von Berufs- und Hochschulbildung in Deutschland, Österreich und der Schweiz. In *Differenzierung im Hochschulsystem. Nationale und internationale Entwicklungen und Herausforderungen*, eds. Ulf Banscherus, Ole Engel, Anne Mindt, Anna Spexard and Andrä Wolter, 163–76. Münster: Waxmann.

Graf, Lukas, Justin J. W. Powell, Johann Fortwengel, and Nadine Bernhard. 2014. *Duale Studiengänge im globalen Kontext: Internationalisierung in Deutschland und Transfer nach Brasilien, Frankreich, Katar, Mexiko und in die USA*. Bonn: DAAD.

Hemkes, Barbara, Karl Wilbers, Michael Heister. 2019. *Durchlässigkeit zwischen beruflicher und hochschulischer Bildung*. Bonn: BIBB.

Jackson, Denise. 2015. Employability skill development in work-integrated learning: Barriers and best practice. *Studies in Higher Education* 40 (2): 350–67.

Jacobs, George M., and Willy A. Renandya. 2019. *Student centered cooperative learning: Linking concepts in education to promote student learning*. Singapore: Springer.

Johannsen, Thies, and Thorsten Philipp. 2021. Duales Studium. In *Handbuch Transdisziplinäre Didaktik*, eds. Tobias Schmohl and Thorsten Philipp, 79–92. Bielefeld: transcript.

Kaiser, Dennis. 2020. Die Modelle der berufsbegleitenden Promotion. In *Berufsbegleitend promovieren in den Wirtschaftswissenschaften: Ein Leitfaden für Berufstätige*, 49–82. Berlin: Springer Gabler.

Kennedy, Monica, Stephen Billett, Silvia Gherardi, and Laurie Grealish, eds. 2015. *Practice-based learning in higher education*. Dordrecht: Springer.

Kolb, David A. 2015. *Experiential learning: Experience as the source of learning and development*. Upper Saddle River, NJ: Pearson Education.

Lešukov Oleg Valer'evič, Natal'ja Vital'evna Isaeva, and Lÿdmila Aleksandrovna Evstratova. 2018. *Proektnoe obučenie: praktiki vnedrenija v universitetah*. Izdatel'skij dom NIU VŠÈ.

Maassen, Peter, Zacharias Andreadakis, Magnus Gulbrandsen, and Bjørn Stensaker. 2019. *The place of universities in society*. Hamburg: Koerber.

Mazzucato, Mariana. 2018. Mission-oriented innovation policies: Challenges and opportunities. *Industrial and Corporate Change* 27 (5): 803–15.

Milley, Peter. 2016. Commercializing higher learning through the discourse of skills in university cooperative education. *Canadian Journal of Educational Administration and Policy* (180): 99–134.

Newhook, Rebecca. 2016. Are university co-operative education students safe? Perceptions of risk to students on work terms. *Canadian Journal of Higher Education* 46 (1): 121–37.

Powell, Justin J. W., Lukas Graf, Nadine Bernhard, Laurence Coutrot, and Annick Kieffer. 2012. The shifting relationship between vocational and higher education in France and Germany: Towards convergence? *European Journal of Education* 47 (3): 405–23.

Schiller, Benjamin, and Liudvika Leisyte. 2020. Study program innovation in the triple helix context: The case of cooperative study programs at a German university of applied sciences. *Triple Helix* 7 (2–3): 160–88.

Schmees, Johannes K., Tatyana Popkova, and Dietmar Frommberger. 2019. *Das Higher Apprenticeship in England*. Gütersloh: Bertelsmann-Stiftung.

Shen, Jinlu, Tuoyu Li, and Mingchuan Wu. 2020. The new engineering education in China. *Procedia Computer Science* (172): 886–95.

Stern, David G. 2003. The practical turn. In *The Blackwell guide to the philosophy of the social sciences*, eds. Paul A. Roth and Stephen Turner, 185–206. Malden, MA: Blackwell.

Tanaka, Yasushi. 2015. *The economics of cooperative education: A practitioners guide to the theoretical framework and empirical assessment of cooperative education*. Milton Park: Routledge.

Tanaka, Yasushi, and Karsten E. Zegwaard, eds. 2019. *Cooperative and work-integrated education in Asia: History, present and future issues.* Abingdon: Routledge.

Webster-Wright, Ann. 2009. Reframing professional development through understanding authentic professional learning. *Review of Educational Research* 79 (2): 702–39.

Zamora-Torres, América-Ivonne, and Luise Thalheim. 2020. El Modelo Mexicano de Formación Dual como modelo educativo en pro de la inserción laboral de los jóvenes en México. *Revista Iberoamericana de Educación Superior* 11 (31): 48–67.

Zegwaard, Karsten E., Ferns Sonia J., and Anna D. Rowe. 2022. Contemporary insights into the practice of work-integrated learning in Australia. In *Advances in research, theory and practice in work-integrated learning: Enhancing employability for a sustainable future,* eds. Sonia J. Ferns, Anna D. Rowe, and Karsten E. Zegwaard, 1–14. London: Taylor & Francis.

Critical Thinking

Philip Barth and Jonas Pfister

Definition

The term critical has its origin in the Ancient Greek word κρίνειν (krinein), which means "discerning, judging". Critical thinking is therefore to be understood as thinking that aims at reasonable judgment. There is consensus about the basic concept as careful reflective thinking (Allen et al. 2020; Hitchcock 2018). To be groundlessly skeptical about anything or to blindly follow a rule is not critical thinking.

The basic concept can be found in different formulations. In a consensus paper known as the Delphi Report, by a panel of 46 experts with different scientific backgrounds ranging from philosophy to physics, zoology, psychology, social sciences and economics, critical thinking is defined as "purposeful, self-regulatory judgment which results in interpretation, analysis, evaluation, and inference, as well as explanation of the evidential, conceptual, methodological, criteriological, or contextual considerations upon which that judgment is based" (Facione 1990, 3). Ennis (1993, 180) characterizes it as "reasonable reflective thinking about what to believe or do". Bailin et al. (1999, 287) add that it is thinking for the purpose of making up one's mind about what to believe or do and, in doing so, trying to fulfill certain standards of adequacy and accuracy of thinking. Critical thinking is rational thinking in the following sense: "To be a critical thinker is to be appropriately moved by reasons" (Siegel 1988, 32). Reasons are related to criteria and standards (Lipman 1987).

To educate for critical thinking means to aim for the ideal of a critical thinker. The ideal critical thinker is one who is "habitually inquisitive, well-informed, trustful of reason, open-minded, flexible, fair-minded in evaluation, honest in facing personal biases, prudent in making judgments, willing to reconsider, clear about issues, orderly in complex matters, diligent in seeking relevant information, reasonable in the selection of criteria, focused in inquiry, and persistent in seeking results which are as precise as the subject and the circumstances of inquiry permit" (Facione 1990, 3). Hence, to be a critical thinker requires knowledge, skills, and dispositions (habits). Knowledge of the critical thinker includes, among others, the concepts of observation and inference, of conclusive and defeasible inference, of necessary and sufficient conditions, of hypothesis and prediction, of argument, premise, and conclusion (Hitchcock 2018).

The skills of the critical thinker can be categorized under the headings of (1) interpretation, (2) analysis, (3) evaluation, (4) inference, (5) explanation, and (6) self-regulation (Facione 1990, 8). The last one includes the recognition of the influence of emotions. It also involves the use of strategies for dealing with cognitive biases (Stanovich and Stanovich 2010). Skills are not sufficient because one can have them and not use them, and hence not be a critical thinker; therefore, certain additional dispositions are needed. The relevant dispositions, understood as habits and attitudes that contribute to being a critical thinker, include attentiveness, the attitude of inquiry, self-confidence, courage, open-mindedness, the willingness to suspend judgment, trust in reason, and the search for truth (Allen et al. 2020; Facione 1990, 25; Hitchcock 2018).

Many subjects are becoming increasingly complex in theory as well as in their practical application and require the integration of multiple disciplines and stakeholders. Critical thinking solidifies as a necessary competence for doing and interpreting research. It is a central educational goal and serves as basis for (higher) education policymaking.

Background

Critical thinking is considered to be one of the central skills in higher education and research and also increasingly in the public sphere (post-truth era, social media, and the use of artificial intelligence in different fields, notably in text production). Its origin lies in philosophy and finds application in all fields, from the humanities to social sciences and natural sciences. Teaching critical thinking involves crossing the boundaries of traditional disciplines and helping to develop general thinking skills such as conducting an inquiry (which involves, among others, the ability to formulate a research question, to effectively search for relevant information, to assess the epistemic justifications of claims made and the truthfulness of sources, to assess the usefulness of the information for answering the research question, to apply methods of investigation, etc.). Such skills are central to transdisciplinary learning and research.

The concept and educational goal of critical thinking has a longer history than its name. An older, more rationalist conception of critical thinking goes back to the Age of Enlightenment, in particular to Kant, who in a famous article pleaded for the courage to use the faculty of reason (Kant 1784, 481). Another, more empiricist conception goes back to Dewey, who, inspired by Bacon, Locke, and Mill, described what he called “reflective thought” as follows: “Active, persistent, and careful consideration of any belief or supposed form of knowledge in the light of the grounds that support it, and the further conclusions to which it tends, constitutes reflective thought” (Dewey 1910, 6). Dewey (1916) stressed the relation between critical thinking and democratic participation. Both conceptions aim at teaching autonomous thinking.

In the 1930s, Dewey’s ideas of teaching thinking through inquiry were put into practice at several schools in an eight-year study of the Progressive Education Association. Glaser (1941) developed a test for measuring critical thinking abilities, and showed that they can be improved through education. Later, in the 1970s, Lipman (2003) took up Dewey’s idea of inquiry and applied it to philosophy with children. Critical thinking abilities were included in Bloom’s influential taxonomy of cognitive learning objectives (Bloom et al. 1956). A landmark in the research on critical thinking was Ennis (1962), who proposed a list of 12 aspects of critical thinking as a basis for research and evaluation. The movement of critical thinking and the movement of informal logic influenced each other. They led, from the late 1970s and 1980s, to the creation of associations, to international conferences, and to the creation of scientific journals (Lipman 2003).

Debate and criticism

Critical thinking is a central educational goal. What is the justification for it? There are several fundamental reasons: to respect the students as persons, which means, among others, to recognize their right to ask for reasons and justifications, to empower them to control their own destiny, to initiate them into the rational traditions, and to prepare them for democratic living (Siegel 1988, 55–61). Education in critical thinking is opposed by those who see the fundamental aim of education as the preservation and tradition of given values or the indoctrination of an ideology.

Critical thinking has been criticized for favoring certain kinds of thinking and knowledge over others, in particular reason over emotion, imagination, intuition, and collaborative inquiry (e.g. Thayer-Bacon 2000). However, such criticism presupposes too narrow a concept (for further references and discussion, see Allen et al. 2020; Hitchcock 2018). A basic concern is the relation between general and domain-specific skills, both conceptually – are there general critical thinking skills? – and didactically – can general thinking skills be taught as an independent subject? McPeck (1981) argued that there are no general (critical) thinking skills since all thinking is about some subject matter. There are obvious counterexamples to this, e.g. the abilities to identify assumptions, to reason from premises to conclusions, or to recognize a confusion of necessary and sufficient conditions

(Siegel 1988, 20). McPeck used the argument also to support the claim that critical thinking skills can only be taught in subject-specific courses. This didactical claim is more controversial. Lipman argues that the existence of philosophy and logic, which are concerned with specifying what good thinking ought to be, shows that it is wrong (Lipman 2003, 44). But the question remains how critical thinking can be learned.

Arguing against teaching critical thinking in a separate course focusing on informal logic (which has become common in college education in the United States), it has been noted that the focus on single arguments taken out of context does not help much in acquiring critical thinking abilities since these require knowledge of the area, knowledge of methodological principles and norms of practices specific to the area (Bailin et al. 1999). Instead, according to Bailin and her colleagues, students should be engaged in tasks pertaining to complex issues calling for reasoned judgment, and they should be supported in developing the abilities to do so. On the other hand, learning general thinking skills cannot take place only within disciplines, as this would ignore those aspects of argumentation that transcend the boundaries of the disciplines, such as the procedures for conducting an inquiry, the analysis of arguments, and the evaluation of sources (Battersby and Bailin 2015).

Ennis (1989) distinguishes four types of approaches to teaching critical thinking: general, infusion, immersion, and mixed. (1) The general approach teaches principles of critical thinking explicitly and independent of a subject matter. (2) The infusion approach combines subject matter teaching with making principles of critical thinking explicit. (3) The immersion approach encourages students to think critically in a subject without making the principles of critical thinking explicit. Finally, (4) the mixed approach combines the general approach with either the infusion or the immersion approach. Abrami et al. (2015) found in a meta-study that all four types of approaches lead to a significant increase in critical thinking skills, and no big differences between the approaches could be observed. Additionally, they used a classification into the following categories: dialogue (learning through discussion), anchored instruction (learning through analysis of real-world problems), and mentoring (one-to-one interaction with a mentor). They figured out that all three categories had positive effects on the critical thinking outcomes. The largest effect in terms of a gain in the ability to think critically was achieved in courses where a combination of all three aspects was present.

The result of any study purporting the measurement of gains in critical thinking ability faces the criticism of how to measure such gains. First, it "requires that we be clear about what we are trying to assess" (Ennis 1993, 179). This means that a definition of critical thinking is needed. Standardized test inventories exist for different definitions (for an overview see Ennis 1993 and the supplement on assessment in Hitchcock 2018). While such tests are used for the assessment of the knowledge and abilities aspects of critical thinking, it is much more difficult to evaluate gains in attitudes, especially since these might only develop slowly, calling for more long-term research in this area.

Jahn (2012) distinguishes four standards of thought: analysis, perspectivity, ideological criticism, and constructivity. Analysis is concerned with the assessment of arguments. Perspectivity means to take on multiple perspectives and contrast them with each other. As human beings, we are biased in favor of our own perspective, sometimes including views from our own (often small) in-group. Critical thinking requires grasping multiple perspectives, sharpening one's own perspective, and assessing reasons to hold it. This is especially true when normative questions are in focus. To be a critical thinker sometimes requires ideological criticism which involves analysis of the (hidden) power structures. Such structures can be found not only in political contexts but also in institutions such as a university or committee that decides on the distribution of funds for research in different fields. Finally, constructivity means developing new solutions to a problem. It can also mean developing action plans that help improve practice.

Current forms of implementation in higher education

Critical thinking is implemented as an educational goal in educational systems and institutions across the world. The teaching is often done in a course devoted to critical thinking. Useful material for such courses can be found, for example, on the University of Hong Kong's Critical Thinking Web (Lau and Chan 2023). However, education in critical thinking is not restricted to such courses.

Identifying successful didactical approaches helps to elucidate factors that are conducive for learning critical thinking (for a study about practices in institutions of higher education in the global south, see Okolie et al. 2022). Significantly better outcomes are found in forms of teaching that involve some kind of mentoring along with dialogue-based instructions and the use of authentic tasks. While using real-world examples clearly increases the relevance of the topic (for the students), and hence fosters students' intrinsic motivation, the findings that dialogue-based instruction has even greater effects highlight the importance of cognitively activating methods. This effect is especially strong when roleplaying is used (being a combination of real-world tasks and dialogue-based instruction, Abrami et al. 2015, 299). The boost in effect when adding elements of mentoring, in turn, emphasize that feedback and some form of guidance are beneficial. Indeed, teaching critical thinking often starts with an initial irritation. This can be in the form of a real-world problem, presented as a text, a video, research data, etc., which triggers in the students the urge to resolve it.

The meta-study by Abrami et al. (2015) cuts across all disciplines. Different aspects of teaching critical thinking can be illustrated by a course, designed in the context of the critical thinking initiative at ETH Zurich, titled Scientific Concepts and Methods (Sieroka et al. 2018). It is a compulsory, one-week course in the Master's program of the pharmaceutical sciences curriculum. Each day, the students receive input on a certain aspect of the philosophy of science (e.g. scientific reasoning, use of images) and input on a thematically aligned scientific method in modern biomedicine (e.g. reproducibility models in drug development, biomedical imaging), leading up to an interdisciplinary discussion of each day's topic together with the philosopher of science and the biomedical expert.

When planning to integrate critical thinking as an intended learning outcome, it is necessary to clarify which standards of thought should be considered. One can follow the four standards by Jahn (see above). Analysis can be taught in a separate course but might as well be integrated into any subject matter that deals with arguments, such as devising formal proofs in mathematics, interpreting experimental results in the natural sciences, debating arguments in political sciences, or justifying value judgments in economics. Classical debates can be one means to integrate the analysis of arguments into a course. Writing an argument based on a scaffolded structure and then giving feedback to and receiving feedback from course peers can be another (Kölbel and Jentges 2017). In order to achieve the standards of perspectivity, students need to be exposed to multiple perspectives on the topic or issue. However, mere exposure is not enough. It is essential also to take on the perspective of others, as in the examples before or in role play, which has a rather high impact in terms of gains in critical thinking (Abrami et al. 2015). Adding transdisciplinary elements into a course can be a great opportunity for students to take on new and different perspectives, either from experts in a different field or from people who are unfamiliar with the research field in question. This can feel rather odd for teachers who are used to passing on knowledge to students. If critical thinking is one of the aims of the course, the teacher should act as a role model of a scientist in the field and share not only the standard canon of knowledge, but also its uncertainties and boundaries, as well as the scientific attitude of dealing with such uncertainties. Including a historical perspective in the course may help to shed light on the process of scientific development – a useful text here is still Kuhn's The Structure of Scientific Revolutions (1962).

Including aspects of ideological criticism in a course can add to the curiosity of students and hone their critical thinking skills. At the level of constructivity, students are asked to find constructive solutions to existing problems and invited to integrate their newly acquired knowledge into their everyday (professional) lives. It is therefore important to allow the students to relate any subject matter and critical discussion to their own context, ideally to some work or produced output of their own. Adding elements of the scientific process to a course can lead to high engagement of students and the acquisition of critical thinking skills. These can involve the writing (and peer reviewing) of grant proposals, article abstracts, scientific presentations, and posters. The use of mentoring will increase the likelihood of students successfully learning to think critically.

The intended learning outcome of a single course cannot be to become a critical thinker because this requires much more time. It is rather to be seen as part of the aims of education in general (or even of lifelong learning). For a particular course, it is helpful to specify which aspects of critical thinking will be in focus and what the teacher expects the students to learn (for example, to be able to reconstruct an argument, to be able to apply their critical thinking skills in real-world problems or to acquire the habit of a critical thinker to be open to new information and to objections). A very effective way is to offer learning activities to students where they can practice, for skills can only be acquired by practicing.

Generally, aligning the learning activities and the assessment to the intended learning outcomes is a basic requirement of good teaching (Biggs and Tang 2011). And a learning and teaching culture that de-emphasizes teacher-centered or solution-focused teaching approaches are likely to be conducive to learning critical thinking (Okolie et al. 2022).

But how should the assessment be done? In order to answer this question, one should first be clear about the purposes of the assessment: Is the primary purpose to diagnose the individual (or collective) level of critical thinking as information for the teacher (or for some institution)? Is it to give feedback to the students so that they know what they are good at and where they need to improve? Is it to motivate students by letting them know what they have learned?

The methods to be used in assessment can vary. One can either use one of the commercially available tests in English (see Norris and Ennis 1989 for guidelines choosing among them) or devise one's own multiple-choice test (see Norris and Ennis 1989 for recommendations). However, devising a valid multiple-choice test is challenging and time-consuming. Since it is inherently difficult to comprehensively assess critical thinking using closed questions, Ennis (1993, 184) recommends asking for a brief written defense of the chosen answer, and generally implementing open assessment formats, especially when going beyond the diagnostic aspects of the exam. Open-ended techniques include short-answer tests, argumentative essays, and individual interviews (Norris and Ennis 1989). Other, rather open forms of assessment can also be used, for example doing a case study, designing a poster, or taking part in a panel discussion.

Due to a widespread bias, one aspect in teaching is often underestimated: the expert blind spot. The time it takes to learn to think critically (or to learn anything for that matter) must not be underestimated. Learning does take time, and so does learning of critical thinking.

References

Abrami, Philip C., Robert M. Bernard, Eugene Borokhovski, David I. Waddington, C. Anne Wade, and Tonje Persson. 2015. Strategies for teaching students to think critically: A meta-analysis. Review of Educational Research 85 (2): 275–314.

Allen, Derek, Sharon Bailin, Mark Battersby, and James B. Freeman. 2020. Critical thinking. Oxford research encyclopedia of education. New York: Oxford University Press.

Bailin, Sharon, Roland Case, Jerrold R. Coombs, and Leroi B. Daniels. 1999. Conceptualizing critical thinking. Journal of Curriculum Studies 31 (3): 285–302.

Battersby, Mark, and Sharon Bailin. 2015. Teaching critical thinking as inquiry in higher education. In Palgrave handbook of critical thinking in higher education, eds. Martin Davies and Ronald Barnett, 123–38. New York: Palgrave Macmillan.

Biggs, John B., and Catherine Tang. 2011. Teaching for quality learning at university. 4th edition. Buckingham: Open University Press.

Bloom, Benjamin S., Max D. Engelhart, Edward J. Furst, Walker H. Hill, and David R. Krathwohl. 1956. Taxonomy of educational objectives. Handbook I: Cognitive domain. New York: David McKay.

Dewey, John. 1910. How we think. Boston: D. C. Heath.

Dewey, John. 1916. Democracy and education: An introduction to the philosophy of education. New York: Macmillan.

Ennis, Robert H. 1962. A concept of critical thinking: A proposed basis for research on the teaching and evaluation of critical thinking ability. Harvard Educational Review 32 (1): 81–111.

Ennis, Robert H. 1989. Critical thinking and subject specificity: Clarification and needed research. Educational Researcher 18 (3): 13–16.

Ennis, Robert H. 1993. Critical thinking assessment. Theory into Practice 32 (3): 179–86.

Facione, Peter A. 1990. Critical thinking: A statement of expert consensus for purposes of educational assessment and instruction, research findings and recommendations prepared for the Committee on Pre-College Philosophy of the American Philosophical Association. Available from https://eric.ed.gov/?id=ED315423.

Glaser, Edward M. 1941. An experiment in the development of critical thinking. New York: Teachers college, Columbia University. Contributions to education no. 843.

Hitchcock, David. 2018. Critical thinking. The Stanford Encyclopedia of Philosophy. Available from https://plato.stanford.edu/entries/critical-thinking.

Jahn, Dirk. 2012. Kritisches Denken fördern können. Entwicklung eines didaktischen Designs zur Qualifizierung pädagogischer Professionals. Aachen: Shaker.

Kant, Immanuel. 1784. Beantwortung der Frage: Was ist Aufklärung? Berlinische Monatsschrift 4: 481–94.

Kölbel, Julian, and Erik Jentges. 2017. The six-sentence argument: Training critical thinking skills using peer review. Management Teaching Review 3 (2): 118–28.

Kuhn, Thomas. 1962. The structure of scientific revolutions. Chicago: University of Chicago Press.

Lau, Joe, and Jonathan Chan. 2023. Welcome to critical thinking web. Available from https://philosophy.hku.hk/think.

Lipman, Matthew. 1987. Critical thinking – What can it be? Educational Leadership 45: 38–43.

Lipman, Matthew. 2003. Thinking in education. 2nd edition. New York: Cambridge University Press.

McPeck, John E. 1981. Critical thinking and education. New York: St. Martin's Press.

Norris, Stephen P., and Robert H. Ennis. 1989. Evaluating critical thinking. Pacific Grove, CA: Midwest.

Okolie, Ugochukwu Chinonso, Paul Agu Igwe, Ifeanyi Kalu Mong, Hyginus Emeka Nwosu, Clementina Kanu, and Chidiebere C. Ojemuyide. 2022. Enhancing students' critical thinking skills through engagement with innovative pedagogical practices in Global South. Higher Education Research & Development, 41 (4): 1184–98.

Siegel, Harvey. 1988. Educating reason: Rationality, critical thinking, and education. New York: Routledge.

Sieroka, Norman, Vivianne I. Otto, and Gerd Folkers. 2018. Critical thinking in education and research – Why and how? Angewandte Chemie International Edition 57 (51): 16574–75.

Stanovich, Keith E. and Paula J. Stanovich. 2010. A framework for critical thinking, rational thinking, and intelligence. In Innovations in educational psychology: Perspectives on learning, teaching and human development, eds. David D. Preiss and Robert J. Sternberg, 195–237. New York, Springer.

Thayer-Bacon, Barbara. 2000. Transforming critical thinking: Thinking constructively. New York: Teachers College Press.

Data Literacy

Valentin Unger, Michael Beck, and Vera Husfeldt

Definition

The competence to deal with data sovereignly, or "Data Literacy" is one of the fundamental skills of modern society. To define the term data literacy, its two semantic constituents data and literacy are examined in more detail.

1. Data are signs or symbols based on observations of the world and are constantly collected by or through us – both consciously and unconsciously (Cambridge University Press 2023b; Zins 2007, 482). Etymologically, the term data goes back to the plural of the Latin datum, which can be translated as "that which is given, a present" (Glare 2012, 532) and has been associated with computer processes since the mid-1940s (Harper 2021). Early on, data was associated with a prospective meaning: data have an impact on the future.
2. In 2000, the OECD defined literacy as follows: "The ability to understand and employ printed information in daily activities, at home, at work and in the community – to achieve one's goals and to develop one's knowledge and potential" (OECD 2000, x). This definition of literacy was criticized for several reasons, first, the focus on standardized testing. By the late 20th century, it was pointed out that a measurable continuum cannot reflect the diversity of ways in which people interact through spoken and written language (St. Clair 2012, 771). Second, the OECD's definition of literacy is also driven by the idea that the measured areas of reading, numeracy, and problem solving are directly related to economic and social progression in modern societies. By excluding other factors and focusing on one's own culturally shaped world, the model remains very simplistic and limited according to some critics (e.g. Guadalupe 2017, 334; Perry et al. 2020, 12; Sellar and Lingard 2014, 922). The increasing use of digital technologies over the past two decades has also made it necessary to refine the OECD's concept of literacy, which originally focused on printed information (OECD 2021, 5). Eventually, the original concept of

literacy changed to a multiplicity of knowledge, skills, and values relevant to society's success (Vincent 2003, 342; Ware et al. 2016, 307).

The term data literacy emerged around the turn of the millennium and is often used in the context of digital competencies (Schüller 2020, 11–12). It is, therefore, socially relevant knowledge, skills, and attitudes necessary for dealing with and handling (digital) data sensitively, in a society characterized by permanent multifaceted information (Ridsdale et al. 2015, 4).

The model of the data–information–knowledge–wisdom pyramid states that data must be transformed into usable knowledge by sequencing its hierarchical ascent from data to information, knowledge, and wisdom. Each level can be seen as a precursor to the next, with unorganized and therefore useless data being the lowest. Data becomes usable information only by ascending the levels, which can then be used to build knowledge (Rowley 2007, 163–68). The highest level is the so-called "wisdom", which can be described as integrated, usable knowledge (Rowley 2007, 174). Appropriate use of data includes the insightful perception and production of data. Data literacy also includes knowledge, skills, and attitudes for insightful and ethical data collection, processing, and interpretation. It is also relevant when decisions are made based on the data that affect other people (Schüller 2020, 11). Schüller (2020, 23–40) has described six steps that need to be taken when encoding and decoding information: (1) "Establish a data culture", (2) "provide data", (3) "evaluate data", (4) "interpret results", (5) "interpret data", and (6) "derive actions".

Background

Data literacy is increasingly more relevant at work and in private life, which is why research is also increasingly addressing this phenomenon. "Data literacy is a hot topic" (Van Audenhove et al. 2020, 2) for various reasons:

As societies become more digital, people receive information from around the world – almost in real time (Hai et al. 2021, 25–28). The Covid-19 pandemic gave a significant boost to digitization. Many organizations had to switch to a remote mode of operation almost overnight and adapt their ways of communicating and working (Amankwah-Amoah et al. 2021, 605). Of course, this adaptation has not happened equally everywhere, but it is astonishing how quickly and widely digitalization has been adopted in order to avoid face-to-face contact. According to a study by Dingel and Neiman (2020, 1), 37 percent of jobs in the US can be done entirely from home. Schools around the world have also had to adapt to the new circumstances (Unger et al. 2022, 174).

Aside from the advantages of the increase in constantly updated information (accessibility of general education to almost everyone, worldwide information can help people form opinions and generate knowledge, the world is moving closer together, etc.), there are also problematic aspects. For example, digitalization increasingly leads to poor quality data, i.e. deliberately misinterpreted or misleadingly prepared information that can be disseminated unfiltered – especially via the internet (Schüller 2020, 11). New ethical challenges also arise, such as the increased need for awareness about privacy and data protection (Faraj et al. 2021, 4). Furthermore, the ability to detect misinformation has become more important (Nguyen 2021, 212).

"The world's most valuable resource is no longer oil, but data" (The Economist 2017). Referencing this, the slogan "data is the new oil" became a popular refrain, claiming data, especially "big data", was becoming increasingly valuable. Big data can be defined as "very large sets of data ... that can only be stored, understood, and used with the help of special tools and methods" (Cambridge University Press 2023a). Data is collected and used commercially everywhere and at all times – in smartphones, online shopping, mobility services, etc. Based on big data, numerous decisions are made that directly or indirectly affect citizens. While there are positive effects attributed to big data such as improved public safety or cancer diagnostics optimized by machine learning (Bhagespur 2019; Chang 2021, 1–10), the conclusions drawn from big data are often criticized – especially in the context of screening people for economic reasons under questionable data protection rules (Brayne 2017, 980; Zuboff 2019). Nonetheless, it is relevant that people are familiar with the basics of big data mining and how big data is interpreted and acted upon.

Understanding decisions on the basis of data is becoming more complex (Van Audenhove et al. 2020, 3), while data analysis is automated via algorithms and artificial intelligence. Data literacy can enable people to understand and challenge these automated data analyses and the decisions based on them through relevant knowledge, skills, and values. Furthermore, people should also be able to recognize and evaluate algorithmic bias (Baker and Hawn 2021, 1083).

Another argument for data literacy is "misunderstandings, misperceptions, mistrust and misgivings" (Wallmann 1993, 1) regarding statistics: Data is often aggregated in order to generalize information as far as possible. However, this aggregation inevitably means that information is lost, making individual inferences very difficult (Holderness 2016, 9). One must be aware of this to interpret the omnipresent data in an informed way. Furthermore, there are often misconceptions in society, which can be traced back to a simplified or simply incorrect view of statistical terms, statistical evaluation procedures, and how they can be interpreted. For example, some people misunderstand the mean value as the highest expression of the values in a population (Yilmaz 2013, 22). Problems can also arise from improperly drawing causal conclusions from correlations (Pearl 2000, 1) due to a lack of knowledge about (1) how data are generated, (2) what conclusions are even possible, (3) and by what process of data generation (observational or experimental) or (4) statistical analysis procedures (traditional statistical inference or modern causal inference).

This shows that data literacy subsumes competencies that help to address challenges systematically by enabling the confident handling of data. It is a transdisciplinary key competence for 21st-century society (Ludwig and Thiemann 2020, 436; Schüller 2020, 44), which is relevant in several fields. Data literacy can be an important skill when assessing the nature, quality, and credibility of different sources of information, which can vary greatly between different disciplines. Accordingly, data literacy can be adapted anywhere in a specific subject context with corresponding learning cultures. For example, data literacy is relevant for dealing with "big data" in large companies as it provides the basis for making future decisions based on this data, but also for understanding major societal issues such as climate change, mass migration, or pandemics.

Debate and criticism

The evolution of the term data literacy itself as distinct from other closely related and sometimes synonymous terms, such as statistical literacy or information literacy, is worth highlighting. As early as 2004, Schield (2004, 8) presented an integrative model in which data literacy is seen as one of the core competencies for critical thinking. He emphasizes the technical component of data literacy, especially the use of suitable software solutions (e.g. databases, analysis, and presentation software); the term is closely related to practical aspects of data collection and analysis. In his model, data literacy can be seen as an essential sub-competence of statistical literacy, which is an essential sub-competence of information literacy.

Given this, it is surprising that Schüller (2020, 11) notes that even today, definitions and competency frameworks do not distinguish between data, statistical, and information literacy. The delineation of the term, as well as its classification as a super- and sub-category of other forms of literacy, turns out to be fundamentally problematic, as Gould (2017) demonstrates. Partly under the influence of a workshop (Oceans of Data Institute 2016), he calls for broadening the concept of statistical literacy to include aspects of data literacy (and thus follows Schield's 2004 model), although he immediately notes that definitions of data literacy include statistical literacy but also go far beyond it. According to his observations, "this is because the notions of SL [statistical literacy] that have arisen from the statistics education community are perceived by those who work in data science as falling short of what is required" (Gould 2017, 23). It remains unclear whether data literacy is a subset of statistical literacy, or vice versa. The quote suggests that data literacy may also be statistical literacy in a new (more digital) guise. This is supported by earlier definitions of statistical literacy that are indistinguishable from current definitions of data literacy (e.g. Gal 2002, 3–4; Wallmann 1993, 1). A closer look at the levels of associated skills and competencies often reveals little difference (apart from a stronger focus on the creation and collection of data and the inclusion of buzzwords such as "big data" and "machine learning" in the term data literacy).

Schield's (2004) article gives an opportunity to explore the evolution of the term data literacy and its relative importance in the context of critical thinking, information literacy, and statistical literacy. His search of the ERIC database (Institute of Education Sciences n.d.) at the time yielded over 10,000 hits for "critical thinking" and just under 1,500 for "information literacy", whereas "quantitative literacy", "statistical literacy", and "data literacy" only received a maximum of 65 hits each (Shield does not provide individual figures for the last three aspects of competence). In 2022, on the other hand, "critical thinking", "information literacy", and "data literacy" have between 12,000 and 15,000 more entries, whereas "statistical literacy" (just under 3,000 more entries) and "quantitative literacy" (just over 1,600 more entries) have become much less established as terms (see Figure 1).

Figure 1. Occurrence of different terms in Education Resources Information Center 2022 (Institute of Education Sciences n.d.)

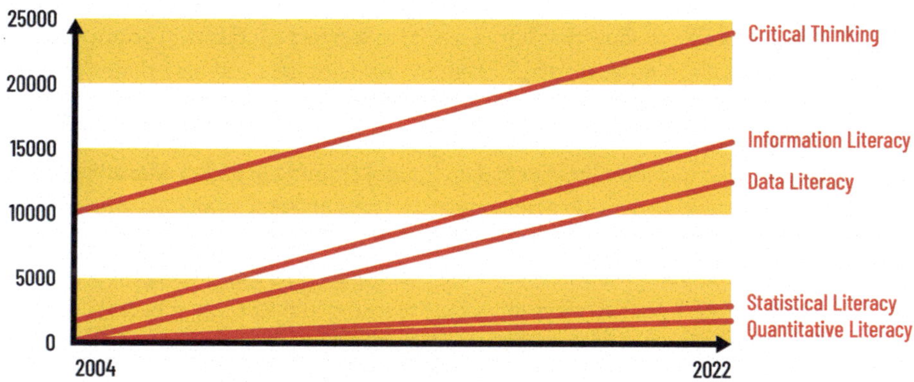

thinking, has grown substantially, whereas statistical literacy has seen a comparatively small increase in hits. It seems that statistical literacy is now seen as part of data literacy. Information literacy, on the other hand, continues to be considered quite separately. However, it would undoubtedly be instructive to break down the frequency of use of the terms by discipline to consider the extent to which the different terminology is simply a consequence of subject-specific differences in terminology.

However, based on the definitions and competence frameworks in use today, it remains unclear to what extent competencies in data literacy differ from statistical literacy, information literacy, or problem-solving competencies in general, especially in comparison to Schield (2004). It seems problematic that most definitions tend not to focus explicitly on information technologies, but data literacy without reference to digitalization seems to have no added value.

A critical problem in defining the term seems to be that in the course of digitalization, the distinction between "data" and "information" now seems obsolete (Schüller 2020, 11), as information that cannot be stored in the form of data has become almost inconceivable. Schüller (2020, 11) makes a claim for the "triumph" of data literacy: she assumes that it is a fashionable term resulting from the fact that jobs with the title "data engineer" or "data scientist" pay higher salaries than comparable titles such as "statistician" or "IT specialist" (unfortunately, she does not support this assumption with actual salary analyses).

Current forms of implementation in higher education

Data literacy has significant role in today's educational landscape. When it comes to curricula for schools or universities, the significance of digital literacy and data literacy, in particular, is undeniable (Bandtel et al. 2021, 396).

As part of a European Erasmus+-funded program, an international research group was able to show that it is essential to support educators in acquiring data literacy skills in order to enable them to deal with data and base their decisions on data (Papamitsiou et al. 2021, 21). Using an extensive literature review of international databases from academic publishers, the authors identified different courses that include educational data literacy training in higher education or professional development (Papamitsiou et al. 2021, 8). Some of them are briefly presented here. The Norwegian University of Science and Technology offers a course on digital literacy and smart learning that aims to develop a thoughtful relation to the use of digital services in various teaching and learning processes (Norwegian University of Science and Technology 2020). A 15-week interactive hands-on course on data literacy is given online by the RETAIN Center of Excellence at Newberry College for educators from South Carolina (n.d.). Edx, a consortium of universities from all over the world that offer online courses, provides various courses for teachers. One of them is a six-week course on learning analytics that addresses schoolteachers who want to improve their teaching through valuable data-driven insights during three to four hours a week (Edx 2023a). Another course focuses on big data in education (Edx 2023b), and again another one focuses on wise action in connection with data (Edx 2023c).

In addition, there are other examples from around the world. In Tanzania, the Tanzania data lab (dLab), is fostering data literacy to engage communities to address issues like the HIV/AIDS epidemic but also gender inequality and economic growth by using data. Together with the University of Dar es Salaam, College of Information and Communication Technology (UDSM CoICT) they helped to start the first East African Masters in Data Science (dLab 2021).

As part of the "Swiss Digital Skills Academy", a project funded by Swissuniversities, the association of higher education institutions in Switzerland, led by the École polytechnique fédérale de Lausanne, the "Develop Data Literacy" project was launched in 2021, with four universities of applied sciences participating as a core group. At the University of Applied Sciences of the Grisons, two courses were held where information science students first studied the theoretical foundations of data literacy and subsequently developed Open Educational Resources on various topics of data literacy. Open Educational Resources are freely accessible learning opportunities that are mostly made available digitally (UNESCO & Commonwealth of Learning 2011, v). At the St.Gallen University of Teacher Education, an Open Educational Resource is currently being developed in cooperation with educational researchers and in-service teachers as part of the same project, which will enable in-service and prospective teachers to receive further training regarding computer-based educational technologies as an example.

The contribution shows that data literacy – despite existing terminological imprecision – is a transdisciplinary key competence of our time, which should be fostered more actively in university teaching.

References

Amankwah-Amoah, Joseph, Zaheer Khan, Geoffrey Wood, and Gary Knight. 2021. COVID-19 and digitalization: The great acceleration. Journal of Business Research 136: 602–11.

Baker, Ryan S., and Aaron Hawn. 2021. Algorithmic bias in education. International Journal of Artificial Intelligence in Education 32: 1052–92.

Bandtel, Matthias, Leonie Kauz, and Natalia Weißker. 2021. Data Literacy Education für Studierende aller Fächer. Kompetenzziele, curriculare Integration und didaktische Ausgestaltung interdisziplinärer Lehr-Lern-Angebote. In Digitalisierung in Studium und Lehre gemeinsam gestalten, eds. Geschäftsstelle beim Stifterverband, 395–412. Wiesbaden: Springer.

Bhagespur, Kiran. 2019. Data is the new oil – And that's a good thing. Forbes. Available from https://www.forbes.com/sites/forbestechcouncil/2019/11/15/data-is-the-new-oil-and-thats-a-good-thing/?sh=796640cb7304.

Brayne, Sarah. 2017. Big data surveillance: The case of policing. American Sociological Review 82 (5): 977–1008.

Cambridge University Press, ed. 2023a. Big data. Available from https://dictionary.cambridge.org/dictionary/english/big-data.

Cambridge University Press, ed. 2023b. Data. Available from https://dictionary.cambridge.org/dictionary/english/data.

Chang, Victor. 2021. An ethical framework for big data and smart cities. Technological Forecasting & Social Change 165 (120559): 1–11.

Dingel, Jonathan, and Brent Neiman. 2020. How many jobs can be done at home? Journal of Public Economics 189 (104235): 1–8.

dLab. 2021. The Tanzania data lab (dLab) and the UDSM. Available from https://dlab.or.tz/the-tanzania-data-lab-dlab-and-the-university-of-dar-es-salaam-launch-the-first-masters-of-science-in-data-science.

Edx. 2023a. Analytics for the classroom teacher. Available from https://www.edx.org/course/analytics-for-the-classroom-teacher.

Edx. 2023b. Big data and education. Available from https://www.edx.org/course/big-data-and-education.

Edx. 2023c. Introduction to data wise: A collaborative process to improve learning & teaching. Available from https://www.edx.org/course/introduction-to-data-wise-a-collaborative-process.

Faraj, Samer, Wadih Renno, and Anand Bhardwaj. 2021. Unto the breach: What the COVID-19 pandemic exposes about digitalization. Information and Organization 31 (1): 1–7.

Gal, Iddo. 2002. Adults' statistical literacy: Meanings, components, responsibilities. International Statistical Review 70 (1): 1–51.

Glare, Peter G. W., ed. 2012. Oxford Latin dictionary. 2nd edition. Vol. A–L. Oxford: Oxford University Press.

Gould, Robert. 2017. Data literacy is statistical literacy. Statistics Education Research Journal 16 (1): 22–25.

Guadalupe, César. 2017. Standardisation and diversity in international assessments: Barking up the wrong tree? Critical Studies in Education 58 (3): 326–40.

Hai, Thanh Nguyen, Quang Nguyen Van, and Mai Nguyen Thi Tuyet. 2021. Digital transformation: Opportunities and challenges for leaders in the emerging countries in response to Covid-19 pandemic. Emerging Science Journal 5: 21–36.

Harper, Douglas. 2021. Data. Available from https://www.etymonline.com/word/data.

Holderness, Clifford G. 2016. Problems using aggregate data to infer individual behavior: Evidence from law, finance, and ownership concentration. Critical Finance Review 5 (1): 1–40.

Institute of Education Sciences. n.d. ERIC. Available from https://eric.ed.gov.

Ludwig, Thomas, and Hannes Thiemann. 2020. Datenkompetenz – Data Literacy. Informatik Spektrum 43 (6): 436–39.

Nguyen, Dennis. 2021. Mediatisation and datafication in the global COVID-19 pandemic: On the urgency of data literacy. Media International Australia 178 (1): 210–14.

Norwegian University of Science and Technology. 2020. Digital literacy – Smart learning. Available from https://www.ntnu.edu/studies/courses/SOS2020/2017/1#tab=omEmnet.

Oceans of Data Institute, ed. 2016. Building global interest in data literacy: A dialogue. Workshop report. Available from http://oceansofdata.org/our-work/building-global-interest-data-literacy-dialogue-workshop-report.

OECD [Organisation for Economic Cooperation and Development], ed. 2000. Literacy in the information age. Final report of the international adult literacy survey. Paris: OECD Publications Service. Available from https://www.oecd.org/education/skills-beyond-school/41529765.pdf.

OECD [Organisation for Economic Cooperation and Development], ed. 2021. 21st-century readers: Developing literacy skills in a digital world. Paris: OECD Publishing. Available from https://www.oecd-ilibrary.org/education/21st-century-readers_a83d84cb-en.

Papamitsiou, Zacharoula, Michail E. Filippakis, Marilena Poulou, Demetrios Sampson, Dirk Ifenthaler, and Michail Giannakos. 2021. Towards an educational data literacy framework: Enhancing the profiles of instructional designers and e-tutors of online and blended courses with new competences. Smart Learning Environments 8 (18): 1–26.

Pearl, Judea. 2000. Causality: Models, reasoning, and inference. Cambridge: Cambridge University Press.

Perry, Kristen, Donita Shaw, and Sara Saberimoghaddam. 2020. Literacy practices and the Programme for the International Assessment of Adult Competencies (PIAAC): A conceptual critique. International Review of Education 66 (1): 9–28.

RETAIN Center of Excellence at Newberry College. n.d. Data literacy 01. Available from http://artofeducating.com/online-courses.

Ridsdale, Chantel, James Rothwell, Mike Smit, Hossam Ali-Hassan, Michael Bliemel, Dean Irvine, Daniel Kelley, Stan Matwin, and Brad Wuetherick. 2015. Strategies and best practices for data literacy education knowledge synthesis report. Halifax: Dalhousie University. Available from http://rgdoi.net/10.13140/RG.2.1.1922.5044.

Rowley, Jennifer. 2007. The wisdom hierarchy: Representations of the DIKW hierarchy. Journal of Information Science 33 (2): 163–80.

Schield, Milo. 2004. Information literacy, statistical literacy and data literacy. IASSIST Quarterly 28 (2/3): 6–11.

Schüller, Katharina. 2020. Future skills: A framework for data literacy. Competence framework and research report. Working Paper No. 53. Berlin: Hochschulforum Digitalisierung.

Sellar, Sam, and Bob Lingard. 2014. The OECD and the expansion of PISA: New global modes of governance in education. British Educational Research Journal 40 (6): 917–36.

St. Clair, Ralf. 2012. The limits of levels: Understanding the International Adult Literacy Surveys (IALS). International Review of Education 58 (6): 759–76.

The Economist, ed. 2017. The world's most valuable resource is no longer oil, but data. The Economist. Available from https://www.economist.com/leaders/2017/05/06/the-worlds-most-valuable-resource-is-no-longer-oil-but-data.

UNESCO and Commonwealth of Learning, ed. 2011. Guidelines for open educational resources (OER) in higher education. Paris: UNSECO.

Unger, Valentin, Laura Schmidberger, and Albrecht Wacker. 2022. "Ja, der Fernunterricht hat sich deutlich verbessert…!" Befunde einer Schüler*innenbefragung zum schulisch angeleiteten Lernen zu Hause während der zweiten Schulschliessungen in Deutschland. In Schule in Distanz – Kindheit in Krise. Auswirkungen der Covid-19 Pandemie auf Wohlbefinden und Lebensbedingungen von Kindern und Jugendlichen, eds. Jürgen Budde, Drorit Lengyel, Caroline Böning, Carolina Claus, Nora Weuster, Katharina Doden, and Tobias Schroedler, 173–96. Wiesbaden: Springer VS.

Van Audenhove, Leo, Wendy Van den Broeck, and Ilse Mariën. 2020. Data literacy and education: Introduction and the challenges for our field. Journal of Media Literacy Education 12 (3): 1–5.

Vincent, David. 2003. Literacy literacy. Interchange 34 (2): 341–57.

Wallmann, Katherine K. 1993. Enhancing statistical literacy: Enriching our society. Journal of the American Statistical Association 88 (421): 1–8.

Ware, Paige, Richard Kern, and Mark Warschauer. 2016. The development of digital literacies. In Handbook of second and foreign language writing, eds. Rosa M. Manchón and Paul Kei Matsuda. Boston: De Gruyter, 307–28.

Yilmaz, Zuhal. 2013. Usage of tinker plots to address and remediate 6th grade students' misconceptions about mean and median. The Anthropologist 16 (1–2): 21–29.

Zins, Chaim. 2007. Conceptual approaches for defining data, information, and knowledge. Journal of the American Society for Information Science and Technology 58 (4): 479–93.

Zuboff, Shoshana. 2019. The age of surveillance capitalism: The fight for a human future at the new frontier of power. London: Profile Books.

Design Thinking

Sadaf Taimur, Daniela Peukert, and BinBin Pearce

Definition

Design refers to "the plan, project, or working hypothesis which constitutes the 'intention' in intentional operations" (Buchanan 1992, 10). Or as Rittel (1971, 19) formulates it, "a [wo]man designs whenever [s]he has a purpose in mind and devises a scheme to accomplish this purpose". In this chapter, design thinking is defined as a human-centered problem-solving approach which was intended originally to aid in the development of novel products but has more recently been adapted to deal with system-level challenges and "wicked" problems (Buchanan 1992; Dam and Siang 2018; Von Thienen et al. 2014).

The meaning of words changes with time and the term *design* is no exception. In the 14th century, the word originated from the Latin *designare*, which means "to plan" or "to devise". In the 16th century, the Italian *disegnare* developed the meaning "to plot, to draw, to embroider", etc., which was absorbed into the French language in different forms and then passed on to English. In the English-speaking world, the term design covers all creative, planning, and drafting activities, various disciplines of both design and engineering, as well as the designed object itself. Only the noun and verb forms of the term are distinguished. As a noun, design refers to various disciplines and the designed object. As a verb, design encompasses planning and drafting procedures, i.e. process-oriented action.

The term *design thinking* originated and was used in 20th-century literature, e.g. in the *Architect and Engineer Magazine* in 1946, but it has mostly been used by researchers in the field of design since the 1980s (Dorst 2015; Rowe 1987). Starting from the 1990s, various models and understandings of design thinking have emerged, e.g. design thinking as a mindset, toolkit, method, and process (Dorst 2011). Dorst (2011) elaborates that there is a reason why people from diverse fields are interested in design: Designers have been dealing with complex and open problems for years, and they have developed professional practices to do so; therefore, adopting their practices can help solve such problems. Adopting design thinking in other non-design fields led to a demand for definite and clear knowledge about design thinking, which included a definition and methodological guidelines for non-designers to understand and engage with. With a consistent gain in popularity, literature has reported a considerable number of success stories of the application of design thinking in diverse fields (e.g. Brown 2008; Brown and Wyatt 2010; Dorst 2010). However, considerable ambiguity still exists regarding the use of the term and its definition.

Using design thinking in education has a long history, and the use of the term varies in application. For example, Panke (2019) conducted a literature review on applying design thinking in different educational settings and concluded the review with seven themes. These themes identified design thinking as: (1) a method for instructional design for course material development; (2) a technique for curriculum development; (3) a strategy for teaching to achieve specific objectives; (4) a learning goal; (5) a facilitation technique to support students; (6) an approach for leadership and organizational development; (7) a method for product development or process improvement.

Background

Various models of design thinking have been developed over time based on Simon's (1969) design thinking process. Simon's design thinking approach was focused more on creating a way to change or adapt the existing conditions to ones compliant with the current context. Another discourse on design thinking comes from Schön (1983), who emphasized the role of reflection in design thinking and regarded reflection as a core of design thinking. Design thinking as a problem-solving activity is related to Rittel's interpretation and Buchanan's elaboration of the wicked problem's solution approach (Buchanan 1992). According to this approach, the process of design is divided into two segments, i.e. problem definition and problem solution (Buchanan 1992, 15). Another discourse by Krippendorff (2005) elaborates design thinking as an approach to create meaning rather than artifacts, and making meaning becomes the core of the design process while artifacts become the medium of communicating the meaning (Johansson-Skoldberg et al. 2013, 126) The most notable design thinking approach for its application in the field of education is developed by the Hasso Plattner Institute of Design at Stanford University (Hasso Platter Institute 2023; Melles et al. 2015; Plattner et al. 2009). This transferable approach emerges from the human-centered design principles and entails five iterative stages: (1) *Empathize* – building empathy by deeply understanding the problems and realities of people who are facing the problems; (2) *Define* – defining a specific problem by unpacking and synthesizing the findings from the empathize stage; (3) *Ideate* – generating creative and radical ideas to deal with the framed problem; (4) *Prototype* – bringing the ideas into some physical form to eliminate ambiguities and check the feasibility; (5) *Test* – putting low-resolution prototypes into a real-world context and refining the solution idea further by gathering feedback from stakeholders (Hasso Plattner Institute 2023).

The application of design thinking in the field of education is recent and varied. However, design thinking can effectively promote transdisciplinarity by connecting students to the practice via real-world experiences (Pohl et al. 2018). Transdisciplinarity is defined as the process of extended knowledge production that encompasses a variety of actors and requires openness to the different forms of knowledge produced by scientific and lay communities (Mobjörk 2010). Design thinking takes an interdisciplinary approach and put learners into diverse teams which expose them to a plurality of knowledge and perspectives – this set-up encourages transdisciplinary learning as it allows learners to establish a shared understanding of the content while considering multiple perspectives (Taimur et al. 2022). We need to encourage transdisciplinarity (Darbellay 2015; Polk 2015) in order to deal with wicked, real-world problems while collaborating with different actors from society and science to produce robust and action-oriented knowledge (Biberhofer and Rammel, 2017). Design thinking, as a problem-solving approach, aims to deal with wicked problems, and in an educational setting it can encourage real-world, practice-oriented learning experiences – hence, it can be regarded as a transdisciplinary didactic tool for training learners to deal with wicked challenges.

Debate and criticism

There have been elaborate discussions regarding the transdisciplinary nature of design thinking among educators (Buchanan 1992; Cross 2006; Welsh and Dehler 2013), demanding that more research is needed on design thinking in education from a transdisciplinary perspective. Most of the literature focuses on design thinking in education within a specific discipline. For example, in architecture and design (Lloyd 2013), natural sciences, technology (Mubin et al. 2017), entrepreneurship, and business management (Nielsen and Stovang 2015; Von Kortzfleisch et al. 2013), medicine (Deitte and Omary 2019), and engineering (Dym et al. 2005).

However, the application of design thinking in an educational setting requires individuals to consider multiple perspectives from different actors during the empathize phase in order to frame the problem well. If perspectives from a single discipline are considered to frame the problem, it may solve the problem in the context of that discipline but may end up creating problems for other fields. Therefore, educators must pay attention to how design thinking is being used in educational settings to promote transdisciplinarity. Panke (2019) explains that design thinking is being used in educational settings in different forms, and that this pluralism is worth appreciating. However, educators should be aware that dealing with complex problems (in any field), which is the purpose of design thinking, requires individuals to consider multiple perspectives from diverse actors and be exposed to real-world experiences. Koria (2015) argues that deep and diverse disciplinary expertise should be brought into design thinking education to create disciplinary capabilities and promote interdisciplinarity, and that design thinking education should not be limited to teaching skills, but students should be engaged in the application of these skills, learning to collaborate across functions, people, and cultures.

Mobjörk (2010) discusses two types of transdisciplinarity, consulting and participatory. Consulting transdisciplinarity allows learners to consult stakeholders on their projects; however, the participation of actors is limited. On the other hand, participatory transdisciplinarity involves actors fully in the knowledge production process along with learners. Both kinds of transdisciplinarity can be promoted through design thinking, but educators should determine what kind of transdisciplinarity is relevant to their context and to the objectives of the educational program – specifically, when design thinking is applied in formal educational settings with time constraints related to its application. For example, one course is scheduled only for six months; in such a case, consulting transdisciplinarity works well to engage actors in the design thinking process (e.g. Taimur et al. 2022).

The application of design thinking in education also has ethical implications, as educators are required to organize diverse perspectives from different actors, disciplines, and functions. In this regard, educators (taking the role of facilitators) should ensure that they expose learners to all the perspectives (dominant and subservient) without concealing any perspectives based on their personal preferences to implement design thinking for promoting transdisciplinarity. The ethical considerations imply that applying design thinking may be difficult in undemocratic or authoritarian regimes as participants will not be able to encounter diverse perspectives or pluralism to develop the competencies required to frame problems in a real-world setting.

Current forms of implementation in higher education

Multiple cases have been reported in the literature that focuses on applying design thinking in education to foster transdisciplinarity. Cases from Germany, Romania, Japan, and Switzerland identify the essential elements of design thinking for transdisciplinary learning by showcasing how this approach can be adapted to varying contexts.

A Romanian–German case study (Peukert and Vilsmaier 2021) focuses on the application of a specific aspect of design thinking: design prototyping. Design prototyping is a method of constructing small two- and three-dimensional designs to develop and visualize ideas, which can then be discussed and revised. The application of design prototyping in transdisciplinary research processes differs from the application of prototyping in design disciplines in several ways. The role of a designer facilitates the prototyping process, prototyping context is detached from design or a product orientation, and the heterogeneity of the actors in a transdisciplinary team can be far more diverse than those of product development teams. Design prototyping in transdisciplinary research is strongly embedded in the process and is, therefore, in itself only an intermediate step in the overarching frame.

A case study by Taimur et al. (2022) identified the use of design thinking in higher education settings. That article used design thinking to implement a six-month-long, graduate-level field exercise course at a public university in Japan – where learning processes and environment were referred to as pedagogy. The course's overall objective was to deal with sustainability challenges in a specific context (Kashiwa no ha). Five-stage design thinking (adapted from Hasso Plattner Institute 2023) was used for pedagogy and all the stages of design thinking were implemented during the course, as indicated in Figure 2.

Figure 2. Stages of design thinking implemented during the field exercise in the sustainability science course (Taimur et al. 2022)

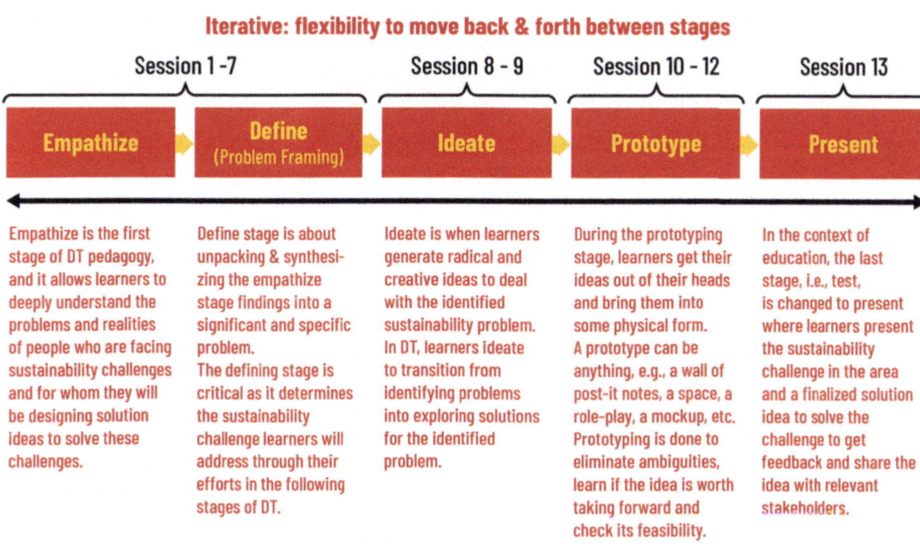

In this case, design thinking promoted consulting transdisciplinarity (Mobjörk 2010), where stakeholders collaborated with the learners to respond and react to their questions, projects, and research. The case study showed that design thinking can support the characteristics of transdisciplinarity as it allows learners: (a)

to understand the sustainability problems (which lack distinct system boundaries) and define these problems; (b) to move back and forth between different stages of design thinking and to iterate their shared understanding openly; (c) to collaborate with other learners and with the stakeholders outside the university settings. In order to establish a shared understanding of the problem and the solutions, it is crucial for learners to collaborate with each other.

Tackling Environmental Problems is a first-year bachelor-level course that is compulsory for students who are part of the Department of Environmental System Science at ETH Zurich. This intensive, year-long course builds on a tradition of using real-world case studies as a basis for group learning but also incorporates new elements of the integration of design thinking with systems thinking. The course provides students with the mindset and tools that allow them to analyze *and* take action on complex problems in the environmental field using the design thinking approach adapted from the process codified by Hasso Plattner Institute of Design at Stanford University (Figure 3).

Figure 3. The structure of design thinking followed in Tackling Environmental Problems (author's work adapted from Hasso Plattner Institute of Design 2023)

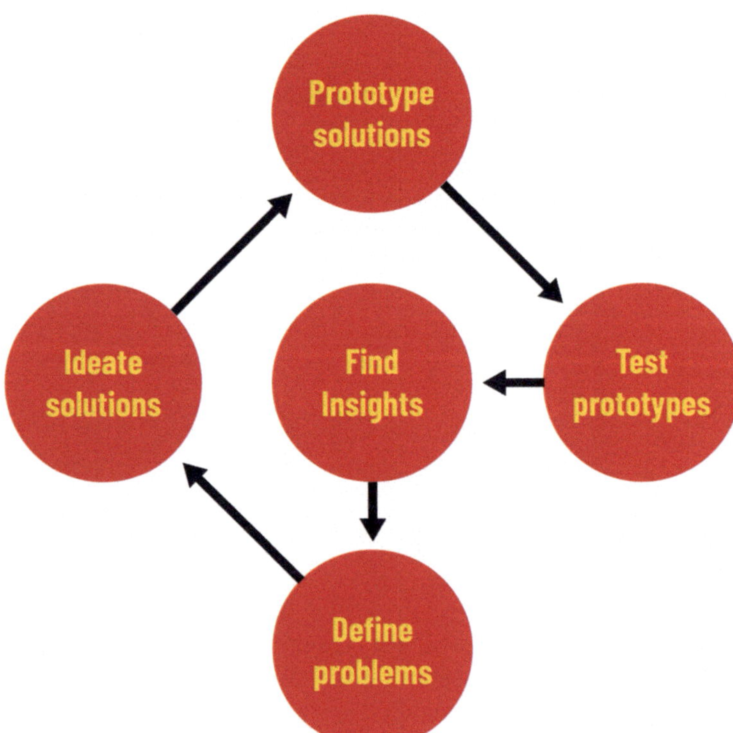

The transdisciplinary approach of this design-based course is exhibited in four ways. First, the course is an inherently interdisciplinary course because it not only introduces first-year students to methodologies or methods specific to a particular discipline, but also to the perspective of systems thinking as a whole. Second, the course makes use of approaches from a variety of disciplines to push students to understand the world through an analytical lens. It also introduces a design perspective to students that enables them to develop a sense of empathy for the people taking part in the systems they are studying and encourages them to identify leverage points in the system where action can be taken. Thus, the methodology interweaves qualitative systems modeling and soft systems methodology with design thinking (Pohl et al. 2021). Third, the course relies on the contribution of real-world stakeholders for the identification of case studies and for providing feedback to students throughout the course. Fourth, collaboration and self-reflection have a particular emphasis in the course.

Another case study described by Ambole (2020), Living Lab at the University of Nairobi, Kenya, highlighted a model for embedding design thinking in the African context. Through the Living Lab, design thinking workshops were facilitated for community engagement and to run problem-based learning programs from urban development. Most notably, workshops were facilitated in urban development for urban actors drawn from five East African countries; the purpose of the workshop was to engage local actors to co-create ideas for local solutions by contextualizing local expertise and knowledge through design thinking. Further, the study placed the design thinking work done by the Living Lab in the context of transdisciplinary research, as these workshops allowed the design teams to contextualize the design thinking tools to tackle local challenges. This contextualization enabled the researchers and learners to engage meaningfully with diverse multiple stakeholders and local actors to realize sustainable solutions. The study highlighted that design education, infused with transdisciplinary lenses, needs to gain prominence in Africa for sustainable development, because in Africa disciplinary boundaries are still concrete and the use of technology has not been rooted in urban development.

The cases presented in this section have identified that design thinking can promote transdisciplinarity by encouraging autonomy, openness to appreciate diverse perspectives, and active learning in the real-world setting. This kind of transdisciplinarity develops learners' ability to solve real-world complex problems. The way design thinking has been implemented in the case studies differs according to the context, indicating that it can be applied to diverse teaching environments, but it has to be contextualized and adapted to the specific learning objectives and context. In all the cases, participants (learners) followed a specific sequence of thinking: (a) building early thoughts and views (how to visualize them, thinking which aspects are important, where feedback is necessary); (b) communicating the early thoughts and views; (c) thinking of how to test them; (d) openly giving and receiving feedback on early thoughts and views (giving appreciative feedback, taking different perspectives); (e) iteratively developing and rejigging the thoughts and ideas. This thinking pattern can also be termed as a design mindset for transdisciplinarity, which can be promoted via design thinking.

Transdisciplinary collaborations also come with many challenges, like managing the diversity in a team, translating different ways of communication, leveling power imbalances, bridging different epistemological approaches, dealing with personal and team issues, and keeping a common focus (Peukert and Vilsmaier 2021). Design thinking as a process and its methods, like design prototyping, do not provide a general cure in overcoming these challenges, but by adding for example a visual and tangible dimension (through working with drawings or prototypes) to the collaborative process, a further cognitive mode is added that complements written text and the spoken word. The common design ideas and objects in design thinking processes can act as boundary objects (Carlile 2002; Heiss 2020; Leigh Star 2010), which are able to improve communication, level power imbalances, bridge epistemic difference, and create a common focus.

References

Ambole, Amollo. 2020. Embedding design in transdisciplinary research: Perspectives from urban Africa. *Design Issues* 36 (2): 28–40.

Biberhofer, Petra, and Christian Rammel. 2017. Transdisciplinary learning and teaching as answers to urban sustainability challenges. *International Journal of Sustainability in Higher Education* 18 (1): 63–83.

Brown, Tim. 2008. Design thinking. *Harvard Business Review* 86 (6): 84.

Brown, Tim, and Jocelyn Wyatt. 2010. Design thinking for social innovation. *Development Outreach* 12 (1): 29–43.

Buchanan, Richard. 1992. Wicked problems in design thinking. *Design Issues* 8 (2): 5–21.

Carlile, Paul R. 2002. A pragmatic view of knowledge and boundaries: Boundary objects in new product development. *Organization Science* 13: 442–55.

Cross, Nigel. 2006. *Designerly ways of knowing*. London: Springer.

Dam, Rikke, and Teo Yu Siang. 2018. *What is design thinking and why is it so popular?* Interaction Design Foundation. Available from https://www.interaction-design.org/literature/article/what-is-design-thinking-and-why-is-it-so-popular.

Darbellay, Frédéric. 2015. Rethinking inter-and transdisciplinarity: Undisciplined knowledge and the emergence of a new thought style. *Futures* 65: 163–74.

Deitte, Lori A., and Reed A. Omary. 2019. The power of design thinking in medical education. *Academic Radiology* 26 (10): 1417–20.

Dorst, Kees. 2010. The nature of design thinking. Available from https://opus.lib.uts.edu.au/handle/10453/16590.

Dorst, Kees. 2011. The core of "design thinking" and its application. *Design Studies* 32 (6): 521–32.

Dorst, Kees. 2015. *Frame innovation: Create new thinking by design*. Cambridge, MA: MIT Press.

Dym, Clive L., Alice M. Agogino, Ozgur Eris, Daniel D. Frey, and Larry J. Leifer. 2005. Engineering design thinking, teaching, and learning. *Journal of Engineering Education* 94 (1): 103–20.

Hasso Plattner Institute of Design, ed. at Stanford University. 2023. *Design thinking bootleg*. Available from https://dschool.stanford.edu/resources/design-thinking-bootleg.

Heiss, Leah. 2020. Iterative prototypes as "boundary objects": Facilitating interdisciplinary collaboration of a modular hearing aid. *Design Journal* 23 (6): 865–83.

Johansson-Sköldberg, Ulla, Jill Woodilla, and Mehves Çetinkaya. 2013. Design thinking: Past, present and possible futures. *Creativity and Innovation Management* 22 (2): 121–46.

Koria, Mikko. 2015. Four dimensions in learning design thinking: Capabilities, constraints, collaboration, and the diffusion of ideas. *Research in Arts and Education* 2: 15–26.

Krippendorff, Klaus. 2005. *The semantic turn: A new foundation for design*. Boca Raton: CRC Press.

Leigh Star, Susan. 2010. This is not a boundary object: Reflections on the origin of a concept. *Science, Technology and Human Values* 35 (5): 601–17.

Lloyd, Peter. 2013. Embedded creativity: Teaching design thinking via distance education. *International Journal of Technology and Design Education* 23 (3): 749–65.

Melles, Gavin, Neil Anderson, Tom Barrett, and Scott Thompson-Whiteside. 2015. Problem finding through design thinking in education. In *Inquiry-based learning for multidisciplinary programs: A conceptual and practical resource for educators*, eds. Patrick Blessinger and John M. Carfora, 191–209. Bingley: Emerald.

Mobjörk, Malin. 2010. Consulting versus participatory transdisciplinarity: A refined classification of transdisciplinary research. *Futures* 42 (8): 866–73.

Mubin, Omar, Mauricio Novoa, and Abdullah Al Mahmud. 2017. Infusing technology driven design thinking in industrial design education: A case study. *Interactive Technology and Smart Education* 14 (3): 216–29.

Nielsen, Suna Løwe, and Pia Stovang. 2015. DesUni: University entrepreneurship education through design thinking. *Education + Training* 57 (8/9): 977–91.

Panke, Stefanie. 2019. Design thinking in education: Perspectives, opportunities and challenges. *Open Education Studies* 1 (1): 281–306.

Peukert, Daniela, and Ulli Vilsmaier. 2021. Collaborative design prototyping in transdisciplinary research: An approach to heterogeneity and unknowns. *Futures* 132: 102808.

Plattner, Hasso, Christoph Meinel, and Ulrich Weinberg. 2009. *Design Thinking: Innovationen lernen- Ideenwelten öffnen*. München: mi-Wirtschaftsbuch.

Pohl, Christian, Pius Krütli, and Michael Stauffacher. 2018. Teaching transdisciplinarity appropriately for students' education level. *GAIA-Ecological Perspectives for Science and Society* 27 (2): 250–52.

Pohl, Johanna, Vivian Frick, Anja Hoefner, Tilman Santarius, and Matthias Finkbeiner. 2021. Environmental saving potentials of a smart home system from a life cycle perspective: How green is the smart home? *Journal of Cleaner Production* 312: 127845.

Polk, Merritt. 2015. Transdisciplinary co-production: Designing and testing a transdisciplinary research framework for societal problem solving. *Futures* 65: 110–22.

Rittel, Horst. 1971. Some principles for the design of an educational system for design. *Journal of Architectural Education* 26 (1–2): 16–27.

Rowe, Peter G. 1987. *Design thinking*. Cambridge, MA: MIT Press.

Schön, Donald. 1983. *The reflective practitioner: How professionals think in action*. New York: Basic Books.

Simon, Herbert A. 1969. *The sciences of the artificial*. 1st edition. Cambridge, MA: MIT Press.

Taimur, Sadaf, Motoharu Onuki, and Huma Mursaleen. 2022. Exploring the transformative potential of design thinking pedagogy in hybrid setting: A case study of field exercise course, Japan. *Asia Pacific Education Review* 23 (4): 571–93.

Von Kortzfleisch, Harald F. O., Dorothée Zerwas, and Ilias Mokanis. 2013. Potentials of entrepreneurial design thinking for entrepreneurship education. *Procedia – Social and Behavioral Sciences* 106: 2080–92.

Von Thienen, Julia, Christoph Meinel, and Claudia Nicolai. 2014. How design thinking tools help to solve wicked problems. In *Design thinking research*, eds. Larry Leifer, Hasso Plattner, and Christoph Meinel, 97–102. New York: Springer.

Welsh, M. Ann, and Gordon E. Dehler. 2013. Combining critical reflection and design thinking to develop integrative learners. *Journal of Management Education* 37 (6): 771–802.

Education for Sustainable Development

Michael Brennan and Lyda Patricia Sabogal-Paz

Definition

The word sustainable derives from the Latin *sustinere*, which means "to hold up, hold upright", or "furnish with means of support" (Stevenson 2010, 4612). The uses and meaning of the word have evolved gradually to include the provision of the necessities of life. This contemporary usage can be understood in two ways: as a broad concept relating to ecosystems, but also in a narrow sense relating to human well-being (Harrington 2016). In universities, sustainability encompasses a wide range of activities and is commonly identified using the term *education for sustainable development* or ESD, defined as "holistic and transformational education which addresses learning content and outcomes, pedagogy, and the learning environment" (UNESCO 2020, 8). Engagement with sustainability and sustainable development in higher education continue to expand in scope since the publication of a comprehensive review in 2016 (Barth et al. 2016). Recent developments have highlighted the significance of transdisciplinary approaches to producing and circulating knowledge, as well as transforming higher education for global sustainability (Parr et al. 2022).

However, each of the ESD words is contested, and there are multiple interpretations in higher education. A sense of the definitional challenges is highlighted in the first volume of the *International Journal of Sustainability in Higher Education* (Leal Filho 2000). The different meanings of the word sustainability in different languages are illustrated, encompassing, for example, the long-term use of resources; how social and economic development takes place; the ethics of development; and the environmental impact of development. Combining sustainability with the different perspectives of the global north and south, post-colonial societies, and post-conflict settings generates additional complexity (Janssens et al. 2022).

How does one make sense of this kaleidoscope of ideas? Three key points can be made. Firstly, how do we define sustainable development? A seminal United Nations publication, *Our Common Future* (also called the Brundtland Report), describes sustainable development as "meet[ing] the needs of the present without compromising the ability of future generations to meet their own needs" (Brundtland 1987, 43). The broad scope of this definition has been questioned and multiple alternative definitions proposed, but the sentiment expressed is enduring: the idea that sustainability encompasses the intergenerational needs of humanity. Nevertheless, the redefinition of sustainable development is ongoing, with, for example, the introduction of new perspectives such as sustainable entrepreneurship within planetary boundaries (Hummels and Argyrou 2021).

Secondly, how do we engage with the idea of sustainable development? UNESCO, the United Nations agency tasked with education relating to sustainable development, emphasizes the nature of ESD as: "a lifelong learning process and an integral part of quality education that enhances cognitive, social and emotional and behavioural dimensions of learning" (UNESCO 2020, 8). However, at least two approaches along a gradient of types of engagement have been recognized in educational settings. A narrow approach conceives of ESD as an addition to, or extension of, conventional courses and taught within established academic disciplines. This contrasts with a broad approach to ESD that is holistic and privileges education reform and transformation (Sterling 2021).

Thirdly, how do we practice sustainable development? A considerable effort has been made over the last 20 years to identify the knowledge, skills, and attitudes (i.e. competencies) that are relevant outcomes for higher education learners. Different frameworks and models have been used, of which the most influential is that proposed by Wiek et al. (2011). This framework is the most accepted among experts in the field and informs the learning objectives for achieving the United Nations Sustainable Development Goals (Redman et al. 2021).

In summary, the multiple definitions, approaches, and frameworks associated with ESD have highlighted the limitations of partial and discipline-based approaches to sustainability. To overcome these limitations, it is argued that transdisciplinarity must become the lens through which the relevance of all disciplinary research and teaching relating to sustainability needs to be understood.

Background

Since the early 1960s the emergence of debate, research, and actions relating to sustainable development can be tracked though a series of historical events triggered by increasing concerns about how human actions impact the environment. Table 1 identifies and includes highlights relating to some of the more influential ESD-related events which predate the introduction of the UN's Sustainable Development Goals in 2015.

The historical development of education for sustainable development outlined in Table 1 has influenced, and is influenced by, transdisciplinary thinking. The seminal work of Erich Jantsch (1970) viewed education as evolving from "training for well-defined, single-track careers and professions [...] towards an education which enables judgement of complex and dynamically changing situations" (Jantsch 1970, 407). Underpinning this evolution was the increasing adoption by universities of transdisciplinary approaches to teaching and research as a means of increasing the capability for innovation. More recently, Scholz (2020) has highlighted the significance of transdisciplinary approaches for transitioning to sustainable development and reiterated the role of universities for the public good. The 50-year period separating the work of Jantsch and Scholz has witnessed a wealth of research and practice reported in dedicated academic journals and handbooks. In addition, interest in the approach is reflected in the emergence of global communities of practice, such as the Network for Transdisciplinary Research (td-net) and the International Center for Transdisciplinary Research (CIRET).

Table 1. History of education for sustainable development (adapted from UNESCO 2020, 65)

Year	Event or publication	UNESCO ESD highlights
1972	United Nations Conference on the Human Environment, Stockholm.	The need for education in environmental matters.
1987	Our Common Future (Brundtland Report).	Sustainable development defined.
1992	The United Nations Conference on Environment and Development (Rio Summit), Rio de Janeiro.	Education is critical for promoting sustainable development.
2002	World Summit on Sustainable Development (Johannesburg Summit), Johannesburg.	Designated 2005–2014 as the United Nations Decade of Education for Sustainable Development.
2005	UN Decade of Education for Sustainable Development (2005–2014).	Reorientation of education globally towards a central goal: to learn to live and work sustainably.
2009	UNESCO World Conference on ESD, Bonn.	Emphasized ESD as a "life-saving measure" for promoting ESD as "an investment in the future".
2012	The United Nations Conference on Sustainable Development (Rio +20), Rio de Janeiro.	The need to integrate sustainable development more actively into education.
2014	UNESCO World Conference on ESD, Aichi-Nagoya.	Launched the Global Action Programme: scaling up action in education to accelerate progress towards sustainable development.

What can be synthesized from the coevolution of sustainability and transdisciplinarity of relevance to education? Arguably, the most profound concept is that of mutually dependent knowledge. This idea is underpinned by a typology consisting of systems, target, and transformation knowledge, which together give meaning to a particular interpretation of a problem area (Brennan and Rondón-Sulbarán 2019; Pohl and Hirsch Hadorn 2007). Systems knowledge relates to a current situation and questions about the interpretations, origins, interactions, and trends relating to a problem. Target knowledge looks to desired future states and questions relating to better ways of operating and behaving. Transformation knowledge examines the means of changing from a current situation to a desired future state. The dynamic relationship between the three types of knowledge is important and mutually dependent. In other words, the knowledge about, and assumptions relating to, a particular challenge in sustainable development are provisional and not absolute. This interdependence creates a particular way of knowing about a challenge: such knowledge is conditional and interpretative (Popper 1959, 79).

The contingent character of knowledge, implicit in sustainable development research, has resulted in an emphasis on key ESD competencies and specific learning outcomes. The reason for this focus is partly due to the continually evolving interpretation of sustainable development, as well as the need to span the different "worlds" of the physical environment, societies, and economies – for example, de Haan's articulation of the concept of *Gestaltungskompetenz* ("shaping competence") relating to the capacity to act and solve problems in a particular setting (de Haan 2010, 318). In addition, several frameworks have been developed with attendant lists of different types of competencies, though this approach has been criticized repeatedly for the "laundry list" manner of articulating such competencies (Brundiers et al. 2021). Wiek's introduction of a model-based framework (Wiek et al. 2011) resulted from a reported convergence in the education literature around a set of key competencies in sustainability: systems thinking; futures or anticipatory thinking; values or normative thinking; strategic and action-oriented thinking; and collaboration or interpersonal approaches. These were subsequently broadly adopted, with additions, by UNESCO in the identification of the learning objectives for achieving the United Nations Sustainable Development Goals or SDGs (UNESCO 2017).

Debate and criticism

The Sustainable Development Goals act as a framework that identifies 17 end-states that are important for human survival on earth. Education relates to all 17 areas and in addition has a dedicated focus within SDG4: Quality Education. SDG4 encompasses a series of targets with associated indicators that explain what is involved in the each. Table 2 highlights key ESD-related events and publications linked to the SDGs.

Table 2. Education for sustainable development and the Sustainable Development Goals

Year	Event or publication	UNESCO ESD highlights
2015	UN introduction of the SDGs and the 2030 Agenda for global transformation.	Global Action Programme on ESD (2015–2019) aimed at promoting concrete actions in ESD.
2017	UNESCO ESD goals: learning objectives identified.	"ESD requires is a shift from teaching to learning ... inter- and transdisciplinarity ... linking of formal and informal learning" (UNESCO 2017, 2).
2018	UNESCO highlights issues and trends in education for sustainable development.	"ESD entails rethinking the learning environment, physical and virtual" (UNESCO 2018, 8).
2019	40th Session of the UNESCO General Conference.	Adoption of a framework for the implementation of ESD beyond 2019 (2020–2030).
2020	Education for Sustainable Development Goals: A Roadmap.	"Often ESD is interpreted with narrow focus on topical issues rather than with a holistic approach" (UNESCO 2020, 9).
2022	Berlin Declaration on Education for Sustainable Development.	"ESD must be based on ... respect for nature, as well as human rights, democracy, the rule of law, non-discrimination, equity and gender equality" (UNESCO 2022, 3).
2022	Knowledge-driven action: transforming higher education for global sustainability.	The imperative need for institutions to become open, fostering epistemic dialogue and integrating other ways of knowing (Parr et al. 2022, 14).

The emerging discourse summarized in Table 2 emphasizes the need for transdisciplinarity to tackle the complexity of sustainability challenges. This need is based on a recognition that individual scientific disciplines can only ever provide partial solutions: challenges can be perceived and interpreted in different ways. This takes place through knowledge integration and a recognition of differing societal and scientific discourses (Jahn et al. 2012). Increasingly, the evolution of ESD approaches is being viewed as a series of phases. An initial orientation phase (1970–1990), with a focus on environmental issues; a secondary transition phase (1990–2000), with the broadening of the debate to include development themes; and finally, the current expansionary phase (2015 onwards), with a focus on sustainability as a key agent of change (Michelsen et al. 2016). However, debate and criticism highlight a concern that the ESD concept is more often described than defined. This is unsurprising, as "no one discipline can claim education for sustainable development" (UNESCO 2005, 31). ESD is typically explained in terms of frameworks of competencies, despite no explicit consensus on a specific framework (Brundiers et al. 2021). More critical debate suggests that international efforts to promote ESD have been hampered by lack of clarity on how to implement this form of education (Vare et al. 2019). Further, there is a need to include other, non-European, ways of knowing, including indigenous perspectives (Rondón-Sulbarán et al. 2021) and experts from Latin America, Middle Eastern, and African higher education with alternative perceptions on development.

Current forms of implementation in higher education

The aspiration is for education for sustainable development in higher education research and teaching to become transdisciplinary in perspective and transformational in practice (Parr et al. 2022). This transformational aspect of education encompasses learning content and outcomes, pedagogy, and the learning environment itself. How this takes place in practice is conditioned, arguably, by the contrasting narrow and broad approaches to education for sustainable development. A narrow approach deconstructs ESD into component parts (students, faculty staff, and institutions) and then looks at novel ways in which learning and teaching engage with sustainable development. In this way discipline-based curricula are modified and redesigned to reflect the sustainability agenda. Faculty staff are encouraged to collaborate with colleagues from different disciplines, and institutions register and promote such initiatives as examples of ESD with relevance to regional or national economies. This approach arguably fails to understand the inherent complexity of sustainability challenges and at worst can be viewed as an optional addition to education practices.

An alternative, broad approach to education for sustainable development recognizes the complexity of sustainable development and the dynamic nature of human actors, social groups, and institutions involved in education. A useful way of understanding the implications of this approach is to view education as an innovation system (Jantsch 1970), consisting of a nested hierarchy of analytic dimensions (Geels 2004). This approach privileges different forms of innovation activities at different levels. A micro-level involves novel configurations or niches that are shaped by an existing education regime in a particular local or regional setting – for example, Utrecht University in the Netherlands with its emphasis on transformative hubs in Future Food; Negative Emissions; Transforming Cities; Water, Climate, and Future Deltas; and Circular Economy and Society (Parr et al. 2022, 40). A second, meso-level is envisioned as patchworks of regimes encompassing sociocultural elements; market networks; policy; science; and technology. These regimes constantly evolve and interact with micro-level learning and teaching innovations. For example, the Quality Assurance Agency for Higher Education in the United Kingdom has produced a guide (QAA/Advance HE 2021) that promotes practical actions for higher education across British universities. The opportunity for ESD to reinforce individual institutional objectives is explicit and includes the promotion of transdisciplinary learning, employability, enterprise, entrepreneurship, and civic engagement. Key competencies are linked to an overarching model of learning which is identified as central to the transformational learning experience. A third, macro-level is conceived as an evolving sociotechnical landscape that is transformed by a patchwork of meso-regimes. The dynamic nature of this evolving macro-landscape creates emerging opportunities. For example, GreenComp, the European competence framework (Bianchi et al. 2022) is an organizing framework consisting of: developing sustainability values, embracing complexity, envisioning sustainable futures, and acting for sustainability. The benefit of such a micro-, meso-, and macro-innovation systems approach is that it provides a more contextualized and dynamic understanding of the increasing numbers of case study examples of ESD in different contexts across the globe, as illustrated in Table 3.

Table 3. Learning and teaching education for sustainable development

Source	University	Education for sustainable development as ...
Baumber 2022	University of Technology Sydney, Australia	A transdisciplinary approach to facilitate transformative learning through a focus on real-world challenges.
Taylor et al. 2021	Tampere University, Finland	Skills and competencies required and effective pedagogic practices that could help educate future professionals.
Cavalcanti-Bandos et al. 2021	Higher education institutions in Peru, Brazil, and Colombia	Critical thinking surrounding rational bases for exploring the environment. Organizational development, supporting culture, and planning for sustainability integration.
Galvão et al. 2020	University of Lisbon, Portugal	Student learning as collaborative experience towards transdisciplinary knowledge creation.
Jia et al. 2019	Tongji University, China	Comprehensive transformation of curricula and pedagogy to bring coordinated innovation at multiple levels.
Awuzie and Emuze 2017	Central University of Technology, South Africa	Implementation drivers such as cost-related, regulations, competitive advantage, and community engagement.

In summary, the idea of education for sustainable development is continually evolving but broadly speaking can be understood in three ways: (1) Education *about* sustainable development with an emphasis on raising awareness. (2) Education *for* sustainable development as a way of widening perspectives with a view to influencing practice. (3) Education *as* sustainable development involving behavioral and paradigm change.

References

Awuzie, Bankole, and Fidelis Emuze. 2017. Promoting sustainable development implementation in higher education: Universities in South Africa. *International Journal of Sustainability in Higher Education* 18 (7): 1176–90.

Barth, Matthias, Gerd Michelsen, Marco Rieckmann, and Ian Thomas, eds. 2016. *Routledge handbook of higher education for sustainable development.* Abingdon: Routledge.

Baumber, Alex. 2022. Transforming sustainability education through transdisciplinary practice. *Environment, Development and Sustainability* 24 (6): 7622–9.

Bianchi, Guia, Ulrike Pisiotis, and Marcelino Cabrera Giraldez. 2022. *GreenComp: The European sustainability competence framework.* Luxembourg: JRC, EU.

Brennan, Michael, and Janeet Rondón-Sulbarán. 2019. Transdisciplinary research: Exploring impact, knowledge, and quality in the early stages of a sustainable development project. *World Development* 122: 481–91.

Brundiers, Katja, Matthias Barth, Gisela Cebrián, Matthew Cohen, Liliana Diaz, Sonya Doucette-Remington, Weston Dripps, Geoffrey Habron, Niki Harré, Meghann Jarchow, Kealalokahi Losch, Jessica Michel, Yoko Mochizuki, Marco Rieckmann, Roderic Parnell, Peter Walker, and Michaela Zint. 2021. Key competencies in sustainability in higher education – Toward an agreed-upon reference framework. *Sustainability Science* 16 (1): 13–29.

Brundtland, Gro Harlem. 1987. *Report of the World Commission on Environment and Development: Our common future.* United Nations General Assembly Document A/42/427.

Cavalcanti-Bandos, Melissa Franchini, Silvia Quispe-Prieto, Alberto Paucar-Caceres, Toni Burrowes-Cromwel, and Héctor Heraldo Rojas-Jiménez. 2021. Provision of education for sustainability development and sustainability literacy in business programs in three higher education institutions in Brazil, Colombia, and Peru. *International Journal of Sustainability in Higher Education* 22(5): 1055–86.

de Haan, Gerhard. 2010. The development of ESD-related competencies in supportive institutional frameworks. *International Review of Education* 56 (2): 315–28.

Galvão, Cecília, Cláudia Faria, Wanda Viegas, Amélia Branco, and Luís Goulão. 2020. Inquiry in higher education for sustainable development: Crossing disciplinary knowledge boundaries. *International Journal of Sustainability in Higher Education* 22 (2): 291–307.

Geels, Frank W. 2004. From sectoral systems of innovation to socio-technical systems: Insights about dynamics and change from sociology and institutional theory. *Research Policy* 33 (6): 897–920.

Harrington, Lisa M. Butler. 2016. Sustainability theory and conceptual considerations: A review of key ideas for sustainability, and the rural context. *Papers in Applied Geography* 2 (4): 365–82.

Hummels, Harry, and Aikaterini Argyrou. 2021. Planetary demands: Redefining sustainable development and sustainable entrepreneurship. *Journal of Cleaner Production* 278: 123804.

Jahn, Thomas, Matthias Bergmann, and Florian Keil. 2012. Transdisciplinarity: Between mainstreaming and marginalization. *Ecological Economics* 79: 1–10.

Janssens, Lise, Tom Kuppens, Ingrid Mulà, Egle Staniskiene, and Anne B. Zimmermann. 2022. Do European quality assurance frameworks support integration of transformative learning for sustainable development in higher education? *International Journal of Sustainability in Higher Education* 23 (8): 148–73.

Jantsch, Erich. 1970. Inter-and transdisciplinary university: A systems approach to education and innovation. *Policy Sciences* 1: 403–28.

Jia, Qian, Ying Wang, and Li Fengting. 2019. Establishing transdisciplinary minor programme as a way to embed sustainable development into higher education system: Case by Tongji University, China. *International Journal of Sustainability in Higher Education* 20 (1): 157–69.

Leal Filho, Walter. 2000. Dealing with misconceptions on the concept of sustainability. *International Journal of Sustainability in Higher Education* 1 (1): 9–19.

Michelsen, Gerd, Maik Adomßent, Pim Martens, and Michael von Hauff. 2016. Sustainable development – Background and context. In *Sustainability science*, 5–29. Dordrecht: Springer.

Parr, Adrian, Agnes Binagwaho, Andy Stirling, Anna Davies, Cheikh Mbow, Dag Olav Hessen, Helena Bonciani Nader, Jamil Salmi, Melody Brown Burkins, Seeram Ramakrishna, Sol Serrano, Sylvia Schmelkes, Tong Shijun, and Tristan McCowan. 2022. *Knowledge-driven actions: Transforming higher education for global sustainability: Independent Expert Group on the Universities and the 2030 Agenda*. Paris: UNESCO Publishing.

Pohl, Christian, and Gertrude Hirsch Hadorn. 2007. *Principles for designing transdisciplinary research*. Munich: oekom.

Popper, Karl Raimund. 1959. *The logic of scientific discovery*. London: Hutchinson [1923].

QAA/Advance HE [Quality Assurance Agency for Higher Education/Advance Higher Education Academy]. 2021. *Education for sustainable development guidance*. Quality Assurance Agency for Higher Education and Advance HE. Gloucester: QAA.

Redman, Aaron, Arnim Wiek, and Matthias Barth. 2021. Current practice of assessing students' sustainability competencies: A review of tools. *Sustainability Science* 16 (1): 117–35.

Rondón-Sulbarán, Janeet, Ian Balam, and Michael Brennan. 2021. A transdisciplinary approach to water access: An exploratory case study in indigenous communities in Chiapas, Mexico. *Water* 13 (13): 1811.

Scholz, Roland W. 2020. Transdisciplinarity: science for and with society in light of the university's roles and functions. *Sustainability Science* 15 (4): 1033–49.

Sterling, Stephen. 2021. Educating for the future we want. Opening essay for GTI Forum: The Pedagogy of Transition. *Great Transition Initiative*. Available from https://greattransition.org/gti-forum/pedagogy-transition-sterling.

Stevenson, Angus, ed. 2010. *Oxford dictionary of English*. Oxford: Oxford University Press.

Taylor, Jonathon, Salla Jokela, Markus Laine, Juho Rajaniemi, Pekka Jokinen, Liisa Häikiö, and Antti Lönnqvist. 2021. Learning and teaching interdisciplinary skills in sustainable urban development – The case of Tampere University, Finland. *Sustainability* 13 (3): 1180.

UNESCO. 2005. *United Nations Decade of Education for Sustainable Development (2005–14)*. Paris: UNESCO.

UNESCO. 2017. *Education for sustainable development goals: Learning objectives*. Paris: UNESCO.

UNESCO. 2018. *Issues and trends in education for sustainable development*. Paris: UNESCO.

UNESCO. 2020. *Education for sustainable development goals: A roadmap*. Paris: UNESCO.

UNESCO. 2022. *Berlin Declaration on Education for Sustainable Development*. Paris: UNESCO.

Vare, Paul, Grete Arro, Andre de Hamer, Giovanna Del Gobbo, Gerben de Vries, Francesca Farioli, Chrysanthi Kadji-Beltran , Mihkel Kangur, Michela Mayer, Rick Millican, Carlien Nijdam, Monika Réti, and Aravella Zachariou. 2019. Devising a competence-based training program for educators of sustainable development: Lessons learned. *Sustainability* 11 (7): 1890.

Wiek, Arnim, Lauren Withycombe, and Charles L. Redman. 2011. Key competencies in sustainability: A reference framework for academic program development. *Sustainability Science* 6 (2): 203–18.

Embodied Learning

Lucy Allen, Susanne Pratt, Bem Le Hunte, Jacqueline Melvold, Barbara Doran, Giedre Kligyte, and Katie Ross

Definition

Within transdisciplinary education and beyond, educators are increasingly incorporating embodied learning as part of an expanded understanding of intelligence and cognition. Embodied learning is grounded in the recognition that experience, perception, and knowledge are shaped through the activity of our body in relation to the world and also shape the world (Dewey 1997; Goldin-Meadow 2009; Lakoff and Johnson 2020; Lindgren and Johnson-Glenberg 2013). The term embodied learning is etymologically based in the Old High German *botah*, meaning "body". In recognizing that "the human mind is a relational and embodied process that regulates flows of energy and information" (Siegel 2010, 52), embodied learning offers an educational approach that integrates physical and sensory learning alongside "the work occurring between the ears" (Henriksen et al. 2015, 8). This is not to suggest that simply engaging the body in physical activity within education constitutes embodied learning. Rather, embodied learning occurs when "the meaning of what is learned is grounded specifically in body movement and perception" (Nathan 2021, 81). In this chapter, embodied learning is applied as an umbrella term to encompass the various ways learning is discussed and realized across different fields, where there is an emphasis on learning that "joins body and mind in physical and mental acts of knowledge construction" (Nguyen and Larson 2015, 332). Embodied learning moves beyond mind–body dualism to embrace learning as a holistic, integrated, and felt act. For this reason, it can be seen as a crucial component of transdisciplinary learning in that it supports "attentive [ness] to the body and its experiences as a way of knowing" (Freiler 2008, 40).

In transdisciplinary education, embodied learning supports students from different disciplinary backgrounds to move across disciplinary boundaries and make sense of complex real-world challenges through explicit engagement of their bodies and sensory systems, including reflexivity, creativity, and complexity through attentiveness to body–mind–environment interactions. It also enables students to develop core transdisciplinary competencies in integrating and implementing diverse perspectives, wisdom, and knowledge. Keeping mind–body–environment relations central to education provides opportunities for students to develop meaningful views of themselves as competent learners within a "community of experience" (Nguyen and Larson 2015, 339). It also encourages students to move back and forth between the background, histories, beliefs, choices, experiences, and expressions that influence their own and others' actions (Satina and Hultgren 2001) to support reflexive transdisciplinary ways of being and knowing.

Background

As an approach for transdisciplinary learning, embodied learning has the potential to effectively support students to integrate their felt and emotional experiences, connect knowledge with action and make sense of complex real-world challenges. The 20th century has brought increased interest from academics across disciplines in the role of the body in education, including transdisciplinary learning.

Embodied learning has established itself as a significant field of research and practice in recent decades (Lindgren and Johnson-Glenberg 2013) as scholars adopt an expanded understanding of cognition and acknowledge the limitations of traditional education models. The body has historically been dismissed in education due to its subjective nature and perceived irrelevance in processes of knowledge construction (Dewey 1997; Henriksen et al. 2015; Johnson 1987). Yet at the turn of the 20th century expanded notions of cognition and intelligence emerged. Most notably, embodied cognition is informed by the work of contemporary philosophers such as John Dewey (1997), Martin Heidegger (1975), and Maurice Merleau-Ponty (1962, 1964). Embodied cognition acknowledges that "the brain is not disconnected from the rest of the body and solely responsible for cognition, but an organ occupied with processing perceptions experienced in the body" (Branscombe 2019, 3). Embodied learning offers a holistic approach to education in which the learner's physical, emotional, mental, and spiritual development is supported, and assumptions about the nature of knowledge are challenged (Forgasz and McDonough 2017). As the body is put back inside the mind (Johnson 1987, 7), learners are supported to engage their felt and bodily experiences as a means to make sense of knowledge and the world in new ways.

Embodied learning has a natural affinity with the goals of transdisciplinary education in striving for an "equilibrium between analytic intelligence, feelings, and the body" (Nicolescu 2012, 15). Both embodied learning and transdisciplinary education have emerged in response to postmodern views around the nature of knowledge that recognize it as not static or rational but complex, indeterminate, interpersonal, and contested. To grapple with and develop complex knowledge, learners require higher-level skills, competencies, and diverse approaches to observing and making sense of complex knowledge and social problems (Murray 2009). Mishra et al. (2011, 25) argue that *embodied thinking* is one of seven transdisciplinary skills that support individuals in facing challenging situations through engagement with empathy and by integrating the "physical, mental and the emotional aspects of how we think and experience the world". Within transdisciplinary education, embodied learning is both an approach to learning and a transdisciplinary competency that should be developed. Despite increasing interest and recognition of the value of holistic, integrative education – such as transdisciplinary education and embodied learning – there are still challenges around bridging the gap between these domains and theory and practice, as the next section highlights.

Debate and criticism

The 20th century has seen great theoretical advancement in the domain of embodied learning, yet further empirical research is required to better understand how embodied learning is applied within educational contexts, including that of transdisciplinary higher education. There is still a stigma around embodiment, which presents challenges, as articulated by Gregory (2006, 316): "students of my generation were taught to view embodiment as a circus sideshow, a vulgar distraction like the fat man and the bearded lady who, we assumed, had nothing in common with the glittering flights of mind exhibited by the intellectual trapeze artists soaring high above the centre ring of the educational circus tent".

Table 1. Overview of debate, criticism and limitations of embodied learning research

Debate, criticism, and limitations	Details	References
Criticism and resistance by learners, practitioners, and the institution.	Embodied learning challenges our understanding of what it means to teach and learn. Students and educators have been socialized to see learning as the act of knowledge transfer (involving sitting, thinking, and repeating back information). As such, embodied learning can be seen as alternative, unnatural, unintellectual, and uncomfortable.	Gregory 2006; Gustafson 1999; Monk et al. 2015; Nguyen and Larson 2015

Debate, criticism, and limitations	Details	References
Limited scholarship addressing embodied learning in transdisciplinary higher education.	Research connecting the domains of embodied learning and transdisciplinary higher education is limited. Existing scholarship is currently focused on the application of embodied learning within Kindergarten to Year 12 in contrast to higher education, and when discussed in relation to higher education, the focus is typically discipline-specific rather than transdisciplinary in focus.	Henriksen et al. 2015; Monk et al. 2015
Embodied learning requires time, resources, training and support, which can be challenging in resource-constrained environments.	Students and practitioners must be supported to develop an understanding of the value of embodied learning. There must also be careful consideration of body politics within the classroom and the role of gender, power, culture, trauma, and emotion within embodied learning. If embodied learning is to be supported in academia, education cultures and structures need to change, not just the educators themselves. Institutions must also recognize and support the labor-intensive nature of embodied approaches.	Fugate et al. 2019; Lipson Lawrence 2012; Macintyre Latta and Buck 2008; Nguyen and Larson 2015; Wagner and Shahjahan 2015
Debates around what constitutes embodied learning and when it is appropriate.	It is easy to assume that bodily engagement is always necessary and effective in learning. Yet scholars seek to dispel this assumption, recognizing that embodied learning does not claim that all movement and bodily engagement supports learning. There should be careful consideration of how and when embodied learning can support meaning-making.	Nathan 2021; Skulmowski and Rey 2018

Current forms of implementation in higher education

There are various ways in which embodied learning is implemented within transdisciplinary higher education. However, the emergent nature of both domains means limited scholarship and few case studies explicitly discuss implementation. Therefore, the following discussion does not offer an exhaustive list but instead highlights the various ways embodied learning is currently being implemented within transdisciplinary higher education around the globe, and the need for institutions and practitioners to make these applications more explicit. Given the emphasis on collaborative, real-world, and integrative learning, transdisciplinary education naturally leans towards and draws on embodied learning. Movement and physical activity are often integrated into the classroom as students work collaboratively and in hands-on ways with one another as part of transdisciplinary teams and engage in real-world contexts where students interact with stakeholders to understand and respond to complex challenges. Furthermore, bodily sensations, feelings, and emotions are required to make sense of and integrate different ways of knowing. As a result, higher education institutions and practitioners may unconsciously engage students in embodied learning when designing and delivering transdisciplinary learning. However, the following examples highlight how embodied learning can be engaged intentionally within transdisciplinary higher education, both as an embedded curricula approach and through specific activities.

Within the undergraduate degree of the Bachelor of Creative Intelligence and Innovation at the University of Technology Sydney, embodied learning forms a core approach to transdisciplinary teaching and learning. The course was launched in 2014 as a combined degree that enables students from 25 different "core degrees" from faculties across the university to undertake transdisciplinary learning alongside another undergraduate degree (e.g. design, science, business, communications, and engineering). Within the Bachelor of Creative Intelligence and Innovation, embodied learning is used to foster engagement with complexity, reflexivity, and creativity to support students from different disciplinary backgrounds to work collaboratively on complex real-world challenges with different people (Allen et al. 2021). For example, in a complexity-focused subject, students respond to an assessment brief asking them to enact a complex system in which their challenge occurs. Students use systems thinking and complexity frameworks, such as Dave Snowden's Cynefin framework (Kurtz and Snowden 2003) and Donella Meadows' (2009) leverage points, to intervene in a system and explore extreme challenges facing humanity today, such as out-of-control bushfires, drought, unsustainable farming practices, and overconsumption. These frameworks are introduced alongside different creative embodied practices, such as role-play, improvisation, and "Complex Systems Tableau" (Allen et al. 2021), drawing on systems thinking and theatre-based practices, such as the drama convention of tableau, in which participants make a frozen scene using their bodies, striking different poses and facial gestures. As a result, students design various embodied enactments of complex systems, from games in which the rest of the cohort participates to poetic imaginings of the system from more-than-human perspectives. This embodied experience contributes to students' capacity to question their assumptions and develop a situated, collective, and relational understanding of the system they are engaging with, including an empathetic sensibility to different stakeholders' values and perspectives. Empathy towards other actors in the system and understanding of our complicity in propagating problems was much harder to grasp for students in the past when more analytical approaches to systems thinking and complexity were adopted.

In other cases, embodied learning is applied within specific subjects as part of a broader course to enhance creativity and support transdisciplinary learning in higher education. At the University of Vermont in the USA, students studying the environmental sciences can participate in subjects that draw on an approach called Kitchen-based Learning (O'Neil 2015), developed by environmental sciences instructor Joy Kcenia O'Neil. Within the subject "Environmental cooking", students are invited to learn about the interdisciplinary field of sustainability through experience and interactions in the kitchen "ecosystem". Creativity is enhanced through the coming together of seemingly disparate domains – sustainability and cooking – and by engaging students' sensory systems and embodied emotional states, feelings, and moods. For example, students take part in sensory-based activities such as "Palate solving", where they describe and reflect on the texture, taste, and smell of their dish to "deeply experience their food by connecting to their senses and what memories might be elicited or insights gained" (O'Neil 2016, 326). In this activity, students use their bodies to access hidden wisdom and understanding, which leads to collective and unanticipated learning, trust, and relationships that create the conditions for deep, enriched learning. O'Neil (2016, 328) also describes the broad range of emotions that are "all over the place", and interactions, both human and non-human, that take place within the kitchen as "students go back and forth from the refrigerator, stove, cupboards, talking to one another, with clanking sounds of chopping and cooking ... laughing, silence and focus". Transdisciplinary embodied experiences, such as kitchen-based learning, value and amplify the emotional and energetic states that inform cognition, perception, and creativity (Cherukunnath and Singh 2022) and play a crucial role in transdisciplinary higher education.

Transdisciplinary embodied learning can also be implemented via specific learning activities rather than a specific curricular approach within a subject or course. The Swiss Academies of Arts and Sciences emphasizes studies in the sustainable sciences and society and focuses on strengthening the exchange across scientific disciplines (SCNAT 2023). Scholars and educators within the academy have developed a range of activities that utilize embodied learning approaches to support transdisciplinary learning. For example, the Actor Constellation is an activity developed by Christian Pohl that draws on role-play and physical mapping to unpack different perspectives and interactions relating to a central challenge question. Participants are invited to represent different scientific and societal actors, positioning themselves physically and spatially in relation to the challenge question and other actors to question and demonstrate their relevance to the challenge (Pohl 2020). Embodied activities such as the Actor Constellation support learners in moving beyond current perceptions and biases they may hold in relation to the challenge and to stakeholders. Furthermore, it supports the integration of different perspectives and ways of thinking. Pohl (2020, 1) argues that "the actor constellation helps to bridge thought styles by making the underlying assumptions of the person that positions the actors explicit. The assumptions become known and open for deliberation and discussion". This is an "embodied reflexivity", generating insights that only become accessible through critical engagement with embodied action (Midgelow 2017, 130) that can "disrupt assumptions, passive learning, and mind/body division" (Nguyen and Larson 2015, 341). This activity has also been successfully adapted and applied within the Bachelor of Creative Intelligence and Innovation to support students from different disciplines to notice (and make sense of) the world with their bodies in reflexive ways (Allen 2021).

In conclusion, embodied learning in transdisciplinary higher education supports learners to engage with reflexivity, creativity, and complexity as they tackle real-world social challenges together. Embodied learning fosters core transdisciplinary competencies around knowledge integration, enabling learners to develop as transdisciplinary practitioners. Although there are challenges in engaging embodied learning within transdisciplinary higher education, it offers a valuable teaching and learning approach that supports a more holistic and integrative educational experience. Embodied learning is often implicitly applied within transdisciplinary higher education contexts; however, it can be integrated with intention and awareness to enhance transdisciplinary learning. Scholars, educators, and institutions are advised to focus on developing and sharing understanding and practice around the critical application of embodied learning within transdisciplinary higher education. This requires recognizing that embodied learning is relational and contextual and that there is no one approach or formula for how embodied learning is best engaged within transdisciplinary learning contexts.

References

Allen, Lucy. 2021. *Empowering students in moments of "stuck"*. Available from https://lx.uts.edu.au/blog/2021/05/04/empowering-students-moments-of-stuck.

Allen, Lucy, Susanne Pratt, Giedre Kligyte, Barbara Doran, Bem Le Hunte, Jacqueline Melvold, and Katie Ross. 2021. *Embodied learning for a complex world: Exploring creative education approaches through a transdisciplinary workshop*. Available from https://opus.lib.uts.edu.au/handle/10453/165563.

Branscombe, Margaret V. 2019. *Teaching through embodied learning: Dramatizing key concepts from informational texts*. Abingdon: Routledge.

Cherukunnath, Deepa, and Anita Puri Singh. 2022. Exploring cognitive processes of knowledge acquisition to upgrade academic practices. *Frontiers in Psychology* 13. Available from https://www.frontiersin.org/articles/10.3389/fpsyg.2022.682628.

Dewey, John. 1997. *Democracy and education: An introduction to the philosophy of education*. New York: Simon and Schuster.

Forgasz, Rachel, and Sharon McDonough. 2017. Struck by the way our bodies conveyed so much: A collaborative self-study of our developing understanding of embodied pedagogies. *Studying Teacher Education* 13 (1): 52–67.

Freiler, Tammy J. 2008. Learning through the body. *New Directions for Adult and Continuing Education* 119: 37–47.

Fugate, Jennifer M. B., Sheila L. Macrine, and Christina Cipriano. 2019. The role of embodied cognition for transforming learning. *International Journal of School & Educational Psychology* 7 (4): 274–88.

Goldin-Meadow, Susan. 2009. How gesture promotes learning throughout childhood. *Child Development Perspectives* 3 (2): 106–11.

Gregory, Marshall W. 2006. From Shakespeare on the page to Shakespeare on the stage: What I learned about teaching in acting class. *Pedagogy* 6 (2): 309–25.

Gustafson, Diana L. 1999. Embodied learning: The body as an epistemological site. In *Meeting the challenge: Innovative feminist pedagogies in action*, eds. Maralee Mayberry and Ellen Cronan Rose, 249–73. New York: Routledge.

Heidegger, Martin. 1975. *The basic problems of phenomenology*. Bloomington: Indiana University Press.

Henriksen, Danah, Jon Good, Punya Mishra, and The Deep Play Research Group. 2015. Embodied thinking as a trans-disciplinary habit of mind. *TechTrends* 59 (1): 6–11.

Johnson, Mark. 1987. *The body in the mind: The bodily basis of meaning, imagination, and reason*. Chicago: University of Chicago Press.

Kurtz, Cynthia F., and David J Snowden. 2003. The new dynamics of strategy: Sense-making in a complex and complicated world. *IBM Systems Journal* 42 (3): 462–83.

Lakoff, George, and Mark Johnson. 2020. The embodied mind. In *Shaping entrepreneurship research*, eds. Saras D. Sarasvathy, Nicholas Dew, and Sankaran Venkataraman, 80–103. Abingdon: Routledge.

Lindgren, Robb, and Mina Johnson-Glenberg. 2013. Emboldened by embodiment: Six precepts for research on embodied learning and mixed reality. *Educational Researcher* 42 (8): 445–52.

Lipson Lawrence, Randee. 2012. Coming full circle: Reclaiming the body. *New Directions for Adult and Continuing Education* 134: 71–78.

Macintyre Latta, Margaret, and Gayle Buck. 2008. Enfleshing embodiment: "Falling into trust" with the body's role in teaching and learning. *Educational Philosophy and Theory* 40(2): 315–29.

Meadows, Donella H. 2009. *Thinking in systems: A primer.* London: Earthscan.

Merleau-Ponty, Maurice. 1962. *Phenomenology of perception*. London: Routledge.

Merleau-Ponty, Maurice. 1964. *Signs*. Evanston, IL: Northwestern University Press.

Midgelow, Vida L. 2017. A new kind of learning: Somatics, dance improvisation and transdisciplinarity. In *Transdisciplinary higher education*, ed. Paul Gibbs, 121–35. Cham: Springer International Publishing.

Mishra, Punya, Matthew J Koehler, and Danah Henriksen. 2011. The seven transdisciplinary habits of mind: Extending the TPACK framework towards 21st century learning. *Educational Technology* 51(2): 22–28.

Monk, Nicholas, Carol Chillington Rutter, Jonothan Neelands, and Jonathan Heron. 2015. *Open-space learning: A study in transdisciplinary pedagogy*. London: Bloomsbury.

Murray, Tom. 2009. What is the integral in integral education? From progressive pedagogy to integral pedagogy. *Integral Review* 5 (1): 96–134.

Nathan, Mitchell J. 2021. *Foundations of embodied learning: A paradigm for education*. New York: Routledge.

Nguyen, David J., and Jay B. Larson. 2015. Don't forget about the body: Exploring the curricular possibilities of embodied pedagogy. *Innovative Higher Education* 40 (4): 331–44.

Nicolescu, Basarab. 2012. The Need for Transdisciplinarity in Higher Education in a Globalized World. *Transdisciplinary Journal of Engineering & Science* 3.

O'Neil, Joy Kcenia. 2015. *"Cooking to learn" while "learning to cook": (Be)coming and (re) membering sustainability*. Available from https://www.proquest.com/openview/e49ea1d768b991be4969da0918a2319d.

O'Neil, Joy Kcenia. 2016. (Be)coming and (re)membering through kitchen based learning as sustainability: An innovative living learning systems model for higher education. In *Handbook of theory and practice of sustainable development in higher education*, World Sustainability Series, eds. Walter Leal Filho, Constantina Skanavis, Arminda do Paco, Judy Rodgers, and Olga Kuznetsova, 317–33. Cham: Springer .

Pohl, Christian. 2020. *Actor constellation*. Available from https://naturalsciences.ch/co-producing-knowledge-explained/methods/td-net_toolbox/actor_constellation_final_.

Satina, Barbara, and Francine Hultgren. 2001. The absent body of girls made visible: Embodiment as the focus in education. *Studies in Philosophy and Education* 20: 521–34.

Siegel, Daniel J. 2010. *Mindsight: The new science of personal transformation*. New York: Random House.

Skulmowski, Alexander, and Günter Daniel Rey. 2018. Embodied learning: Introducing a taxonomy based on bodily engagement and task integration. *Cognitive Research: Principles and Implications* 3 (6): 1–10.

SCNAT [Swiss Academy of Sciences], ed. 2023. Available from https://scnat.ch/en.

Wagner, Anne E., and Riyad A. Shahjahan. 2015. Centering embodied learning in anti-oppressive pedagogy. *Teaching in Higher Education* 20 (3): 244–54.

Engaged Learning

Alexander Chmelka, Mary Griffith, and Hendrik Weiner

Definition

"[T]he concept of Engaged Learning emerges from multiple theoretical frameworks and educational practices" (Swaner 2007, 16). Therefore, different approaches to classification and definition can be identified. Definitions might focus on the learners (Swaner 2007, 19), on the learning process and its products (Schreiner 2009, 43), or on engagement with communities beyond the learning institution itself (Jacob et al. 2015, 1).

Engaged learning is a broad, inclusive umbrella term (Lund and Wright 2017, 652) for the pedagogical approach that enables students to derive learning from meaningful community engagement while working on real-world problems. "We define Engaged Learning as the process where students apply the theory learned at Higher Education Institutions to a context outside of Higher Education Institutions by addressing societal concerns, challenges or needs while producing knowledge in an equitable, mutually beneficial partnership" (Marsh et al. 2021, 23).

Some publications use engaged learning as a derivation from the more common *service learning* (Sachs and Clark 2017, 54). However, the replacement of "service" with "engagement" is accompanied by a fundamental change in meaning. While "service" aims at support, help, or a remunerated offer, "engagement" goes hand in hand with a socially emancipated attachment and a personal dedication to a cause, and is oriented towards the well-being of all actors concerned. Activities of academic teaching and learning are combined with civic engagement (Bandy 2011). Engaged learning, also referred to as community-engaged learning (Bandy 2011; Berard and Ravelli 2021), builds on ideas of service learning, but puts an even stronger emphasis on community involvement. Civic communities should make a significant contribution to the design and implementation of engaged learning initiatives, while independence in the development of solutions and innovations should be preserved to prevent conflicts of interest. This refers to engaged learning's underlying principle of reciprocity, which can be described as a state in which all partners mutually benefit from the actions of the respective counterpart

(Weyer 2014, 49). While service learning has increasingly developed into a distinct teaching method with recognized quality standards (Aramburuzabala et al. 2019) and processes for institutionalization (Bringle and Hatcher 1996), engaged learning takes the opposite approach and tries to gather as many methods and concepts as possible in order to work out their common core concern and educational understanding. This can be seen as an attempt to dissolve and unite the previous parallelism or even competition between approaches that are similar in essence.

At the universities, engaged learning is understood "as part of a third mission" (Chmelka et al. 2020, 11), i.e. as activities of a higher education institution that take place in the context of teaching and research without being teaching or research alone (Henke et al. 2015, 5). In terms of mission, the experiences of people outside academia need to be recognized and included in academic teaching and learning. In this way, engaged learning is an approach to transformative science which not only observes social transformation processes and describes them from the outside, but also initiates and catalyzes change processes itself (Schneidewind 2015).

From a systemic perspective, engaged learning is a transdisciplinary approach. It aims to overcome the increasingly contested boundaries between society and science. At times this means "unlearning" or perhaps just challenging some concepts traditionally taught in secondary education and introducing learners to the world of science from divergent perspectives. Academic teaching in the sense of engaged learning involves actors outside the university in the teaching process and tries to take into account the identified needs of all stakeholders. In this way, the previously separate systems of society and science are blended, preventing the infamous ivory tower of higher education and instead making the university increasingly more relevant to society as a whole.

Background

The term engaged learning is obviously composed of two words – engagement and learning. Both are as familiar as they are loaded with assumptions. It is, therefore, advisable to explore the specific understanding and link between engagement and learning in engaged learning.

Educators think of engagement in four related but different ways (Bowen 2005). The most fundamental is student engagement with the learning process: just getting students actively involved. The second is student engagement with the object of study. Here the emphasis is on the stimulation of students' learning by direct experience of something new. Another is student engagement with contexts of the subject of study. This gives emphasis to the importance of context as it may affect and be affected by the students' primary subject. When social and civic contexts are considered, this inevitably raises ethical issues. Finally, there is student engagement with the human condition, especially in its social, cultural, and civic dimensions. Engaged learning is born from this fourth aspect.

The idea of learning as reflected experiential knowledge, gained through civic commitment in social settings facing actual problems, points to the work of researchers and theorists on learning, as well as on the purposes of education itself. One of the first foundations for what we now call engaged learning is Marx's statement that the purpose of scientific work should not be understanding the world alone, but rather the transformation for the better that goes with it (Marx [1888] 1976). Later, Pasteur revealed the value of practical teaching combining the awakening of a love for science with an interest in the needs of the surrounding community (Vallery-Radot 1906, 75–86), making a case for engaging approaches. John Dewey (1938) revealed the interconnection of education and democracy, where the relationship between education and civil society were highlighted. For Dewey, teaching practice, through interaction, could reveal a better starting point of direct inquiry. In this way new material, factual and conceptual, is disclosed, material which is more relevant, more weighted and confirmed, than were the initial facts which served as the point of departure (Dewey 2008, 145). At the end of the 20th century, educational researchers increasingly questioned the teacher-centered (Henson 2003) and low-activation approach (Kolb 1984) of traditional learning. The concept of engaged learning draws from both Kurt Lewin's (1946) contribution with Action Research and Donald Schön's (1987) observations that the problems that concern people outside academia resist purely research-based, technical attempts to solve them. Engaged learning must therefore be grounded in the respective social reality of life and, at the same time, strive to overcome problematic conditions within it.

Engaged learning has evolved from these philosophical traditions into a higher education agenda, which responds to the recurring demands for greater educational effectiveness, economic efficiency, political relevance, social responsibility, and environmental sustainability of higher education institutions. The normative imperative of engaged learning (Sachs and Clark 2017, 54) is to see higher education institutions not above, but at the same level and as an integral part of the surrounding social life and to make appropriate contributions to living together. This can also be seen as a strategic approach to bring scientific expertise into the public debates and as an attempt to maintain or expand the relevance of higher education institutions in times of great social upheaval and crisis-driven historic turning points.

Debate and criticism

It is obvious that an agenda alone is not enough to overcome local, let alone global challenges. Engaged learning requires cross-sectional partnerships, where all involved must be willing, able, and committed to work together on complex challenges, at least in the medium term. One of the best ways to open the debate and criticism of engaged learning is to understand the potential and challenges to working across the disciplines together with working directly with community partners. For scientists, engaged learning may lead to new research directions or reinforce existing research lines (Van der Windt et al. 2014, 7). In the field of teaching, engaged learning opens up the "opportunity to [co-]educate the next generation of professionals, citizens, board members, policy makers, and donors" (Worrall 2007, 11). But for this to occur, the doors of universities, which form a boundary between the relatively safe and autonomous academic world and the pressing concerns and limitations of the outside world, literally must be opened (Oonk et al. 2020). One of the clear challenges is that partnerships between communities and universities need to be developed, with building sustainable partnerships "often requiring enormous investment in human and social capital to build sufficient trust in relationships to enable the co-creation of possible solutions. This ... requires time and perseverance" (Sachs and Clark 2017, 44), meanwhile taking into account the multiple perspectives and interests of all stakeholders.

Following a current engaged learning toolkit discussion (CaST project 2022), four different groups of interest can be identified: students; teachers and staff at universities; community partners; and policymakers. For students, learning formats should support them in achieving the formal requirements to obtain an academic degree and prepare them for a subsequent professional career, while at the same time developing self-determined and critical personalities. Students who participate in engaged learning initiatives find it gratifying, particularly because teamwork and open communication, which tend to play a subordinate role in the usual course of studies, take on concrete meaning (Nieto-Herman and Viera 2019, 5). For teachers and staff, engaged learning initiatives should be compatible with the multitude of other tasks, like ongoing research and third-party-funded projects. It should be recognized as a valuable teaching approach that is not only fruitful for student learning and development, but also contributes to furthering the university's third mission, as well as rewarding teachers who contribute to it. Community partners highlight the need for sustainable partnerships. These relationships may include associations, municipalities, companies, or agencies. Many desire permanent contacts at the university, and connections with students, whose cooperation helps to address challenges. For this to be viable, expectations must be managed at the initial stages of project development. And finally, for policymakers, the agendas proposed and concepts developed by engaged learning should be practically implemented and evaluated to serve as a basis for new decisions and governance.

Once the various stakeholder positions are explored broadly, the debate moves towards a deeper understanding of collaboration across academic disciplines. The need to "manage the expectations of external partners" (Anderson 2022, 154) is highlighted in recent publications on engaged learning. At the same time, collaboration within the university itself needs to consider discrepancies in teaching and research styles across the disciplines. When disciplines with different learning modes are supposed to work together, this can lead to misunderstandings (Barron 2002). Clearly this is a contextual issue, for complex real-world problems have no disciplinary boundaries. Teaching and learning on those problems is therefore conducive to conflicts across the disciplines, which can provoke new combinations of experience and knowledge that are the basis of innovation (Nonaka and Konno 1998).

By recognizing the diversity of perspectives and interests, the complexity of previously underestimated issues becomes clear. From this point of view, engaged learning has the potential to increase the awareness of inter- and transdisciplinarity as a necessity for tackling the most urgent sustainable development goals.

In addition to discussing issues related to building understanding and stabilizing partnerships, the actual substantive work in engaged learning initiatives is "intensive and time-consuming", while the immediate "benefits for staff are less clear" (Anderson 2022, 28). After all, the investment of time and perseverance does not necessarily pay off for those seeking a long-term career in higher education institutions. Some even argue that excellent research is the only way to gain a reputation in today's academic system and ultimately to advance one's career as a university teacher (Schneidewind 2016, 14). As a result, even such pioneers who can overcome the first hurdles and are highly engaged in building partnerships, as well as implementing initiatives with content, risk missing promotion opportunities due to a delay or lack of presentable outputs which can be turned into publications or grant applications. The loss of even one committed individual often means setbacks or even termination of the initiatives, since their fundamental "[l]ongterm, healthy, sustained partnerships are grounded in personal relationships" (Worrall 2007, 5).

Implementing initiatives at appropriate interfaces between higher education institutions and external partners, as well as ensuring their sustainability, are the two biggest challenges for the realization of the engaged learning agenda and often are not possible without additional external funding (Anderson 2022).

Current forms of implementation in higher education

So far, the term engaged learning has not been theoretically developed or used in a distinctive way in daily speech, across various languages and professional action. In order to function as an umbrella term, it has many overlaps with other concepts such as service learning, science shops, community engagement, outreach, international cooperation and development, third mission, and many more. Because of this, it is not yet possible to make a global statement about the state of implementation in higher education. Here, basic and applied research in international alliances is needed to close the gap and to establish a common understanding of the term or a systematization according to certain characteristics. For the interdisciplinary communities of science and higher education research in particular, it will, indeed, be intriguing as the lofty idea of engaged learning begins to be a matter of concern (Latour 2004) or fact, as it is implemented; a place where good intentions are put into action.

The current state of research on the implementation of engaged learning in higher education is based on general observational findings and experience-based perceptions on the one hand, and on regionally and culturally limited findings on the other. Purposeful movements towards greater social responsibility in universities can be observed around the world. Initiators can be the universities themselves, as well as reforms initiated by educational policy, and municipal and regional communities, together with alliances of committed individuals. Since engaged learning emerged from service learning, which originated in the United States, the dissemination of engaged learning formats has progressed furthest there, while in the Asian region, China and India are identified as drivers of this development (Ma 2018). Science shops that offer independent, participatory research in response to the challenges of surrounding communities open up spaces for accompanying engaged learning. They originated in the Netherlands and have spread from there since the 1970s through global networks such as LivingKnowledge.

The fact that the conceptual understanding of engaged learning varies greatly depending on cultural, political, economic, and social contexts up to local levels of analysis and even in increasingly harmonized higher education spaces is shown by the reports of the European ERASMUS+ funded project Communities and Students Together (CaST). The project examined 28 engaged learning initiatives in six European Union countries in terms of structures, processes, resources, stated aims, outputs, and benefits for participating groups as well as evaluation. It noted enormous variation in approaches, access to resources, and the degree of institutionalization. CaST partners commented: "Each [initiative] varies in structure and approach, as well as size and the availability of resources. The initiatives sit within a diverse range of university disciplines, and tackle an array of societal challenges" (Chmelka et al. 2020, 7). In addition, results showed that "[t]he degree of institutionalization of the programmes also varies substantially, with some having a more structured (and well-funded) approach from a higher, university level while other initiatives are working predominantly at an individual level with little to no funding from the university or elsewhere" (Chmelka et al. 2020, 7).

What all engaged learning initiatives have in common, despite their differences, is that they seek to connect the education of students in terms of both professionally relevant skills and social maturity with current real-world challenges by evoking encounters between local actors from which collaboration and exchange should emerge. To achieve this, all of them initiate transition zones between organizations, institutions, and systems to function as spaces of shared understanding and mutual trust; they serve as intermediaries of relevant information and facilitate joint projects in which participants can both contribute their specific skills and learn from each other (Penfold and Goodman 2011; Urias et al. 2020). Since engaged learning always goes hand in hand with scientific problematization of social conditions, the transformation of these conditions or their causes is always a declared goal of all initiatives – be it the expansion of ecotourism in eastern Sri Lanka, the provision of tutoring services for pupils in the coronavirus pandemic, or the critical examination of infrastructure projects in inner cities. For this reason, any attempt at implementation always means the involvement of higher education institutions in sociopolitical controversies, the settlement of which is accompanied by the establishment of interpretive sovereignty or majorities in decision-making situations, taking opposition into account.

References

Anderson, Lindsey. 2022. *Communities and Students Together (CaST): Piloting new approaches to engaged learning in Europe.* Antwerpen: Maklu.

Aramburuzabala, Pilar, Lorraine McIlrath, and Héctor Opazo, eds. 2019. *Embedding service learning in European higher education: Developing a culture of civic engagement.* London: Routledge.

Bandy, Joe. 2011. *What is service learning or community engagement?* Vanderbilt University Center for Teaching. Available from https://cft.vanderbilt.edu/guides-sub-pages/teaching-through-community-engagement.

Barron, Colin. 2002. Problem-solving and EAP: Themes and issues in collaborative teaching venture. *English for Specific Purposes* 22(3): 297–314.

Berard, Ashley, and Bruce Ravelli. 2021. In their words: What undergraduate sociology students say about community-engaged learning. *Journal of Applied Social Science* 15 (2): 197–210.

Bowen, Stephen. 2005. Engaged learning: Are we all on the same page? *Peer Review* 7 (2): 4–7.

Bringle, Robert G., and Julie A. Hatcher. 1996. Implementing service learning in higher education. *Journal of Higher Education* 67 (2): 221–39.

CaST project, ed. 2022. *Engaged learning toolkit.* Communities and Students Together (CaST) EU's Erasmus + Programme. Available from https://engagedlearningtoolkit.net.

Chmelka, Alexander, Lindsey Anderson, Eleonora Ferraresi, Mary Griffith, Noel Klima and Courtney Marsh, Philipp Pohlenz, Jarkko Rasinkangas, Tom Ritchie, Sampo Ruoppila, and Elina Sutela. 2020. *Communities and Students Together (CaST): A state-of-the-art review of engaged learning in Belgium, Finland, Germany, Italy, Spain and the United Kingdom.* Available from https://www.cast-euproject.eu.

Dewey, John. 1938. *Experience and education*. New York: Simon & Schuster.

Dewey, John. 2008. *The later works of John Dewey, 1925–1953: 1938, Logic – The theory of inquiry*, Vol. 12, ed. Jo Ann Boydston. Carbondale: Southern Illinois Press.

Henke, Justus, Peer Pasternack, and Sarah Schmid. 2015. Viele Stimmen, kein Kanon. Konzept und Kommunikation der Third Mission von Hochschulen. Available from https://www.hof.uni-halle.de/publikation/hof-ab-2-15-viele-stimmen-kein-kanon.

Henson, Kenneth T. 2003. Foundations for learner-centered education: A knowledge base. *Education* 124 (1): 5–16.

Jacob, W. James, Stewart E. Sutin, John C. Weidman, and John L. Yeager, eds. 2015. *Community engagement in higher education: Policy reforms and practice*. Rotterdam: Sense.

Kolb, David A. 1984. *Experiential learning: Experience as the source of learning and development*. Englewood Cliffs, NJ: Prentice-Hall.

Latour, Bruno. 2004. Why has critique run out of steam? From matters of fact to matters of concern. *Critical Inquiry* 30 (2): 225–48.

Lewin, Kurt. 1946. Action research and minority problems. *Journal of Social Issues* 2 (4): 34–46

Lund Dean, Kathy, and Sarah Wright. 2017. Embedding engaged learning in high enrollment lecture-based classes. *Higher Education* 74 (4): 651–68.

Ma, Carol Hok Ka. 2018. *Community engagement and service-learning in Asia*. Available from https://www.researchgate.net/publication/328306763_Community_Enagagement_and_Service-Learning_in_Asia.

Marsh, Courtney, Lindsey Anderson, and Noel Klima, eds. 2021. *Engaged learning in Europe*. Antwerpen: Maklu.

Marx, Karl. [1888] 1976. Theses on Feuerbach. In *Ludwig Feuerbach and the end of classical German philosophy*, ed. Frederick Engels, 61–65. Peking: Foreign Languages.

Nieto-Herman, Aidee, and Nayagara Viera. 2019. A global service-learning program: A case study at one dental school. *Dentistry & Dental Practises Journal* 2 (2).

Nonaka, Ikujiro, and Noboru Konno. 1998. The concept of "Ba": Building a foundation for knowledge creation. *California Management Review* 40 (3): 40–54.

Oonk, Carla, Judith T. M. Gulikers, Perry J. den Brok, Renate Wesselink, Pieter-Jelle Beers, and Martin Mulder. 2020. Teachers as brokers: Adding a university–society perspective to higher education teacher competence profiles. *Higher Education* 80 (4): 701–18.

Penfold, Erika, and Suki Goodman. 2011. *An evaluation of a knowledge partnership – A review of the literature*. Available from http://www.knowledgeco-op.uct.ac.za/usr/knowledgeco-op/downloads/Penfold&Goodman_Lit%20Review_2011.pdf.

Sachs, Judyth, and Lindie Clark, eds. 2017. *Learning through community engagement*. Singapore: Springer.

Schneidewind, Uwe. 2015. Transformative Wissenschaft – Motor für gute Wissenschaft und lebendige Demokratie. *GAIA* 24 (2): 17–20.

Schneidewind, Uwe. 2016. Die "Third Mission" zur "First Mission" machen? *Die Hochschule* (1): 14–22.

Schön, Donald. A. 1987. *Educating the reflective practitioner.* San Francisco: Jossey-Bass.

Schreiner, Christopher S. 2009. *Handbook of research on assessment technologies, methods, and applications in higher education*. Hershey, PA: IGI Global.

Swaner, Lynn E. 2007. Linking engaged learning, and well-being, and civic: A review of the literature. *Liberal Education* 93 (1): 16–25.

Urias, Eduardo, Floor Vogels, Seda Yalcin, Rosina Malagrida, Norbert Steinhaus, and Marjolein Zweekhorst. 2020. A framework for science shop processes: Results of a modified Delphi study. *Futures* (123): 1–18.

Vallery-Radot, René. 1906. *The life of Pasteur.* New York: McClure Phillips.

Van der Windt, Henny, Pádraig Murphy, Diana Smith, and Andrea Vargiu. 2014. Evaluating projects of public engagement with research and research engagement with society. *Final report on PERARES Work Package 9: Monitoring and evaluation* Available from http://www.academia.edu/16349590/Evaluating_Projects_of_Public_Engagement_with_Research_and_Research_Engagement_with_Society.

Weyer, Johannes, ed. 2014. *Soziale Netzwerke: Konzepte und Methoden der sozialwissenschaftlichen Netzwerkforschung*. München: Oldenbourg.

Worrall Laurie. 2007. Asking the community: A case study of community partner perspectives. *Michigan Journal of Community Service Learning* 14 (1): 5–17.

Entrepreneurship Education

Ewald Mittelstädt, Olena Mykolenko, and Claudia Wiepcke (authors contributed equally)

Definition

The term *entrepreneur* originates in the 13th century to describe traders who wanted to make arbitrage profits at their own risk. It is derived from the French verb *entreprendre* and means "to undertake something" or "to do something" (Hekman 2005). For many centuries the concept of the entrepreneur has been interpreted differently, but in a narrow economic perspective (Mittelstädt and Wiepcke 2013, 87). As a representative of classical economic theory, Jan Baptiste Say (1803) placed the function of coordinating resources at the center of entrepreneurial activity. Knight (2014), as a neoclassicist, described the entrepreneur as a taker of unavoidable risk. Schumpeter (1934), as a representative of the modernist phase, sees the entrepreneur as a catalyst, an innovator. In economics, theorizing is important, but it was Sarasvathy, a cognitive scientist who devised a transdisciplinary approach. Sarasvathy sought a deeper understanding of real-life entrepreneurship and engaged practitioners. Her approach emphasizes the application of entrepreneurial thinking and action and understands it as a set of principles, decision-making logic, and techniques that anyone can learn to a certain extent (Sarasvathy and Venkataraman 2011, 115). A closer scholarly examination reveals that entrepreneurship is understood as creating value for others, whether by founding an innovative organization (Gartner 1989, 51) or within an existing organization (Shane and Venkataraman 2000).

Entrepreneurship education, i.e. the promotion of entrepreneurial thinking and action, is defined very differently. A narrow definition refers to encouraging students to become self-employed (becoming an entrepreneur). A broad version (Lackeus 2015) focuses on empowering students to be more creative, opportunity-oriented, proactive, and innovative for all walks of life (becoming entrepreneurial). The narrow version is regularly justified regarding the economic benefits of entrepreneurship and is therefore often controversial. From a humanistic point of view, such a utilitarian view is incompatible with the public education mandate. The broad version is compatible with it, but an overextension of entrepreneurship as a soft skill bears the danger of arbitrariness of entrepreneurial education and leads to demarcation problems. Is all learning by doing entrepreneurial education? And are problem-based and project-oriented learning – which recur in learning by doing – entrepreneurial education, too? A minimum delimitation can be seen in the recourse to subject-specific definitions of entrepreneurship as "learning by creating value for others" (Lackeus 2015). According to the European Commission (2012, 5)

> "Entrepreneurship education" is about learners developing the skills and mindset to turn creative ideas into entrepreneurial action. This competence is crucial for all learners, supporting personal development, active citizenship, social inclusion, and employability. It is relevant across the lifelong learning process, in all disciplines of learning, and to all forms of education and training (formal, non-formal, and informal) that contribute to an entrepreneurial spirit or behaviour, with or without a commercial objective.

The European Commission's definition emphasizes entrepreneurial education as the acquisition of key life skills in a broad version with an entrepreneurial core ("entrepreneurial action") and is directly applied in the education system in about half of the EU members and associated countries. The European definition emerged from Anglo-Saxon approaches as they can be found in the Global Entrepreneurship Monitor Consortium (Amorós et al. 2013). In Asia, especially Japan and South Korea, so-called Humane Entrepreneurship offers a new perspective, combining entrepreneurship with leadership and human resource management (Kim et al. 2018). In Africa, e.g. Nigeria, entrepreneurship education is strongly connected to improving small business management as the lack of basic business knowledge is often the main reason for entrepreneurial failure (FATE Institute 2021).

Background

Entrepreneurship emerges from the interaction of person and context, from the active pursuit of an opportunity and its success or failure. Nevertheless, what does it all come down to? A variety of factors can be identified that seem to have an influence, such as personality traits (achievement orientation, control beliefs, or willingness to take risks), experience, culture, or other demographics (age, socioeconomic status, etc.). On the other hand, the pertinent question is: what can be influenced in educational institutions, and how effectively?

The theory of planned behavior (Ajzen 1991) contributes significantly to answering this question. In entrepreneurship education, the theory of planned behavior argues that an entrepreneurial attitude (*I can or will …*) is first established before entrepreneurial intentions (*I plan …*) and entrepreneurial actions (*I am entrepreneurial ...*) occur. The entrepreneurial attitude is formed through self-efficacy and sufficient knowledge, skills, and experience. Therefore, educational institutions' task is to provide learning arrangements that set entrepreneurial tasks or challenges. Tasks stimulate emotions, situations, and activities, e.g. interaction with the outside world, enduring uncertainty, teamwork, and presentations to others. Mastering these tasks and overcoming their inherent obstacles fosters entrepreneurial competencies and increases self-efficacy. Therefore, at the heart of effective entrepreneurial education is the promotion of self-efficacy (Bandura 1977; Boyd and Vozikis 1994) in adolescents and young adults. Only if they believe they can change the world, will they try it.

The three most internationally established scales for measuring entrepreneurial self-efficacy (Moberg 2014) are based on Chen et al. (1998), DeNoble et al. (1999), and McGee et al. (2009). They have four dimensions in common: (1) *Action*, e.g. the capability to manage time or budgets in projects; (2) *Context* and Outward Orientation, e.g. the capacity to establish contact with others and exchange information; (3) *Creativity*, e.g. the ability to think around corners; and (4) *Mindset*, e.g. the asset to deal with unexpected change.

Effective entrepreneurship education promotes competencies in these four dimensions by using transdisciplinary approaches (Mittelstädt et al. 2019, 56). Particularly the entrepreneurial action and outward orientation (context) enable students to deal with the plurality of knowledge. At the same time, entrepreneurial working itself contributes to the plurality of knowledge resources and enlarges the community of knowledge producers. Students have to actively create their own, individual knowledge path and educational biography (creativity), and to manage time and budgets (action). They interact with others (context) and have to deal with unexpected change (mindset). Transdisciplinary learning thereby fosters self-efficacy and ambiguity tolerance, wherefore transdisciplinary entrepreneurship education offers a promising approach (Martínez and Muñoz 2021).

Debate and criticism

Entrepreneurship can be criticized in many ways. Firstly, entrepreneurship in theory and practice has been shaped by *Euro-American perspectives*. These concepts originate in trader guilt and stem from the emergence of the bourgeoisie in cities like Amsterdam, London, or New York. Trader guilt – amongst others – pursued economic, but also political freedom and supported the creation of liberal societies at home (Gelderblom 2010, 156). However, it was also European traders who facilitated the transatlantic slave trade (Williams 1990, 199). As these historical dimensions show, entrepreneurship education needs an ethical foundation, as entrepreneurs are confronted with complex moral problems (Hannafey 2003).

Secondly, entrepreneurship involves a *market- and technology-optimistic view* and a growth paradigm that has helped entrepreneurial societies achieve great prosperity (Naudé 2007, 7). While entrepreneurship has optimized the use of scarce resources, free resources such as clean air, clean water, biodiversity, and a stable climate have been squandered. The relevant question is whether entrepreneurship and entrepreneurship education also play a relevant part in the solution of these global challenges. In fact, entrepreneurship presupposes sustainable thinking and acting (United Nations 2022, 44).

Thirdly, *religion* plays an important role in entrepreneurial behavior, too. The Euro-American Christian perspective misses Confucian (not competing for profit but for excellence) and Islamic (Shariah-compliance) aspects of entrepreneurship (Dodd and Gotsis 2007, 93). As a consequence, entrepreneurship education has to embrace diversity.

Fourthly, Euro-American entrepreneurship created the central, socially shared *stereotype of entrepreneurs* being mostly male heroes who manage to build up a company based on innate character traits, alone and against all odds (see Ogbor 2000, 607). This stereotype is just not suitable for entrepreneurship education. It has an exclusionary effect in particular on women (Leffler 2012, 39), suppresses responsibility for one's personal development, negates the potential of teamwork (Drnovsek et al. 2009, 201), and reproduces oversimplified and distorted images of entrepreneurship (Jones 2012, 252). Again, entrepreneurship has to embrace diversity.

Fifthly, entrepreneurship education focuses too much on content and not enough on how to teach it (Ebbers 2019, 43). How can this be changed? Current understandings of entrepreneurship education put the individual to be qualified primarily in the center (subject orientation) and focus on the acquisition of competencies with the emphasis on (1) the core of entrepreneurial action, (2) enabling the individual to cope with future challenges concerning expected careers, and (3) entrepreneurial action in society to initiate social change (see Halbfas and Liszt-Rohlf 2019, 17–18).

As not all university teachers are optimistic about the introduction of entrepreneurship education, it is argued that entrepreneurship education is primarily about economic interests and not about goals of personal maturity and emancipation (Eichhorn and Erlacher 2022, 101).

Regarding the motivation of entrepreneurs, it is necessary to consider that not only financial rewards or purely economic interests are a main driver of entrepreneurship (Shepherd and Patzelt 2018). Passion, a strong inclination toward an entrepreneurial activity in order to gain self-determination, contributes, and the goal of preserving natural and communal environments, generating economic and non-economic gains for disadvantaged others, or strong beliefs in values are also relevant. People with health-related limitations or who are underprivileged often freely choose entrepreneurial careers (Pagán 2009, 219). Ebbers (2019, 209) states that educational institutions are more likely to open up to entrepreneurship education if they include the holistic approach of different motivators. Social entrepreneurship education, or sustainable entrepreneurship education, increasingly finds its way into academic teaching. In addition to enabling students to think and act entrepreneurially, it also focuses on the assumption of social or ecological responsibility. Entrepreneurship can address the sustainability challenge (Villar and Miralles 2019, 104) and social issues (Austin et al. 2006, 6). In turn, social or sustainable entrepreneurship education encompasses all educational measures that address, for example, social, cultural, or environmental problems based on innovative problem-solving processes, and focuses on developing the learners' competencies. The development of competencies is not limited to entrepreneurial thinking and action but also to the ability to solve social, cultural, and environmental problems, among others.

In addition to professional competence (basic entrepreneurial knowledge) and methodological competence, humanistic competence (curiosity, creativity, critical thinking, value-based motivation), and social competence (social sensitivity, empathy, ability to act in solidarity or with environmental awareness) come to the fore (see Wiepcke 2019).

According to Schwarz (2014, 230), design competence is considered central in social entrepreneurship education, which aims to actively shape society and participate in the development processes of civil society. Creating ideas with future potential is also oriented toward sustainable business. Since design competence is also central to other educational approaches such as Education for Sustainable Development (Strachan 2018), Service Learning (Delano-Oriaran et al. 2015) or Transformative Learning (Ramsgaard 2018, 8), Social as well as Sustainable Entrepreneurship Education can be inter- and transdisciplinarily linked together with other subject areas. Thus, complex problems of other subjects such as geography, biology, or politics can be experienced under social, environmental, and entrepreneurial aspects, and students with different disciplinary backgrounds can jointly develop solutions.

Current forms of implementation in higher education

While the concepts and models of entrepreneurship and entrepreneurship education are rather Western oriented, the implementation occurs globally. As universities are considered to be a key institution for opportunity-oriented entrepreneurship, they are urged to provide all students with facilities that promote these competencies.

University activities can be divided into three categories: Learning about, for, and through entrepreneurship (Hannon 2005, 105); *Learning about entrepreneurship* encompasses the content-based theoretical approach to entrepreneurship to enable a general understanding of the phenomenon. It focuses on object orientation and is currently the predominant form in higher education institutions, as it corresponds to traditional knowledge-based approaches. Typically, this is done within existing modules like General Management or Small Business Management. *Learning for entrepreneurship* means to prepare students to consider becoming an entrepreneur as a career option and focuses on promoting required competencies (subject orientation). Entrepreneurial courses are mainly offered in extracurricular schedules at universities, in seminars like Business Planning, Marketing for Entrepreneurs, Finance for Entrepreneurs, etc. They have a little systematic effect (Lackeus and Williams-Middleton 2018, 39).

Learning through entrepreneurship as a gold standard puts the process character of entrepreneurship projects in the foreground and includes process-oriented learning and simulative approaches to real entrepreneurial situations in teaching. Process-oriented learning creates incidents to provide feedback and embark on methodological teaching. In order to achieve it, universities use plural, activating, and action-oriented methods in addition to subject content – and thereby creatively practice one form of transdisciplinary learning. Methods such as entrepreneurial project work, practice firm or mini-companies, entrepreneurial case studies, business model development, design thinking, idea competitions, and role models are used for practice-oriented entrepreneurship education at universities (Kirchner and Loerwald 2014; Neck et al. 2014).

Due to such worldwide organizations as the United Nations (e.g. UNESCO-UNEVOC) and their initiatives on promoting entrepreneurship, best practices from the International Council for Small Business, and the Global Entrepreneurship Monitor consortium are shared globally. The Global Entrepreneurship Week, a yearly event in over 170 countries, attempts to sensitize different stakeholders in society, like universities or schools, for entrepreneurship education – e.g. entrepreneurs come to class, and entrepreneurial challenges or pitch events are conducted across different institutions.

Entrepreneurship competencies can only be developed to a limited extent through individual measures. According to Ashmore (2006) and Bacigalupo et al. (2016), continuous entrepreneurship education achieves two goals: on the one hand, it supports the development of autonomy and responsibility in the process of implementing ideas; on the other hand, it strengthens the ability to create value in simple and predictable contexts to complex, constantly changing environments.

For entrepreneurship education to be sustained in higher education institutions, universities require a stable ecosystem (see Progression Model for Entrepreneurship Education Ecosystems in Europe; McCoshan 2010) to be anchored at three levels: (1) *Macro level*: National or state-specific strategy for entrepreneurial education at universities by the Ministry of Science; (2) *Meso level*: University anchoring of an entrepreneurship strategy, financial support, regional embedding, and local partnerships; (3) *Micro level*: Quantity (number of learners participating in entrepreneurship projects) and quality of measures.

It should not be underestimated that entrepreneurship education is still a young field at universities. With its application and action orientation, it has a highly innovative impact on academic teaching and learning. It is promising, yet still marginalized, as it is at odds with traditional knowledge-based and academic disciplinary approaches. Universities still have a long way to go in order to teach entrepreneurship in a way that is both effective and efficient.

References

Amorós, José Ernesto, Niels Bosma, and Jonathan Levie. 2013. Ten years of global entrepreneurship monitor: Accomplishments and prospects. *International Journal of Entrepreneurial Venturing* 5: 120–52.

Ajzen, Icek. 1991. The theory of planned behavior. *Organizational Behavior and Human Decision Processes* 50 (2): 179–211.

Ashmore, Catherine M. 2006. *Entrepreneurship everywhere: The case for entrepreneurship education.* Columbus, OH: Consortium for Entrepreneurship Education.

Austin, James E., Howard Stevenson, and Jane Wei-Skillern. 2006. Social and commercial entrepreneurship: Same, different, or both? *Entrepreneurship Theory and Practice* 30: 1–22.

Bacigalupo, Margherita, Panagiotis Kampylis, Yves Punie, and Lieve Van den Brande. 2016. *EntreComp: The entrepreneurship competence framework, JRC Science for Policy Report.* Brussels: European Commission.

Bandura, Albert. 1977. Self-efficacy: Toward a unifying theory of behavioral change. *Psychological Review* 84: 191–215.

Boyd, Nancy G., and George S. Vozikis. 1994. The influence of self-efficacy on the development of entrepreneurial intentions and actions. *Entrepreneurship Theory and Practice* 18 (4): 63–77.

Chen, Chao C., Patricia G. Greene, and Ann Crick. 1998. Does entrepreneurial self-efficacy distinguish entrepreneurs from managers? *Journal of Business Venturing* 13: 295–316.

Delano-Oriaran, Omobolade W., Marguerite Penick-Parks, and Suzanne Fondrite. 2015. *The SAGE sourcebook of service-learning and civic engagement.* London: Sage.

DeNoble, Alex F., Dong Jung, and Sanford B. Ehrlich. 1999. Entrepreneurial self-efficacy: The development of a measure and its relationship to entrepreneurial action. In *Frontiers of entrepreneurship research*, eds. Paul D. Reynolds, William Bygrave, Sophie Manigart, Colin M. Mason, G. Dale Meyer, Harry J. Sapienza, and Kelly Shaver, 73–87. Wellesley, MA: Babson College.

Dodd, Sarah, and George Gotsis. 2007. The interrelationships between entrepreneurship and religion. *International Journal of Entrepreneurship and Innovation* 8 (2): 93–104.

Drnovsek, Mateja, Melissa S. Cardon, and Charles Y. Murnieks. 2009. Collective passion in entrepreneurial teams. In *Understanding the entrepreneurial mind*, eds. Alan L. Carsrud and Malin Brännback, 191–215. New York: Springer.

Ebbers, Ilona. 2019. Entrepreneurship Education als Möglichkeits- und Ermöglichungsraum – eine erste theoretische Annäherung aus fachdidaktischer Perspektive. In *Entrepreneurship Education. Begriff – Theorie – Verständnis*, eds. Teita Bijedic, Ilona Ebbers, and Brigitte Halbfas, 43–61. Wiesbaden: Springer Gabler.

Eichhorn, Susanne, and Irene Erlacher. 2022. Entrepreneurship Education – Überfachliches Thema im neuen Lehrplan. *Journal für Elementar- und Primarbildung* 1 (1): 100–06.

European Commission. 2012. Entrepreneurship EDUCATION AT SCHOOL in Europe – National strategies, curricula and learning outcomes. Education, Audiovisual and Culture Executive Agency Brussels. Available from https://eunec.eu/sites/www.eunec.eu/files/attachment/files/135en.pdf.

FATE Institute, ed. 2021. *State of entrepreneurship in Nigeria report 2021*. Lagos: FATE.

Gartner, William B. 1989. "Who is an entrepreneur?" is the wrong question. *Entrepreneurship Theory and Practice* 13 (4): 47–68.

Gelderblom, Oscar. 2010. The golden age of the Dutch republic. In *The invention of enterprise*, eds. David S. Landes, Joel Mokyr, and William J. Baumol, 156–82, Princeton, NJ: Princeton University Press.

Halbfas, Brigitte, and Verena Liszt-Rohlf. 2019. Entwicklungslinien und Perspektiven der Entrepreneurship Education – eine Analyse von Definitionen. In *Entrepreneurship Education. Begriff – Theorie – Verständnis*, eds. Teita Bijedic, Ilona Ebbers, and Brigitte Halbfas, 3–20. Wiesbaden: Springer Gabler.

Hannafey, Francis. 2003. Entrepreneurship and ethics: A literature review. *Journal of Business Ethics* 46: 99–110.

Hannon, Paul D. 2005. Philosophies of enterprise and entrepreneurship education and challenges for higher education in the UK. *International Journal of Entrepreneurship and Innovation* 6 (2): 105–14.

Hekman, Björn. 2005. *Entrepreneurship education in Europa*. Available from https://kups.ub.uni-koeln.de/1744/1/Bjoern-Hekman-Dissertation.pdf.

Jones, Sally. 2012. Gendered discourses of entrepreneurship in UK higher education: The fictive entrepreneur and the fictive student. *International Small Business Journal* 32 (3): 237–58.

Kim, Ki-Chan, Ayman ElTarabishy, and Zong-Tae Bae. 2018. Humane entrepreneurship: How focusing on people can drive a new era of wealth and quality job creation in a sustainable world. *Journal of Small Business Management* 56 (4/5): 10–29.

Kirchner, Vera, and Dirk Loerwald. 2014. *Entrepreneurship Education in der ökonomischen Bildung*. Hamburg: Joachim Herz Stiftung.

Knight, Frank H. [1921] 2014. *Risk, uncertainty and profit*. New York: Martino Fine.

Lackeus, Martin. 2015. *Entrepreneurship in education: What, why, when, how*. Available from https://www.oecd.org/cfe/leed/BGP_Entrepreneurship-in-Education.pdf.

Lackeus, Martin, and Karen Williams-Middleton. 2018. Assessing experiential entrepreneurship education: Key insights from five methods in use at a venture creation programme. In *Experiential learning for entrepreneurship: Theoretical and practical perspectives on enterprise education*, eds. Denis Hyams-Ssekasi and Elizabeth B. Caldwell, 19–49. Cham: Palgrave Macmillan.

Leffler, Eva. 2012. Entrepreneurship in schools and the invisible of gender: A Swedish context. In *Entrepreneurship: Gender, geographies and social context*, ed. Thierry Burger-Helmchen, 31–52. London: IntechOpen.

Martínez, Lizbeth, and Jaime Muñoz. 2021. Are andragogy and heutagogy the secret recipe for transdisciplinary entrepreneurship education? *European Business Review* 33 (6): 957–74.

McCoshan, Andrew. 2010. *Towards greater cooperation and coherence in entrepreneurship education: Report and evaluation of the pilot action high level reflection panels on entrepreneurship education initiated by DG Enterprise and Industry and DG Education and Culture*. Brussels: Publications Office of the European Union.

McGee, Jeffrey E., Mark Peterson, Stephen L. Mueller, and Jennifer M. Sequeira. 2009. Entrepreneurial self-efficacy: Refining the measure. *Entrepreneurship Theory and Practice* 33 (4): 965–88.

Mittelstädt, Ewald, Maya Ivanova, and Judith Michels. 2019. *Unternehmergeist in die Schulen – aktuelle Trends und Entwicklungen, Nachhaltigkeit der Projekte, Transparenz und Erfolgsindikatoren*. Berlin: BMWi.

Mittelstädt, Ewald, and Claudia Wiepcke. 2013. Entrepreneurship Education – Unternehmertum und Werte. *Wirtschaftspolitische Blätter* 60 (1): 87–97.

Moberg, Kare. 2014. *Assessing the impact of entrepreneurship education: From ABC to PhD*. Copenhagen: Copenhagen Business School.

Naudé, Wim. 2007. *Peace, prosperity and pro-growth entrepreneurship, Discussion paper 2007/002*. Helsinki: UNU-WIDER.

Neck, Heidi, Patricia Greene, and Candida Brush. 2014. *Teaching entrepreneurship, a practice-based-approach*. Cheltenham: Edward Elgar.

Ogbor, John O. 2000. Mythicizing and reification in entrepreneurial discourse: Ideology-critique of entrepreneurial studies. *Journal of Management Studies* 37 (5): 605–35.

Pagán, Ricardo. 2009. Self-employment among people with disabilities: Evidence for Europe. *Disability and Society* 24 (2): 217–29.

Ramsgaard, Michael B. 2018. Experiential learning philosophies of enterprise and entrepreneurship education. In *Experiential learning for entrepreneurship: Theo-*

retical and practical perspectives on enterprise education, eds. Denis Hyams-Ssekasi and Elizabeth B. Caldwell, 3–18. Cham: Palgrave Macmillan.

Sarasvathy, Saras D., and Sankaran Venkatamaran. 2011. Entrepreneurship as method – Open questions for an entrepreneurial future. *Entrepreneurship Theory and Practice* 35: 113–35.

Say, Jean-Baptiste. 1803. *Traité d'*économie *politique*. Paris: Crapelet.

Schumpeter, Joseph A. 1934. *The theory of economic development: An inquiry into profits*. Cambridge, MA: Harvard University Press.

Schwarz, Sabine. 2014. *Social Entrepreneurship Projekte – Unternehmerische Konzepte als innovativer Beitrag zur Gestaltung einer sozialen Gesellschaft*. Wiesbaden: Springer.

Shane, Scott, and Sankaran Venkataraman. 2000. The promise of entrepreneurship as a field of research. *Academy of Management Review* 25 (1): 217–26.

Shepherd, Dean, and Holger Patzelt. 2018. *Entrepreneurial cognition: Exploring the mindset of entrepreneurs*. Cham: Palgrave Macmillan.

Strachan, Glenn. 2018. Can education for sustainable development change entrepreneurship education to deliver a sustainable future? *Discourse an Communication for Sustainable Communication* 9 (1): 36–49.

United Nations. 2022. *The sustainable development goals report* 2022. New York: United Nations Publications.

Villar, Eula Bianca, and Fracesc Miralles. 2019. Sustainable entrepreneurship in response to grand challenges: What do we know and how do we move forward? *DLSU Business & Economics Review* 28 (3): 102–11.

Wiepcke, Claudia. 2019. Social Entrepreneurship Education zur Förderung von Inklusion. In *Entrepreneurship Education. Begriff – Theorie – Verständnis*, eds. Teita Bijedic, Ilona Ebbers, and Brigitte Halbfas, 193–212. Berlin: Springer.

Williams, Eric. 1990. *Capitalism and slavery*. Chapel Hill: University of North Carolina.

Experiment

Christina West, Bernd Böttger, and Wing Shing Tang

Definition

The (scientific) experiment (lat. *expĕrīmentum* "proof, trial, test" and "experience", Lewis and Short 2020, 403) is a creative, useful, fundamental, systematic empirical method to gain knowledge. With the Enlightenment, experiments were used in the classical experimental sciences of chemistry and physics (Franklin and Perovic 2021). Since then, they have evolved into an essential source of scientific knowledge in almost all disciplines. The experiment is a tool for exploring the new. But unlike mere observation, it involves an intended manipulation of the observed system. As a connecting element, it further allows integrating different research approaches of various disciplines into a methodological context, thus promoting inter- and transdisciplinary processes (West 2021).

Steinle (1997) distinguishes hypothesis-driven or confirmatory from exploratory experimentation. While the former designates concrete hypotheses as true or false and reflects the dominant scientific model of critical rationalism (Popper 2013), explorative experiments are more suitable for generating hypotheses in unknown contexts and deriving rules inductively. Their epistemological importance is often underestimated, which changes with the "experimental turn" (Overdevest et al. 2010; West 2018) of the humanities and social sciences: experiments are now valued as a scientific method even under uncertain, uncontrollable conditions, recognizing complexity and conflict between knowledge and ignorance. For art, architecture, and design, explorative experiments have always been fundamental. They often become an open-ended creative-artistic intervention, generating astonishment, amazement, or even snubs in society with the aim of triggering reflections and (re)action. In the social sciences, psychology, and medicine, distinction is made between laboratory and field experiments. In quantitative studies the limited reproducibility of the less controllable field trials is often compensated by statistical methods.

The thought experiment has a long tradition in philosophy and natural sciences. By making counterfactual assumptions that are difficult or impossible to establish in reality, researchers check whether a theory is consistent or leads to paradoxical situations. "Schrödinger's cat" (Schrödinger 1935, 812, 827; Trimmer 1980, 323–38) – hypothetically placed in a special box – would be dead and alive at the same time according to the laws of quantum mechanics, until the observer opens the box. Exploratory thought experiments can help refine or illustrate theories (Steinle 1997) and utopias (West 2019, 2021; West and Kück 2019). Virtual experiments (Pasemann 2017, 9) in theoretical physics and mathematics often represent numerical computer simulations under model assumptions which could hardly be achieved in reality and are popular in natural and engineering sciences (Böttger et al. 2019), architecture, and social sciences. Explorative thought experiments are gaining importance in transdisciplinary contexts. Distinctions between "qualitative" and "quantitative" experiments in social sciences (Kleining and Witt 2001) refer back to the qualitative and quantitative methods in empirical social research.

In research practice uniform categories of experimentation exist neither in general nor within individual disciplines. The narrative of a clearly defined procedure of experimentation has to be understood as a consequence of canonized reporting, which in publications and textbooks intentionally ignores the temporal and causal de facto course of the research process, obscuring the actual research strategy *a posteriori*. In teaching, dealing with uncertain evidence and the problem of stabilizing experimental systems are largely avoided, for example in physics lessons (Ruhrig and Höttecke 2015). Rheinberger (1998, 287–88) comes to a similar conclusion, considering the research process as a system. He designs experimental cultures as an epistemology of modern experimentation, based on experimental systems as the smallest functional research units. The underlying concept of culture implies that scientific knowledge itself is not independent of its own history. Less planning and control, more improvisation and coincidence characterize everyday research.

Exploratory field experiments as well as real-world experimentations, e.g. designed as interventions in urban public space, even make it possible to cross the boundaries of science into society in a transdisciplinary manner: not just as an artistic or scientific act, but as a process in which everyday and specialist knowledge are negotiated on an equal footing. With "real-world experiments", a category of transdisciplinary field experiments has recently emerged which do not take place in scientific laboratories, but in and with society (Ehnert 2023; Gross and Hoffmann-Riem 2005; Scholl et al. 2018), where conditions can change in an unforeseeable way.

The orientation on uncertain knowledge is central to more recent approaches, including the concept of real-world experiments which can structure how we deal with uncertainty and ignorance. Due to its grounding in real life and open-ended results, the distinctions between research, teaching, learning, and (inter)action blur: the aim of the experiment then follows not only research logics, but also logics and demands of the "real-world".

The term *real-world experiment* is particularly used in the context of real-world laboratories, which, as an experimental format, have a transdisciplinary setting of institutional, civil society, and scientific actors and address a transdisciplinary objective jointly selected through processes of co-design of the questions, co-production of knowledge for transformation at eye level, and co-evaluation (West and Kück 2019). They differ from related methods of participatory action research (Chevalier and Buckles 2019), intervention research, or transdisciplinary cooperative research (case study approach, research-based learning) by their "twofold transdisciplinary ambition": Firstly, the phases of conceptual development require a co-design of a thought experiment in a transdisciplinary team. Secondly, the implementation and observation of the transformative experiment itself creates application knowledge in a second transdisciplinary context. Due to their orientation on uncertainty, transdisciplinary experiments create spaces for improvisation and reflection, following a reasoning which is transversal, creating "transtopias" (West 2019, 2023).

Background

The future is glocal – global and local at the same time: Global challenges like anthropogenic warming, loss of biodiversity, etc. require local answers and overall societal future-oriented, sustainable transformation and innovation processes, right up to a new social contract. In this sense, the Paris Agreement on climate change (UNFCCC 2015) should not only function as a climate agreement, but also as the basis for a new, integrative, and inclusive agenda of development – the 21st century's social contract. These challenges cannot be mastered only through technical–technological innovations (West 2023): In addition to changes in institutional, regulatory, and political structures, new inter- and transdisciplinary actor constellations and transgressive processes are required, which arise from a transversal individual basic orientation and a transition between different forms of action and reality constellations, across prevailing categories and differences (West 2019).

Research and teaching play a key role in the justification of these transformation processes, as passing from the industrial to the knowledge society, the mode of learning, knowledge generation, and regime changes: knowledge is increasingly generated de-centrally, new places and actors are added, which enable production and recombination of different types of knowledge in new, temporary, experimental constellations and cooperation. Thus, concepts of knowledge change: in addition to individual, solitary recognition, social, jointly generated, inter- and transdisciplinary knowledge is gaining importance. Knowledge, the recognition of ignorance, and uncertainty become transformative energy. Individual and collective innovation and transformation processes are often triggered less by cognitive knowledge than by life-world-motivated needs for change and reform, which can be communicated and implemented in an exploratory, improvisational, experimental way (Epstein 1994, 711; WBGU 2011, 242; West 2019, 2023), whereby reflexivity increases. The (future) *modus operandi* of knowledge societies can then be understood as continuous experimentation, in which scientifically and socially defined problems can hardly be separated (Välimaa and Hoffman 2008; Vilsmaier et al. 2017; West 2023).

The experimental turn in the knowledge society marks the moment when transdisciplinary knowledge generated through experiments becomes the guiding principle for action. Hereby scientific and everyday knowledge are addressed. At the same time, existing knowledge, patterns and schemes of meaning, interpretation and hegemonic practices, social structures and schemes of order are questioned, reflecting on ways of life and seeking options for action and solutions to "wicked problems" (West 2018, 330). This is also reflected in the discussion on "real-world experiments". As a hybrid form of experiment, they oscillate between modes of "knowledge generation", "knowledge application", "controlled" and "situation-specific" boundary conditions that cannot be fully controlled (Schneidewind 2014, 2). Experiments initiate a transition from explorative-improvised action to cognitive insights, if all participants can bear complexity and a temporary loss of control. With the experimental turn, science changes towards *Mode 3*, emphasizing the coexistence and co-development of diverse knowledge and innovation modes (Carayannis et al. 2012), and opens up new perspectives, places, and spaces for public science. By integrating different bodies and systems of knowledge, multiple transdisciplinary teaching and learning processes are released that generate socially robust knowledge for transformation, increase the reflexivity of the actors, and change social rationalities and cultural and disciplinary practices. In experiments, the connection between scientific and everyday learning changes. The knowledge society is an experimental society.

Debate and criticism

Doubts are expressed whether future-oriented sustainable development can be addressed at all through transdisciplinary experiments in science, research, and teaching. The added value that can be generated is also controversial. Criticism primarily concerns (1) *methodological requirements* students and lecturers face, (2) problems of *integrating* transdisciplinary experiments into disciplinary teaching and research curricula, (3) a strong *focus on the Western or European cultural and political context*, and (4) the fundamental scientific *self-conception* on which transformative science is based.

A high degree of openness to results is inherent in transdisciplinary experiments, as in a transdisciplinary environment, project or learning content and corresponding methods cannot be anticipated precisely and in detail. Considering the lack of acceptance of transdisciplinary methods within the disciplines, missing evaluation methods and structures, transdisciplinary research and teaching often do not fit well into academic curricular structures. The "twofold transformation ambition", which is directed outwards (society) and inwards (the scientific system), holds the risk of overstraining research activity (Jaeger-Erben et al. 2018, 117). Whether curricular goals can be achieved at all becomes a problem when academic qualification is envisaged within transdisciplinary projects.

Criticism arises about the cultural context: Developed in the Western hemisphere, transdisciplinary research and experimentation have only recently been applied to other regions. Even if in line with the objective of transdisciplinarity, problems have been encountered in application. Sim et al. (2019) focus on *guanxi* (relationships) in the local context of China – pointing to the importance of state adjudication in knowledge production. In the Western context, "relationship" has been discussed as "complicity" (Ziemer 2016) in experimental space appropriation processes. Schmidt and Neuburger (2017) allude to the importance of North–South imbalances in power relations in the application of transdisciplinary research and thus to processes in and through which space is appropriated, constituted, and manifested (West 2019).

Regarding space, most applications have dealt with macro-physical phenomena like earthquakes, climate change, and sustainable land-use management. Seldom has research focused on phenomena such as development in a city, neighborhood, or community (Schmidt and Neuburger 2017; Sim et al. 2019). Thus, the time is ripe to initiate research on understanding the mutual interaction of processes as real-world experiments in a city like Hong Kong. Transdisciplinary transformative formats are often claimed to have a normative character resulting from the specification of social goals, which blurs the scientific view. Grunwald (2018, 114) argues it is sufficient to explicit one's own normativity in the scientific discourse. In contrast, West (2019, 2023) sees sustainability as itself founded in the transdisciplinary experimental format and thus not as a normative goal.

Current forms of implementation in higher education

While the didactic implementation of disciplinary experiments usually follows rules and forms of the respective disciplines, individual attempts at standardization can be seen for transdisciplinary experiments. But efforts towards canonization are still absent. In transdisciplinary experiments, the paradigm of the knowledge society (Castelfranchi 2007; Välimaa and Hoffman 2008) moves towards a Mode 3 science (Carayannis et al. 2012) through collaborative learning, working, designing, and experimenting. The relation between teachers and students changes, since the participants alternately find themselves in the situation of the respective other.

Transdisciplinary experiments oscillate between two work modes (Figures 1 and 2), which results in different didactic challenges. In the mode of integration (Figure 1), questions from different contexts are linked with one experiment, so that both disciplinary and transdisciplinary methods are further developed. In the mode of addition (Figure 2), different experimental elements, techniques, interventions, methods, and formats are used in a common thematic context. Thus, the integration mode is more research-oriented, while the addition mode is rather a tool. In practice, both modes are often intertwined or combined in varying degrees. The transdisciplinary experiment can be divided into two experimental phases. Firstly, the reflection and development of the scientific and everyday context of the experiment can be described as an exploratory thought experiment. Secondly, its implementation and the observation of its effects in society constitute a real-world experiment. The mode of both experimental phases is transdisciplinary, on different levels, which is why all persons involved must be open-minded and engage with the unforeseeable on both levels. This "twofold transdisciplinarity ambition" challenges or overstresses teachers and learners, but also activates unexploited learning potentials.

Possible forms of didactic transdisciplinary experiments can be illustrated by three formats: (1) *Knowledge to go* (West 2018) is an experimental teaching and research format in which transdisciplinary work is conveyed. The conception phase in the seminar room becomes an explorative thought experiment. The scientific and social conditions, basic hypotheses, and implications of possible interventions are developed in different context levels in the mode of integration (Figure 1) together with transdisciplinary partners. With the real-world experiment, the hypotheses are checked and the results from the various thematic and spatial context levels are brought together and discussed publicly with experts and transdisciplinary audiences. Learning and teaching are understood as active, self-controlled, situational, communicative, improvisational processes – a new transversal learning culture is implemented (West 2019). Similar, comparable concepts of using real-world experiments as a transdisciplinary teaching format have recently been developed in the field of sustainable mobility culture (Baum et al. 2021). (2) The *UrbanUtopiaLAB|Experimenting Utopia: Past ... Present ... Future* (West and Kück 2019) is an iterative multi-phase format for research and teaching that structures and supports comprehensive transdisciplinary transformation processes from utopia to policy advice. Experiments are combined in the mode of addition (Figure 2). Exploratory thought experiments are supported by techniques such as emo-action mapping (West and Kück 2019, 268) and combined with real-world experiments. In this way, joint intentional utopias emerge which are iteratively developed and evaluated in subsequent phases of observation, analysis, intervention, and political participation.

Figure 1. Mode of integration: integrative transdisciplinary experimentation

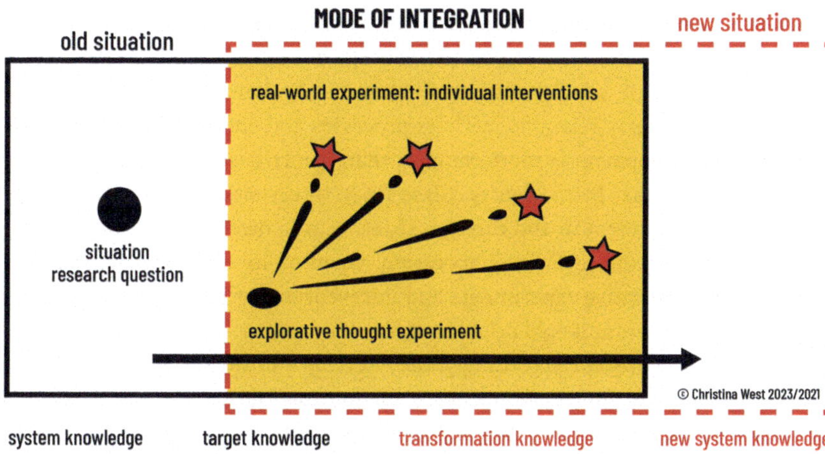

Figure 2. Mode of addition: additive transdisciplinary experimentation

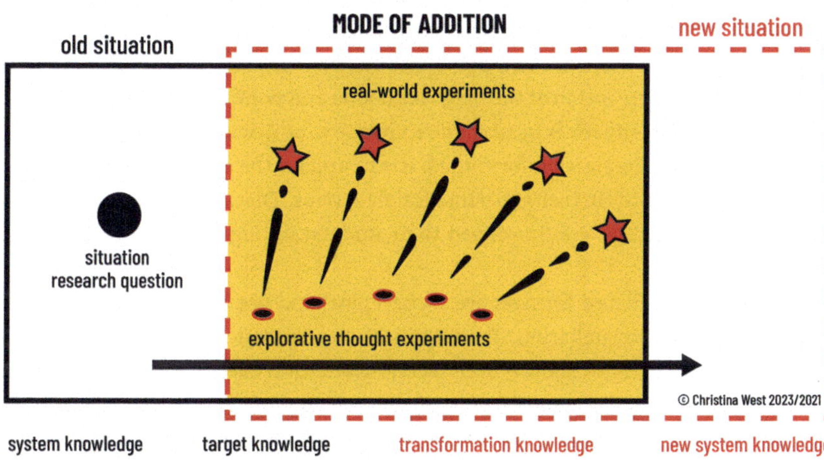

Design: Christina West, based on West and Kück 2019, 272; West 2021.

(3) In Hong Kong, similar formats attempt to put *sustainability as a process* (West 2019), for which cultural specificities must be recognized: Hong Kong, a British colony for 150 years before returning to the sovereignty of China, is still intertwined with modernity and nationalism, and embedded in the Chinese land question. This applies to the understanding of and attitude towards imagination and practicing sustainable development and transformative transdisciplinary experimentation. Utility, efficiency, and productivity have been emphasized at the expense of equality, equity, and environmental sustainability. Relevant for

sustainable development is opening up the mindset of students in a transversal, sustainable way. Two considerations on transdisciplinary transformative experiments emerge under these conditions: The first relates to contents of urban sustainability. Tang (2017) coined the concept of hegemonic-cum-alienated urban redevelopment, which considers the historical colonial base of Hong Kong's landed and property relations, with its processes and concretized patterns. The effect of redevelopment is more complex than "gentrification" – a displacement of the working class by the gentry (Glass 1964). Interventions in Hong Kong as a "real-world experiment in and through space" aim to dwell on its subtleties with a more accurate and inclusive elaboration. Opening up mindsets by transdisciplinary transformative experiments and interventions effectively and efficiently, field trips to different neighborhoods enable students to explore and learn in an experimental approach. Accordingly, they develop a critical attitude towards society and the environment. Sustainability is an experimental process that is to be achieved as people live their lives. A first step is to get acquainted with the neighborhood. Library research enhances their understanding of the forces producing the neighborhood's historical development. This is the prerequisite for reflecting on the issue of development of the city as a whole. At the end of the field trip, in a group discussion, students are requested to reflect on and talk about their experience. It appears that once equipped with the briefing on the ground, they start to change their perception of the neighborhoods and ponder about Hong Kong urban society. All feedback is used at the university for discussing ways to improve. The results of this reflection form the basis to update the experimental approach and organization of further experimental field trips. Discussion among students, teachers, instructors, etc. improved their understanding of Hong Kong's development.

The three discussed formats are archetypical and not limited to any regional context or actor constellation. Their application in practice with regional as well as international groups show – even if conducted under very different conditions and restrictions – that experiments, if they are applied in a genuine transdisciplinary setting, can indeed enable gaps to be bridged between disciplines and between different cultures and fundamental ways of thinking: By oscillating in different modes between thought and real experiment (Figures 1 and 2), building up new common epistemologies will be fostered (and even required!), and transversal thinking will emerge.

References

Baum, Martina, Hanna Noller, and Sebastian Klawiter. 2021. *Realexperimente – On the search for new possibilities.* Available from https://elib.uni-stuttgart.de/bitstream/11682/11774/1/Reallabor-RNM-Zines-C-SI_RE-engl-2021.pdf.

Böttger, Bernd, Bernd Daniels, Lisa Dankl, Thomas Göhler, and Torsten Jokisch. 2019. Systematic phase-field study on microstructure formation during brazing of Mar-M247 with a Si-based AMS4782 filler. *Metallurgical and Materials Transactions A* 50: 1732–47.

Carayannis, Elias G., Thorsten D. Barth, and David F. J. Campbell. 2012. The Quintuple Helix innovation model: Global warming as a challenge and driver for innovation. *Journal of Innovation and Entrepreneurship* 1: 2.

Castelfranchi, Cristiano. 2007. Six critical remarks on science and the construction of the knowledge society. *Journal of Science Communication* 6 (4): 1–3.

Chevalier, Jacques M., and Daniel J. Buckles. 2019. *Participatory action research: Theory and methods for engaged inquiry.* London: Routledge.

Ehnert, Franziska. 2023. Review of research into urban experimentation in the fields of sustainability transitions and environmental governance. *European Planning Studies* 31 (1): 76–102.

Epstein, Seymor. 1994. Integration of the cognitive and the psychodynamic unconscious. *American Psychology* 49 (8): 709–24.

Franklin, Allan, and Slobodan Perovic. 2021. *Experiment in physics: The Stanford encyclopedia of philosophy.* Available from: https://plato.stanford.edu/archives/sum2021/entries/physics-experiment.

Glass, Ruth. 1964. *London: Aspects of change.* London: MacGibbon & Kee.

Gross, Matthias, and Holger Hoffmann-Riem. 2005. Ecological restoration as a real-world experiment: Designing robust implementation strategies in an urban environment. *Public Understanding of Science* 14 (3): 269–84.

Grunwald, Armin. 2018. Transformative Wissenschaft als honest broker? Das passt! *GAIA* 27 (1): 113–16.

Jaeger-Erben, Melanie, Emilia Nagy, Martina Schäfer, Elisabeth Süßbauer, and Jana Zscheischler. 2018. Von der Programmatik zur Praxis: Plädoyer für eine Grounded Theory transformationsorientierter Forschung. *GAIA* 27 (1): 117–21.

Kleining, Gerhard, and Harald Witt. 2001. Discovery as basic methodology of qualitative and quantitative research. *Forum: Qualitative Social Research* 2(1).

Lewis, Charlton T., and Charles Short. 2020. *A Latin dictionary: Founded on Andrews' edition of Freund's Latin dictionary.* Chapel-en-le-Frith: Nigel Gourlay.

Overdevest, Christine, Alena Beicher, and Matthias Groß. 2010. The experimental turn in environmental sociology: Pragmatism and new forms of governance. In *Environmental sociology: European perspectives and interdisciplinary challenges,* eds. Matthias Groß and Harald Heinrichs, 279–94. Heidelberg: Springer.

Pasemann, Frank. 2017. Event-based simulations: Is there a need for new physical theories? In *Interferences and events: On epistemic shifts in physics through computer simulations*, eds. Anne Dippel and Martin Warnke, 75–92. Lüneburg: meson.

Popper, Karl. [1945] 2013. *The open society and its enemies.* Princeton: Princeton University Press.

Rheinberger, Hans-Jörg. 1998. Experimental systems – graphematic spaces. *Inscribing science. Scientific texts and the materiality of communication*, ed. Timothy Lenoir, 285–303. Stanford, CA: Stanford University Press.

Ruhrig, Jan, and Dietmar Höttecke. 2015. Components of science teachers' professional competence and their orientational frameworks when dealing with uncertain evidence in science teaching. *International Journal of Science and Mathematics Education* 13: 447–65.

Scholl, Christian, Joop de Kraker, Thomas Hoeflehner, Mette Agger Eriksen, Petra Wlasak, and Thomas Drage. 2018. Transitioning urban experiments: Reflections on doing action research with urban labs. *GAIA* 27 (1): 78–84.

Schmidt, Laura, and Marina Neuburger. 2017. Trapped between privileges and preciousness: Tracing transdisciplinary research in a postcolonial setting. *Future* 93: 54–67.

Schneidewind, Uwe. 2014. Urbane Reallabore – ein Blick in die aktuelle Forschungswerkstadt. *Pnd online* 3: 1–7.

Schrödinger, Erwin. 1935. Die gegenwärtige Situation in der Quantenmechanik. *Naturwissenschaften* 23: 807–12, 823–28, 844–49.

Sim, Timothy, John Young, Jocelyn Lau, and Ke Cui. 2019. Initiating transdisciplinary research in China: A case study. *International Journal of Environmental Sciences & Natural Resources* 22 (1): 34–43.

Steinle, Friedrich. 1997. Entering new fields: Exploratory uses of experimentation. *Philosophy of Science* 64: 65–74.

Tang, Wing-Shing. 2017: Beyond gentrification: Hegemonic redevelopment in Hong Kong. *International Journal of Urban and Regional Research* 41 (3): 487–99.

Trimmer, John D. 1980. The present situation in quantum mechanics: A translation of Schrödinger's "Cat paradox" paper. *Proceedings of the American Philosophical Society* 124 (5): 323–38.

UNFCCC [United Nations Framework Convention on Climate Change]. 2015: *Adoption of the Paris Agreement.* Paris. Available from https://unfccc.int/resource/docs/2015/cop21/eng/l09r01.pdf.

Välimaa, Jussi, and David Hoffman. 2008. Knowledge society discourse and higher education. *Higher Education* 56: 265–85.

Vilsmaier, Ulli, Vera Brandner, and Moritz Engbers. 2017. Research in-between: The constitutive role of cultural differences in transdisciplinarity. *Transdisciplinary Journal of Engineering & Science*, 8 (1): 169-179.

WBGU [German Advisory Council on Global Change]. 2011. *World in transition: A social contract for sustainability.* Berlin: WBGU.
West, Christina. 2018. "Wissen to Go" – transdisziplinär-transformative Lehre als "Reallabor im Kleinen". In *Transdisziplinär und transformativ forschen. Eine Methodensammlung,* eds. Rico Defila and Antonietta Di Giulio, 329–73. Wiesbaden: Springer.
West, Christina. 2019. Transversal city and transtopia – reflecting and analysing migration, the city, and "the urban" after the postmigrant city. *Geographica Helvetica,* 74: 261–72.
West, Christina. 2021. Experiment. In *Handbuch Transdisziplinäre Didaktik,* eds. Thorsten Philipp and Tobias Schmohl, 93–105. Bielefeld: transcript.
West, Christina. 2023. Reallabore und Innovationen – transdisziplinäre Reallabore als (regionale) Innovationsökosysteme. In *Innovationsökosysteme – Netzwerke nutzen und Innovationskraft steigern,* eds. Klaus-Michael Ahrend and Katrin Redmann, 157–71. Stuttgart: Schäfer-Poeschel.
West, Christina, and Svenja Kück. 2019. "UrbanUtopiaLAB" – einen Möglichkeitsraum zur Produktion von Transformationswissen schaffen. In *Transdisziplinär und transformativ forschen. Eine Methodensammlung,* eds. Rico Defila und Antonietta Di Giulio, 259–91. Wiesbaden: Springer.
Ziemer, Gesa. 2016. *Complicity: New perspectives on collectivity.* Bielefeld: transcript.

Fab Lab

Bonny Brandenburger, Gameli Adzaho, Manon Mostert - van der Sar, Maximilian Voigt, and Peter Troxler (authors contributed equally)

Definition

Fab labs are shared workshops, open to the public and equipped with modern as well as traditional tools and machines. A typical setup may consist of an electronics laboratory, 3D printers, laser cutters, and other computer-controlled machines, as well as classic hand tools and machines for wood-, textile, and metalworking – in line with the motto *make (almost) everything*. The term *fab lab* was coined by Neil Gershenfeld, meaning a "lab for fabrication or simply a fabulous laboratory" (Gershenfeld 2005, 12). *Fab* goes back to the English fabrication, with its etymological origin in the Latin *fabrica*, which in a narrower sense denotes the workshop of an artist working with hard materials (Lewis and Short 2020, 414). *Lab*, short for "laboratory", stems from the Latin *lăbor*, meaning "work" but also "toil, effort, drudgery", which can also be translated as "fruit of labor" (Gershenfeld 2005, 594). The core of its concept is to bring people from different backgrounds – design, engineering, architecture, urban planning, biology, crafts, software development, art, and education – together to create. The common mission is to develop, share, and transform knowledge and to create technologies with practical relevance in everyday life. This turns fab labs into spaces of transdisciplinary learning and working.

On the one hand, some of these facilities present themselves as sites of decentralized and distributed manufacturing or of economically oriented innovation (Kohtala et al. 2020). Concrete goals are often deliberately avoided. The focus is on providing a freely usable infrastructure. Fab labs allow individualized one-offs to be produced, or spare parts that are no longer available on the market (*rapid manufacturing*). The actors in the workshops – also called *makers* – define themselves as part of a grassroots movement that empowers people to deal competently with technology and move from being passive consumers to self-confident producers (Smith et al. 2017). Hepp (2018) challenges this perspective and characterizes makers as a pioneer community that is partly created by media hype and sponsored by corporations. On the other hand, fab labs collaborate with educational institutions (including schools and universities) and provide (in)formal education. Such facilities position themselves as places of learning and portray their activities as *maker education*.

Background

Fab labs became known through an initiative at the Massachusetts Institute of Technology, where in 1998 Gershenfeld offered an experimental course entitled "How to Make (Almost) Anything". The course, designed for a small group of physics and computing students, attracted a broad audience from all backgrounds, including design and architecture. It was the starting point for numerous other activities, including the foundation of the first fab labs as part of an outreach program (Gershenfeld 2005, 12). Similar developments preceded this (Kohtala et al. 2020; Sipos and Franzl 2020; Smith 2014); some of them followed the popular concept from the US. Today, fab labs exist in many large cities around the world (Smith et al. 2017).

Maker education refers to an experience-based and hands-on approach to learning that engages participants in subjects and learning activities at the intersection of computer science, design, art, and engineering, among others (see Brandenburger and Vladova 2020). Mastering subjects in science, technology, engineering, arts, and mathematics (STEAM education) is supposed to prepare students for the challenges of a highly technological and digital society: This is where maker education comes in. It benefits from easy access to digital fabrication and shared software, hardware, and designs, which is seen as democratizing the access to technology and understanding technology. More important than democratizing access, however, are the opportunities maker education creates to empower students and raise consciousness (Blikstein 2013; Halverson and Sheridan 2014).

Fab labs hold a versatile educational potential that has been discussed in various academic studies within and outside the higher education sector (Mostert - van der Sar et al. 2013; Troxler et al. 2014; Rosenbaum and Hartmann 2020). Thus, fab labs can be seen as a key innovation for the tertiary education sector. Due to their open, project-based, and cooperative learning character, fab labs bridge higher education, industry, and society (Pernía-Espinoza et al. 2017). They enable a technology-based environment for knowledge transfer between and beyond academic disciplinary boundaries. This is precisely why they are an ideal opportunity for implementing transdisciplinary learning.

In the higher education context, learning in fab labs has been shown to have a positive impact on team communication, self-efficacy, individual understanding of learning, and overall student outcomes (Andrews and Roberts 2017; Hilton et al. 2018; Tomko et al. 2018). Moreover, studies show that such open learning spaces foster so-called *21st century skills* (skills, abilities, and learning dispositions that have been identified as being required for success in 21st century society and workplaces), including critical thinking and problem-solving skills (Rayna and Striukova 2021). Next to teaching specialized knowledge, it is these skills that are increasingly of interest for new educational concepts, as they prepare students for the future. Overall, students are encouraged to participate and take control and responsibility for their own learning (Martinez and Stager 2013, 81).

Gauntlett's description of "making as connecting" (Gauntlett 2011) offers a possible starting point for such transdisciplinary learning processes: connecting things (materials and ideas), connecting people, and "connecting with our social and physical environments" (Gauntlett 2011, 2). Making is a counterpoint to the "sit back and be told" (Gauntlett 2011, 8) culture inside and outside of school education, it resonates with the idea of learning as a co-creation of knowledge.

For this reason, fab labs that explicitly aim at promoting participation and self-organization are particularly relevant for transdisciplinary learning. They are meeting places and communication spaces, not just providers of manufacturing infrastructure. The focus is on them being open – not limited to access, but encompassing a multi-layered philosophy: participation in governance, in determining the institutional structure and rules, and in the development of the place itself. This refers to the willingness to adapt organization and infrastructure to the needs of the community, the network of people who feel a sense of belonging to each other through a shared practice and place. This includes the existence of formats, sets of rules, or institutional forms of participation through which community members can become active in a process of adaptation and transformation. Fab labs are not only places of learning but also places of transformation that afford to rethink existing structures.

Debate and criticism

Places that emphasize participation are particularly relevant for transdisciplinary learning. Nonetheless, such open educational practices create tensions with established systems and approaches in higher education. Three juxtapositions show the inherent tensions:

(1) *Rigidity versus fluidity*: Self-organization and participant agency in transdisciplinary fab labs go against the planned and streamlined systems of formal education. They have an air of anarchy and chaos, versus the rules and hierarchies in higher education that stifle creativity and innovation. Nagle (2021) identified five specific challenges academic institutions experienced when installing fab labs, challenges which equally hold for non-library fab labs: staffing, shifting culture, policies and procedures, and demonstrating impact. To develop a balance between order and chaos requires extra attention in transdisciplinary education (Mostert - van der Sar and Troxler 2022).

(2) *Science versus arts*: In a fab lab, the focus is not simply on learning to use tools, developing skills, and creating tangible end products. Key experiences for students are iteration, teamwork, accepting failure as part of learning or failure-positivity (Martin 2015), feedback-literacy, and self-efficacy, among others (Rosenbaum and Hartmann 2020). These elements of learning are valuable and also applicable beyond science and engineering disciplines (Halverson and Sheridan 2014; Lande and Jordan 2014). However, a lack of literature on teaching and learning in the arts and social sciences, let alone transdisciplinary education, hinders students from detecting these contexts (for an outstanding example, see Mizeret et al. 2022). Moreover, the way in which groups that differ – for instance, in age or discipline – interact, which often is a core feature in fab labs, has to be reflected more strongly in transdisciplinary educational research.

(3) *Replication versus repurposing*: Fab labs in educational institutions in the Global South enhance quality education through hands-on activities (Ben Rejeb and Roussel 2018) and collaborations with communities (Oladele-Emmanuel et al. 2018). Serving as community labs or innovation centers, they promote citizen science and social innovation (Schonwetter and Van Wiele 2018, 8–23) through research, making, and cultural activities to address local needs. While this approach is embraced in the STEM fields (Buchele and Dafla 2015), funding, infrastructure, and human capital constraints (Herrera and Juárez 2013) are raised as key challenges to implement maker education more broad adaptable, solutions can be found for different contexts, as long as adequate planning and resource mobilization are in place.

While building a communicative, inclusive, and participatory atmosphere is a challenge in general, it is even more so when inclusion needs to address colonial and indigenous cultures and thought systems. Valuable experiences include the fab lab in Wellington, New Zealand (Neale and Hobern 2017) and the integration of digital technology and indigenous culture in Peru (Gonzales Arnao 2016).

Current forms of implementation in higher education

To facilitate transdisciplinary learning in a fab lab, four international principles have been found useful in practice (Troxler and Mostert - van der Sar 2019a).

1. The project *1:1* is for peer instruction (Mazur 1997). Teachers are trained at the lab to actively encourage peer learning by redirecting questions to the group instead of immediately answering themselves. Thus, they can activate the collective knowledge and capacity of the group. This ties in with the idea of the zone of proximal development (Vygotsky 1978, 84–91), where students achieve a level of potential development through problem-solving under the guidance

of more experienced peers. Even when an answer is not readily available in the group, it is part of peer instruction to develop a solution collectively.
2. The initiative *20-60-20* focuses on how people split time between different learning activities – 20 percent for instruction or lecturing, 60 percent for making and experimenting, and 20 percent for reflection with peers. Again, teachers are trained at the lab to design their lessons according to this principle, as they notoriously tend to stretch lecturing to the detriment of the experimenting, where peer instruction can take place (Troxler and Mostert - van der Sar 2019b). Reflection makes room for productive failure (Kapur 2008; Persaud et al. 2022).
3. The idea of *3i* is to foster imitation, iteration, and improvisation (El-Zanfaly 2015), a three-step approach to appropriate technology. Imitation is the basis for learning a particular technology. In iteration, students add changes and modifications, and in improvisation they use the technology for their own ends.
4. The project *4 all* is for lessons at the lab that are open to peripheral participation (Lave and Wenger 2003), i.e. not exclusively to students attending class. Combined with the principle of peer instruction, it can create powerful moments of transdisciplinary learning.

Fab labs have found diverse forms of implementation internationally, as the following examples show. *Vigyan Ashram*, established in 1983, is a center for ancient Indian philosophy in Pabal, India, engaging rural youth in learning rural technologies and entrepreneurship. Since 2002, Vigyan Ashram has been home to a fab lab – the first outside MIT – used by rural youth, often school dropouts. The school awards a diploma in basic rural technology. The pedagogy involves students in "Socially useful productive work" of various domains, focusing on agriculture and fabrication. They also offer services to the community in areas such as biogas, solar energy, food processing, and machine repair. These efforts have transformed the village of Pabal into a hub of innovation, creating opportunities for rural youth to learn and become self-sufficient. The center helps young entrepreneurs start their enterprises, and disseminate technology among rural communities (Kulkarni 2016).

Fab Lab Wgtn is located at a design school in Wellington, New Zealand. It works from a perspective of inclusiveness and of integrating indigenous perspectives into the ecosystem. To do so, Fab Lab Wgtn developed a code of conduct that begins with a *whakatauki*, a proverb written in both te Reo Maori, the first language spoken in Aotearoa New Zealand, and English. The *whakatuki* represents the lab's ethos: "*He waka eke noa* – We are all in this together." It is followed by a statement about honoring indigenous perspectives and acknowledging that diverse approaches enrich the culture of the lab. Innovation begins with inclusion, which leads to the explanation about what is not considered discrimination, such as reasonable communication of boundaries. The code also lists characteristics outside the dominant paradigm, usually subject to discrimination – e.g. ethnicity, age, gender, but also profession and technical ability – which are protected within the Fab Lab Wgtn ecosystem (Neale and Hobern 2017).

Learning by doing at Ashesi University (Ghana): The Introduction to Engineering course at Ashesi University (Beem 2021) teaches students about engineering through a transdisciplinary approach that includes lectures, lab sessions, and real-world projects. The course focuses on technology mastery, critical thinking, problem-solving, leadership, and collaboration. The curriculum is project-based, promoting hands-on learning, and students experience the full product development cycle. Teaching methods include in-person and online lectures, guest presentations, and lab sessions. The course is three credits and evaluates students through class participation, quizzes, and a final project. Beyond the course, students can engage with the community through extracurricular activities and projects led by the Design Lab (Ambole 2020) and projects such as Agbogbloshie Makerspace Platform (Potter et al. 2019).

In conclusion, fab labs can be seen as a seminal step for learning communities where students, teachers, staff, and experts work together in co-creation, adding value for all parties involved within the ecosystem. Technology is used in all areas of life and decisively shapes social development. Fab labs offer a promising hub by linking technology, formal education, and civic and entrepreneurial engagement. At the same time, computer science and engineering components find their way into other subject areas through a technology-oriented learning environment. They allow for cooperative and contemporary learning in an applied and real-world environment characterized by a high degree of exchange, participation, and openness. Institutional and disciplinary boundaries give way to pluralistic project work oriented towards issues affecting society as a whole.

References

Ambole, Amollo. 2020. Rethinking design making and design thinking in Africa. *Design and Culture* 12 (3): 331–50.

Andrews, Deborah, and Dustyn Roberts. 2017. Academic makerspaces: Contexts for research on interdisciplinary collaborative communication. In *Proceedings of the 35th ACM International Conference on the Design of Communication (August 11, 2017)*, 1–7. SIGDOC '17. New York: Association for Computing Machinery. Available from https://www.dl.acm.org/doi/10.1145/3121113.3121230.

Beem, Heather R. 2021. *Exploring the role of project-based learning in building self-efficacy in first year African engineering students. 2021 ASEE Virtual Annual Conference Content Access.* Available from https://www.peer.asee.org/exploring-the-role-of-project-based-learning-in-building-self-efficacy-in-first-year-african-engineering-students.

Ben Rejeb, Helmi, and Benoît Roussel. 2018. Design and innovation learning: Case study in North African engineering universities using creativity workshops and fabrication laboratories. *Procedia CIRP*, Volume 70. 28th CIRP Design Conference 2018: 331–37. Available from https://www.sciencedirect.com/science/article/pii/S2212827118304347.

Blikstein, Paulo. 2013. Digital fabrication and "making" in education: The democratization of invention. In *FabLab: Of machines, makers and inventors*, eds. Julia Walter-Herrmann and Corinne Büching. 203–22. Bielefeld: transcript.

Brandenburger, Bonny, and Gergana Vladova. 2020. Technology-enhanced learning in higher education: Insights from a qualitative study on university-integrated makerspaces in six European countries. In *Seamless Learning – lebenslanges, durchgängiges Lernen ermöglichen*, eds. Claude Müller Werder and Jennifer Erlemann, 27–38. Münster: Waxmann.

Brandenburger, Bonny, and Maximilian Voigt. 2021. FabLab. In *Handbuch Transdisziplinäre Didaktik*, eds. Tobias Schmohl and Thorsten Philipp, 107–17. Bielefeld: transcript.

Buchele, Suzanne Fox, and Aelaf Dafla. 2015. *Crafting a 21st century undergraduate engineering programme for Sub-Saharan Africa*. Available from https://www.ieeexplore.ieee.org/document/7177906.

El-Zanfaly, Dina. 2015. [I3] imitation, iteration and improvisation: Embodied interaction in making and learning. *Design Studies* 41: 79–109.

Gauntlett, David. 2011. *Making is connecting: The social meaning of creativity, from DIY and knitting to YouTube and Web 2.0*. Cambridge: Polity Press.

Gershenfeld, Neil. 2005. *Fab: The coming revolution on your desktop, from personal computers to personal fabrication*. New York: Basic.

Gonzales Arnao, Walter Hector. 2016. Computing systems of the Incas. Software development and hardware for application to teaching. In *Proceedings of the Fab12 Research Papers Stream. Shenzhen: International Fab Lab Association*. Available from https://www.archive.org/details/Fab12GonzalesArnao/mode/2up.

Halverson, Erica Rosenfeld, and Kimberly Sheridan. 2014. The maker movement in education. *Harvard Educational Review* 84 (4): 495–504.

Hepp, Andreas. 2018. What makes a maker? Curating a pioneer community through franchising. *Nordisk Tidsskrift for Informationsvidenskab ok Kulturformidling* 7 (2): 3–18.

Herrera, Pablo C., and Benito Juárez. 2013. *Fabrication laboratories: Problems and possibilities of implementation in Latin America*. Available from https://www.repositorioacademico.upc.edu.pe/bitstream/handle/10757/605215/Pablo%20Herrera-Juarez.pdf.

Hilton, Ethan C., Robert L. Nagel, and Julie S. Linsey. 2018. *Makerspace involvement and academic success in mechanical engineering*. Available from https://www.ieeexplore.ieee.org/document/8658875.

Kapur, Manu. 2008. Productive failure. *Cognition and Instruction* 26 (3): 379–424.

Kohtala, Cindy, Yana Boeva, and Peter Troxler. 2020. Alternative histories in DIY cultures and maker utopias. *Digital Culture & Society* 6 (1): 5–34.

Kulkarni, Yogesh. 2016. *Fab Lab 0.0 to Fab Lab 0.4 – Learning from running a lab in an Indian village*. Available from Bitte https://www.archive.org/details/Fab12Kulkarni.

Lande, Micah, and Shawn Jordan. 2014. *Making it together, locally: A making community learning ecology in the Southwest USA*. Available from https://www.computer.org/csdl/proceedings-article/fie/2014/07044394/12OmNzy7uVh.

Lave, Jean, and Etienne Wenger. 2003. *Situated learning: Legitimate peripheral participation*. Cambridge: Cambridge University Press.

Lewis, Charlton T., and Charles Short. 2020. *A Latin dictionary: Founded on Andrews' edition of Freund's Latin dictionary*. Chapel-en-le-Frith: Nigel Gourlay.

Martin, Lee. 2015. The promise of the maker movement for education. *Journal of Pre-College Engineering Education Research (J-PEER)* 5 (1): 30–39.

Martinez, Sylvia Libow, and Gary Stager. 2013. *Invent to learn: Making, tinkering, and engineering in the classroom*. 2nd edition. Torrance, CA: Constructing Modern Knowledge Press.

Mazur, Eric. 1997. *Peer instruction: A user's manual*. Upper Saddle River, NJ: Prentice Hall.

Mizeret, Jérôme, Nathalie Nyffeler, Sylvie Ray-Kaeser, Noémie Délèze, and Mélanie Thomas. 2022. Case study: The contributions of a FabLab to a Bachelor cursus in Occupational Therapy. *ITM Web of Conferences* 41: 03003.

Mostert - van der Sar, Manon, and Peter Troxler. 2022. Chaos and order in maker coaching: Towards a pathway for library makerspaces. In *6th FabLearn Europe/MakeEd Conference 2022*, 1–4. New York: Association for Computing Machinery.

Mostert - van der Sar, Manon, Ingrid J. Mulder, Leo Remijn, and Peter Troxler. 2013. Fablabs in design education. In *Proceedings of E&PDE 2013, International Conference on Engineering and Product Design Education*, 629–34. Glasgow: The Design Society.

Nagle, Sarah B. 2021. Maker services in academic libraries: A review of case studies. *New Review of Academic Librarianship* 27 (2): 184–200.

Neale, Wendy, and Craig Hobern. 2017. *Innovation begins with inclusion. Integrating the fab charter into a code of conduct at Fab Lab Wgtn*. Available from https://www.archive.org/details/Fab13NealeHobern.

Oladele-Emmanuel, Babasile Daniel, Helmi Ben Rejeb, and Tobias Redlich. 2018. *Strategic management: SWOT analysis of the African digital fabrication laboratories*. Available from https://www.ieeexplore.ieee.org/document/8436269.

Pernía-Espinoza, Alpha, Enrique Sodupe-Ortega, Sergio Peciña-Marqueta, Sergio Martínez-Bañares, Andres Sanz-Garcia, and Julio Blanco-Fernandez. 2017. *Makerspaces in higher education: The UR-Maker experience at the University of La Rioja*. Available from https://www.ocs.editorial.upv.es/index.php/HEAD/HEAD17/paper/view/5400.

Persaud, Stefan, Bas Flipsen, and Erik Thomassen. 2022. Productive failure in action. In *DS 117: Proceedings of the 24th International Conference on Engineering and Product Design Education (E&PDE 2022)*. Glasgow: The Design Society.

Potter, Cher, D. K. Osseo-Asare, and Mugendi K M'Rithaa. 2019. Crafting spaces between design and futures: The case of the Agbogbloshie Makerspace Platform. *Journal of Futures Studies* 23 (3): 39–56.

Rayna, Thierry, and Ludmila Striukova. 2021. Fostering skills for the 21st century: The role of Fab labs and makerspaces. *Technological Forecasting and Social Change* 164: 120391.

Rosenbaum, Leah F., and Björn Hartmann. 2020. *Where be dragons? Charting the known (and not so known) areas of research on academic makerspaces*. Available from https://www.ijamm.pubpub.org/pub/2952q2fi/release/1.

Schonwetter, Tobias, and Bram Van Wiele. 2018. *3D printing: Enabler of social entrepreneurship in Africa? The roles of fablabs and low-cost 3D printers*. Cape Town: Open African Innovation Research.

Sipos, Regina, and Kerstin Franzl. 2020. Tracing the history of DIY and maker culture in Germany's open workshops. *Digital Culture & Society* 6 (1): 109–20.

Smith, Adrian. 2014. *Technology networks for socially useful production*. Available from http://peerproduction.net/issues/issue-5-shared-machine-shops/peer-reviewed-articles/technology-networks-for-socially-useful-production/.

Smith, Adrian, Mariano Fressoli, Dinesh Abrol, Elisa Arond, and Adrian Ely. 2017. *Grassroots innovation movements*. London: Routledge.

Tomko, Megan, Robert Nagel, Melissa Alemán, Wendy Newstetter, and Julie Linsey. 2018. *Learning in academic makerspaces: Preliminary case studies of how academic makerspaces afford learning for female students*. Available from https://www.peer.asee.org/board-147-learning-in-academic-makerspaces-preliminary-case-studies-of-how-academic-makerspaces-afford-learning-for-female-students.pdf.

Troxler, Peter, and Manon Mostert - van der Sar. 2019a. *Didactisch Kompas Stadslab Rotterdam*. Rotterdam: Hogeschool Rotterdam. Available from https://www.hbo-kennisbank.nl/details/sharekit_hr:oai:surfsharekit.nl:3a505a01-5201-499b-beaf-df9219ba8782.

Troxler, Peter, and Manon Mostert - van der Sar. 2019b. *Seven years of plenty? Zeven jaar rijkdom?* 2nd edition. Rotterdam: Hogeschool Rotterdam. Available from https://www.hogeschoolrotterdam.nl/onderzoek/projecten-en-publicaties/pub/zeven-jaar-rijkdom-leren-in-met-door-van-en-v/395d3e6d-efbc-4a89-b782-3e7eb2c3d7e5.

Troxler, Peter, Edgar de With, and Manon Mostert - van der Sar. 2014. *DOT O&O: Developing research and design skills in the classroom (Dutch secondary education).* Available from https://www.hbo-kennisbank.nl/details/sharekit_hr:oai:surf-sharekit.nl:735be763-8e85-4938-b514-6e024b3f5537.

Vygotsky, Lev Semyonovich. 1978. *Mind in society: The development of higher psychological processes.* Cambridge, MA: Harvard University Press.

Feedback Literacy

Jennifer Schluer, Olivia Rütti-Joy, and Valentin Unger

Definition

Feedback constitutes an integral part of teaching and is essential for ensuring student achievement (Wisniewski et al. 2020, 13). A key component of contemporary educational feedback is *feedback literacy*. It has become particularly prominent in instructional feedback research since Carless and Boud's (2018) seminal paper and is increasingly recognized for its significance to transdisciplinary research and practice. Indeed, as a specific form of cooperation between heterogeneous social agents of different hierarchical levels, feedback literacy is crucial to disestablish the reproduction of hegemonic power structures, systems of oppression, misunderstandings, and dysfunctions in educational contexts. Successful feedback practices rely on transdisciplinary approaches and practices to collaboratively construct knowledge amongst a plurality of knowledge resources and participants.

A variety of definitions of feedback literacy exist across the disciplines (Nieminen et al. 2022, 99–104). To define and understand feedback literacy amidst this diversity, understanding the etymology of its semantic constituents can be helpful: (1) *feedback* and (2) *literacy*.

1. *Feedback* was originally an open compound noun, consisting of two separate components (Merriam Webster 2023). *Feed* traces back to Old English *fedan* ("to nourish, give food to, sustain, foster"), which roots in the Proto-Germanic *fodjan* ("to feed"). *Back* originates from Old English *bæc* ("backwards, behind, aback", Harper n.d.a). As a closed compound, *feedback* first appeared in the field of cybernetics, systems theory, and the regulation of machines, organisms, and organizations (Boud and Molloy 2013, 700–701). It was later appropriated into the educational realm and defined from behaviorist, cognitivist, and socio-constructivist perspectives. The conceptualization of feedback thus shifted from manipulating learners' stimulus-response associations (behaviorist, Boud and Molloy 2013, 700) to (unidirectional) information delivery from a more knowledgeable person (cognitivist, Lipnevich and Panadero 2021,

2) to multidirectional, iterative processes of mutual responsibility-sharing and collaborative negotiation of meaning (socio-constructivist, Carless 2022, 145).
2. *Literacy* comes from *literate*, with roots in the Latin expression *litteratus* ("educated, learned, who knows the letters"), a derivative of *littera/litera* (Latin for "alphabetic letter", Harper, n.d.b). Commonly defined as "the ability to read and write" (OECD 2000), the meaning of *literacy* has proliferated starkly in response to the growing diversity of communication modes (The New London Group 1996, 60–61). With the emergence of "new" literacy practices, the original "literacy" concept broadened from a singular skill to multiple literacies and an understanding of "literacy" as complex, dynamic, dialogic, and situated practices that are relevant to succeed in society (e.g. Freire 1985, 17).

Background

Feedback literacy was derived from the constructs *assessment literacy* (Stiggins 1991, 535) and *academic literacies* (skills required by students when transitioning into higher education, Lea and Street 1998). When Sutton (2012, 31–33) deduced feedback literacy from academic literacies, he emphasized the learners' role and responsibilities. Conversely, assessment literacy originally referred to the knowledge and skills assessors need for performing assessment-related actions (Stiggins 1991, 535). The concept was later expanded to include *student assessment literacy* (Carless et al. 2011) because successful assessment processes and interactions rely equally on the assessor and the assessee.

Analogously, the feedback receiver and provider are mutually responsible for successful feedback practices (Carless 2022, 149). As a transdisciplinary construct, feedback literacy is relevant across subjects and disciplines and thus a prerequisite for all feedback participants involved. This sharing of responsibilities is reflected in a recent subdivision of feedback literacy into two dimensions: *student feedback literacy* and *teacher feedback literacy*. Student feedback literacy refers to the knowledge, dispositions (attitudes and willingness), and capacities learners require to seek, understand, and utilize feedback as well as their ability to manage affect (Carless and Boud 2018, 1316–17). This dimension thus emphasizes learners' active role instead of reducing them to mere recipients of feedback information (Carless 2022, 145; Winstone and Carless 2020, 13). The second dimension – teacher feedback literacy – includes teachers' ability to design learning environments that are conducive to students' feedback literacy and the development of their self-regulation skills, to enable effective and multidirectional feedback dialogues as well as successful feedback use by all participants (Boud and Dawson 2023, 158).

Newer conceptualizations of teacher and student feedback literacy refuse to make a clear-cut distinction between the two, and instead capture both constructs in the plural term *feedback literacies* (Tai et al. 2023, 203). This approach is rooted in the belief that teaching and learning are inseparably intertwined. Moreover, it highlights the multiplicity and heterogeneity of contextually shaped feedback practices (Gravett 2022, 266). This reconceptualization also facilitates a paradigm shift towards empowering all feedback agents to negotiate and question hegemonic structures (Tai et al. 2023, 203). With this shift cascading into practice, intertwined *teacher–student feedback literacies* manifest themselves in the "ongoing process of attending, attuning, reading and ... working to modify the conditions which surround feedback" (Tai et al. 2023, 210). Developing and maintaining such literacies requires an openness to the entanglements of the conditions and practices of feedback, as well as a purposeful use across all feedback and learning opportunities, contexts, and disciplines (Chong 2022, 6–7). In line with the underlying principles of transdisciplinarity, a pluralistic perspective on feedback literacy emphasizes responsibility-sharing between all agents in the feedback process and considers the particularities of various disciplines, tasks, and technologies, as well as their impact on the situated and co-constructed feedback processes (Chong 2021, 96; 2022, 3–4).

Debate and criticism

Feedback is a contested, contextualized, and complex practice (Sutton 2012, 31) that has been researched largely separately across the disciplines (Lipnevich and Panadero 2021, 2). While feedback thrives from multidisciplinary influences and is transdisciplinary at its core, its inconsistent conceptualizations result in vivid discussions. One of these debates revolves around the definition of feedback itself and affects the definition of feedback literacy. For example, it is discussed whether one can speak of feedback if it is not acted upon by the recipient or whether it is then "information" only (Boud and Molloy 2013, 701). To emphasize subsequent action, the term "feedforward" was suggested to foreground the transformation of information into action (Reimann et al. 2019, 1279–80). Similarly, feedback has also been defined as an assembly of discursive practices that both reflect and construct reality (Nieminen et al. 2022, 102–4). Other authors, in turn, fear that by focusing (too much) on the subsequent actions, the content of the information would lose its power (Panadero and Lipnevich 2022, 5).

Another field of tension resides in the precise definition of feedback literacy. Three strands dominate this conversation (Nieminen and Carless 2022, 8). The first considers feedback literacy as an *internal psychological construct* and trainable skill that remains consistent across contexts and aims to fabricate feedback-literate and psychologically capable students (skills and capabilities). The second strand conceptualizes feedback as a process that involves an *acculturation of stu-*

dents and teachers to national feedback and academic cultures – cultures that are deeply embedded in the disciplines, institutional structures, and power relations. Therein, feedback literacy is an inconsistent construct that is continually redeveloped and reenacted across contexts and disciplines (socialization) and needs to be approached from a transdisciplinary perspective. The third strand considers feedback literacy as *socially constructed through power and discourse*, thus moving beyond an understanding of feedback literacy as an inherent trait of individuals. Instead, it is inherent to communities and their interactions where feedback-literate participants are critical and political agents who construct and (re)negotiate meaning in feedback processes to change their contexts (Nieminen et al. 2022, 103).

As the abundance of available definitions is rooted in incongruent ideas of learning, scientific inquiry, and feedback, it is not yet clear whether the various approaches can (or should) be reconciled into one overarching definition of feedback literacy (Nieminen et al. 2022, 13–14). On the one hand, this variety, the lack of nuanced conceptualizations, and an inconsistent use of terminology have inhibited clearer insights into the pedagogical designs that restrict or foster learners' agency in the feedback process (Nieminen et al. 2022, 103). On the other hand, as "contexts [enable] and constrain the ways in which individuals can act" (Tai et al. 2023, 203), a reductionist approach is not conducive. Instead, it is necessary to understand feedback practices and processes in a broader social and relational context, i.e. within a particular ecosystem of teaching and learning (Nieminen et al. 2022, 99). Recent conceptualizations highlight the "complex, nuanced, dynamic and situated set of feedback literacies, that are entangled by social, epistemological, material-discursive, spatial and temporal factors" (Gravett 2022, 270). Learners' (and teachers') agency is therein seen as mediated by the interplay of all these factors as well as individual variables (Chong 2021, 96; Gravett 2022, 270–71; Tai et al. 2023, 202). The latter include power, trust, relationships, and emotions. For instance, students are more likely to act on feedback when power is evenly distributed between teachers and students (Dann 2019, 362–63) and when feedback takes place in honest conversations. Overcoming hierarchical barriers and promoting responsibility-sharing as well as inclusive, transdisciplinary teaching and learning requires a changed understanding of teachers' role from information-providing authorities to learning facilitators. Abandoning the separation of teacher and student feedback literacy could contribute to such a change, as a separate treatment builds on predetermined roles of individuals and fails to consider material, discursive, or social dimensions (Tai et al. 2023, 202). The concept of *student–teacher feedback literacies* therefore needs to reach beyond a binary understanding.

While feedback as a discursive process has become prominent in instructional feedback research, it remains a fragile idea in need of more nuance and stronger embedding in transdisciplinary contexts. Indeed, there is hardly any transdisciplinary research on feedback literacy, its development, and its effect on transdisciplinary teaching and learning. Investigating the boundaries between feedback literacy and related concepts (e.g. assessment literacy or digital literacy) as well as embracing a multimodal, transdisciplinary, and critical-transformative approach could promote a more accurate and future-oriented understanding. We thus encourage transdisciplinary research to pursue the development of a multidirectional and multifaceted conceptualization of feedback literacy within the sociopolitical contexts of education (Nieminen and Carless 2022, 13–15). As this form of literacy actively transgresses boundaries, it may establish a space of transdisciplinary participation and collaboration beyond fixed roles and contexts, thereby contributing to the disestablishment of hegemonic power structures within and across disciplines, while recognizing and inviting diversity and a plurality of knowledge sources (Tai et al. 2023, 204). Accordingly, future efforts in research and practice related to feedback need to move away from compartmentalized approaches to understanding feedback literacy within singular (disciplinary) contexts only.

Current forms of implementation in higher education

Feedback literacy is of vital importance to any discipline and is as such a highly contextualized and situated practice. Since conceptualizations are still in their infancy (Chong 2021, 94), pedagogical recommendations for effective feedback designs and institutional implementations are rare (Winstone and Carless 2020), especially regarding digital feedback (Schluer 2022) and transdisciplinary work. The following review establishes an overview of current research and practice in different contexts to derive tentative advice for the development and enactment of feedback literacy for transdisciplinary purposes.

Feedback literacy is increasingly recognized for its transformative power to improve student learning and to foster learners' self-regulation skills (Panadero and Lipnevich 2022, 14). Indeed, a perusal of recent publications in a leading journal in the field (*Assessment & Evaluation in Higher Education*) shows that feedback literacy is of interest to researchers and practitioners from a wide range of geographical contexts, including Europe (e.g. Rovagnati et al. 2022), Australia (Boud and Dawson 2023), Asia (Hsieh and Hill 2022), and the Middle East (Mohammed and Alharbi 2022). Furthermore, studies by multinational teams are increasing, resulting in a growing number of coauthored publications, e.g. from China and the USA (Dong et al. 2023), or Singapore, Australia, and the UK (Hoo et al. 2022). Overall, most published research comes from the UK, the USA, Australia, and Asia, but less so from Africa (e.g. Nieminen et al. 2023, 82).

Research is still ongoing to develop sound and evidence-based theoretical frameworks for teacher feedback literacy (e.g. Boud and Dawson 2023), student feedback literacy (Hoo et al. 2022) and peer feedback literacy (Dong et al. 2023).

The interplay of different feedback sources is investigated, such as peer and teacher feedback (Hsieh and Hill 2022) or self-assessment and peer assessment in academic writing (Cheong et al. 2023). While previous research largely focused on written assignments, studies on oral and multimodal tasks are becoming more prevalent, especially in digital settings (Day et al. 2022). In their linguistic analysis of publications from 2009 to 2019, Winstone, Boud, Dawson, and Heron (2022, 224) observed "a decrease in the use of 'written', 'teacher' and 'detailed' as nouns modifying the term feedback, and an increase in the use of 'peer', 'verbal' and 'video'". This illustrates a shift from teacher to learner orientation and points to a greater relevance of technology-enhanced feedback dialogues across disciplines and contexts in response to the rapid increase of digital educational practices. There is thus a need for developing and investigating pedagogical designs that are conducive to student learning in hybrid or virtual transdisciplinary spaces (Schmidberger et al. 2022, 76–77).

Indeed, technology-enhanced socio-material environments (collaborative documents, online forums, polls, etc.) could reduce power distance and enable equal participation (Schluer 2022, 92–110, 156–63; Tai et al. 2023, 204–207). However, feedback as a process of shared responsibilities has not fully translated into higher education curricula, partly due to students' and educators' persisting expectations of hegemonic power relations (Winstone et al. 2021, 129). For example, Winstone's (2022, 1107) analysis of policy and strategy documents from 134 universities in the United Kingdom showed that most feedback practices center on transmitting feedback information to passive student recipients instead of promoting constructive, learner-focused feedback conversations. Similarly, the analysis of National Qualifications Frameworks from Asia, Africa, Europe, Latin America, North America, and Oceania revealed a strong focus on judgment-making as a graduate outcome, whereas other dimensions of feedback literacy were neglected (Winstone, Baloo, and Carless 2022, 62 and 74). Moreover, feedback literacy is still "widely depicted from a mono-cultural (prevalently Anglophone) perspective that accounts for one literacy rather than multiple literacies" (Rovagnati et al. 2022, 347). This calls for an acknowledgment of individual feedback literacies and "literacy histories" (Rovagnati 2022, 63 and 66) in intercultural and transdisciplinary contexts through open and purposeful dialogues about the underlying principles of feedback practices and responsibilities (Rovagnati 2022, 226).

While concrete institutional practice examples are largely absent and research is still scarce, a review of the literature from different disciplinary fields (e.g. English language teaching and intercultural communication by Schluer and Liu 2023; environmental sustainability by Blythe et al. 2017; as well as Carless and Boud 2018) provides insight into strategies that could contribute to developing feedback literacy in transdisciplinary contexts. Such strategies may include: (1) reflecting and inquiring about all *stakeholders' understanding* of and experiences with feedback literacy; (2) creating a *trustful atmosphere* by openly discussing attitudes, concerns, and challenges regarding transdisciplinary work; (3) *negotiating meanings* and establishing a common ground (third space) to ensure team functionality; (4) critically *inspecting existing communicative modes*, tools, and structures while showing openness to cultivating new interactional norms (especially in linguistically and culturally heterogeneous teams); (5) appreciating and encouraging *diverse perspectives*; and (6) regularly *reflecting on feedback processes and renegotiating them* as needed in dynamically changing transdisciplinary contexts.

Due to the highly situated and dynamic nature of feedback literacy and the scarcity of research and practice in relation to its implementation in transdisciplinary contexts, the application of the above recommendations needs to be reassessed continuously and collaboratively. Indeed, constructive feedback from colleagues in adjacent disciplines holds the potential to uncover new perspectives which may transcend previously rigid disciplinary frames and thinking patterns. While such conversations might cause initial reservations, pursuing feedback processes through critical social reflection and dialogue can lead to transformative reinterpretations and broaden conceptual horizons.

References

Blythe, Jessica, Kirsty Nash, Julian Yates, and Graeme Cumming. 2017. Feedbacks as a bridging concept for advancing transdisciplinary sustainability research. *Current Opinion in Environmental Sustainability* 26–27: 114–19.

Boud, David, and Phillip Dawson. 2023. What feedback literate teachers do: An empirically-derived competency framework. *Assessment & Evaluation in Higher Education* 48(2): 158–71.

Boud, David, and Elizabeth Molloy. 2013. Rethinking models of feedback for learning: The challenge of design. *Assessment & Evaluation in Higher Education* 38 (6): 698–712.

Carless, David. 2022. From teacher transmission of information to student feedback literacy: Activating the learner role in feedback processes. *Active Learning in Higher Education* 23 (2): 143–53.

Carless, David, and David Boud. 2018. The development of student feedback literacy: Enabling uptake of feedback. *Assessment & Evaluation in Higher Education* 43 (8): 1315–25.

Carless, David, Diane Salter, Min Yang, and Joy Lam. 2011. Developing sustainable feedback practices. *Studies in Higher Education* 36 (4): 395–407.

Cheong, Choo Mui, Na Luo, Xinhua Zhu, Qi Lu, and Wei Wei. 2023. Self-assessment complements peer assessment for undergraduate students in an academic writing task. *Assessment & Evaluation in Higher Education* (48) 1: 135–48.

Chong, Sin W. 2021. Reconsidering student feedback literacy from an ecological perspective. *Assessment & Evaluation in Higher Education* 46 (1): 92–104.

Chong, Sin W. 2022. The role of feedback literacy in written corrective feedback research: From feedback information to feedback ecology. *Cogent Education* 9 (1): 1–13.

Dann, Ruth. 2019. Feedback as a relational concept in the classroom. *The Curriculum Journal* 30 (4): 352–74.

Day, Indira N. Z., Nadira Saab, and Wilfried Admiraal. 2022. Online peer feedback on video presentations: Type of feedback and improvement of presentation skills. *Assessment & Evaluation in Higher Education* 47 (2): 183–97.

Dong, Zhe, Ying Gao, and Christian D. Schunn. 2023. Assessing students' peer feedback literacy in writing: Scale development and validation. Available from https://www.tandfonline.com/10.1080/02602938.2023.2175781.

Freire, Paulo. 1985. Resting the world and reading the word: An interview with Paulo Freire. *Language Arts* 62 (1): 15–21.

Gravett, Karen. 2022. Feedback literacies as sociomaterial practice. *Critical Studies in Education* 63 (2): 261–74.

Harper, Douglas. n.d.a. *Etymology of feedback.* Online Etymology Dictionary. Available from https://www.etymonline.com/word/feedback.

Harper, Douglas. n.d.b. *Etymology of literacy.* Online Etymology Dictionary. Available from https://www.etymonline.com/word/literacy.

Hoo, Hui-Teng, Christopher Deneen, and David Boud. 2022. Developing student feedback literacy through self and peer assessment interventions. *Assessment & Evaluation in Higher Education* 47 (3): 444–57.

Hsieh, Yi-Chin, and Christopher Hill. 2022. Reconceptualizing the value of peer and instructor feedback using a sequential structure. *Assessment & Evaluation in Higher Education* 47 (7): 1043–56.

Lea, Mary R., and Brian V. Street. 1998. Student writing in higher education: An academic literacies approach. *Studies in Higher Education* 23 (2): 157–72.

Lipnevich, Anastasiya A., and Ernesto Panadero. 2021. A review of feedback models and theories: Descriptions, definitions, and conclusions. *Frontiers in Education* 31 (6): 1–29.

Merriam Webster. 2023. *Get looped in on "feedback": Its history is more than just noise.* Word History. Available from https://www.merriam-webster.com/words-at-play/the-history-of-feedback.

Mohammed, Murad A. S., and Mohammed A. Alharbi. 2022. Cultivating learners' technology-mediated dialogue of feedback in writing: Processes, potentials and limitations. *Assessment & Evaluation in Higher Education* 47 (6): 942–58.

Nieminen, Juuso H., and David Carless. 2022. *Feedback literacy: A critical review of an emerging concept.* Available from https://link.springer.com/article/10.1007/s10734-022-00895-9.

Nieminen, Juuso H., Joanna Tai, Margaret Bearman, and Michael Henderson. 2022. Student agency in feedback: Beyond the individual. *Assessment & Evaluation in Higher Education* 47 (1): 95–108.

Nieminen, Juuso H., Margaret Bearman, and Joanna Tai. 2023. How is theory used in assessment and feedback research? A critical review. *Assessment & Evaluation in Higher Education* 48 (1): 77–94.

OECD [Organization for Economic Co-operation and Development], ed. 2000. *Literacy in the information age: Final report of the international adult literacy survey.* Available from https://www.oecd.org/education/skills-beyond-school/41529765.pdf.

Panadero, Ernesto, and Anastasiya A. Lipnevich. 2022. A review of feedback models and typologies: Towards an integrative model of feedback elements. *Educational Research Review* 35 (5): 1–22.

Reimann, Nicola, Ian Sadler, and Kay Sambell. 2019. What's in a word? Practices associated with "feedforward" in higher education. *Assessment & Evaluation in Higher Education* 44 (8): 1279–90.

Rovagnati, Veronica. 2022. *Exploring assessment and feedback histories, literacies, and intercultural competence development: Longitudinal narratives of international postgraduate students.* Available from https://kar.kent.ac.uk/96620.

Rovagnati, Veronica, Edd Pitt, and Naomi Winstone. 2022. Feedback cultures, histories and literacies: International postgraduate students' experiences. *Assessment & Evaluation in Higher Education* 47 (3): 347–59.

Schluer, Jennifer. 2022. *Digital feedback methods.* Tübingen: Narr.

Schluer, Jennifer, and Yarong Liu. 2023. Peer feedback in intercultural online communication: Theoretical and practical considerations for English language teaching. In *Language and Interculturality in the Digital World,* eds. Roman Lietz, Milene Mendes de Oliveira, Luisa Conti, and Fergal Lenehan. Berlin: Peter Lang (in press).

Schmidberger, Laura, Valentin Unger, and Albrecht Wacker. 2022. Durch das Internet "gequetschte" Seminare!? Befunde einer explorativen Befragung von Studierenden im ersten Online-Semester. In *Lehren aus der Lehre in Zeiten von Corona. Mediendidaktische Impulse für Schulen und Hochschulen,* eds. Thomas Knaus, Thorsten Junge, and Olga Merz, 65–79. München: kopaed.

Stiggins, Richard J. 1991. Assessment literacy. *Phi Delta Kappan* 72 (7): 534–39.

Sutton, Paul. 2012. Conceptualizing feedback literacy: Knowing, being, and acting. *Innovations in Education and Teaching International* 49 (1): 31–40.

Tai, Joanna, Margaret Bearman, Karen Gravett, and Elizabeth Molloy. 2023. Exploring the notion of teacher feedback literacies through the theory of practice architectures. *Assessment & Evaluation in Higher Education* 48 (2): 201–13.

The New London Group. 1996. A pedagogy of multiliteracies: Designing social futures. *Harvard Educational Review* 66 (1): 60–92.

Winstone, Naomi. 2022. Characterising feedback cultures in higher education: An analysis of strategy documents from 134 UK universities. *Higher Education* 84: 1107–25.

Winstone, Naomi, and David Carless. 2020. *Designing effective feedback processes in higher education: A learning-focused approach*. London: Routledge.

Winstone, Naomi, Kieran Balloo, and David Carless. 2022. Discipline-specific feedback literacies: A framework for curriculum design. *Higher Education* 83: 57–77.

Winstone, Naomi, David Boud, Phillip Dawson, and Marion Heron. 2022. From feedback-as-information to feedback-as-process: A linguistic analysis of the feedback literature. *Assessment & Evaluation in Higher Education* 47 (2): 213–30.

Winstone, Naomi, Edd Pitt, and Robert Nash. 2021. Educators' perceptions of responsibility-sharing in feedback processes. *Assessment & Evaluation in Higher Education* 46 (1): 118–31.

Wisniewski, Benedikt, Klaus Zierer, and John Hattie. 2020. The power of feedback revisited: A meta-analysis of educational feedback research. *Frontiers in Psychology* 10: 1–14.

Global Citizenship Education

Heidi Grobbauer and Mary Whalen (authors contributed equally)

Definition

The globality of everyday life, living conditions, working environments, and economic systems, the crises which have developed in the wake of globalization itself, and the resulting difficulties for individuals are presenting mounting challenges to education systems. International programs, such as the current Global Agenda 2030 (United Nations 2015), highlight the importance of education and science in responding to global challenges and transitioning to an ecologically sustainable and more just world. In this context, Global Citizenship Education has become an internationally established field of education research and practice, most notably within UN General Secretary Ban Ki-Moon's Global Education First Initiative, which called for the promotion of Global Citizenship Education as one of its three focus areas (UNESCO 2014, 11). UNESCO made Global Citizenship Education one of its educational priorities in 2013 (UNESCO 2014, 12), and it is an element of the Sustainable Development Goals, which were adopted by all United Nations member states in 2015 (United Nations 2015).

The complexity of Global Citizenship Education, which can be described more as a field of research and practice than as an educational concept, cannot be captured in one all-encompassing definition. UNESCO describes Global Citizenship Education as a "framing paradigm which encapsulates how education can develop the knowledge, skills, values and attitudes learners need for securing a world which is more just, peaceful, tolerant, inclusive, secure and sustainable" (2014, 9). Global Citizenship Education includes the term citizenship, initially a concept of political philosophy. The idea of citizenship can be traced back to the notions of state and *civitas* in ancient Greece and Rome (Turner 2016, 679). Aristotle points to people's need to belong to a political community such as the πόλις (*polis*, city-state). Citizenship in a Greek city-state was characterized by participation in the political community, those who were qualified as citizens constituted the ruling class, an early form of δημοκρατία (*dēmokratía*, democracy) and δῆμος (*dêmos*). However, citizenship was exclusive and determined by place of residence, descent, gender, age, and socioeconomic status. The Roman concept of citizenship was initially similar; citizens were native free men, the legitimate sons of other native free men. During the expansion of the Roman Empire within Italy and Europe and finally into Africa and Asia, "two important innovations came about. First, the populations of conquered territories were given a version of Roman citizenship while being allowed to retain their own forms of government ... Second, the version of Roman citizenship given was of a legal rather than a political kind – *civitas sine suffragio or citizenship without the vote*. So, the Empire allowed dual citizenship, though it reduced Roman citizenship to a legal status" (Bellamy 2015, 6).

These two classic conceptions form the basis for the later development of various forms of citizenship. Both the historical concept and newer interpretations of citizenship emphasize the following three dimensions. First, a legal status, which defines the relationship between an individual and the state. Citizenship implies membership in a political community (city-state, empire, or later the nation state), which endows a citizen with rights, but also with responsibilities. The legal status is, secondly, closely linked to the right to political participation. Social relationships in society result, thirdly, in a feeling of belonging (Osler and Starkey 2005, 9–16). While membership leads to the formation of a symbolic community, it also implies the exclusion of all those who live within the geographic boundaries of a national state, but have not acquired formal citizenship, and are thus barred from exercising certain rights (for example, the right to vote). In the context of international migration, both the concept of citizenship and democracy, as a concept of political participation of the "demos", are increasingly under pressure.

In response to globalization, the associated transnationalization of political decision-making, and the challenges to citizenship and democracy in the context of growing international migration, various alternative concepts of global citizenship have emerged (Wintersteiner et al. 2015, 16–18). The idea of global citizenship refers above all to belonging to a global community. In the context of this global community, it is also possible to discuss the rights and duties of global citizens, as well as opportunities for political participation.

Background

The informed analysis of global crises, and the discussion of potentials for transformation, involve immense factual and ethical complexity. Fostering an understanding of global developments and potentials for change has thus become the recognized goal of educational approaches such as education for sustainable development (UNESCO 2014), Global Education (Bourn 2021) and Global Citizenship Education (Wintersteiner et al. 2015). Currently, the transformative potential of education is emphasized in Goal 4 of the Sustainable Development Goals. Target

4.7 places the following obligation on all states: "By 2030, ensure that all learners acquire the knowledge and skills needed to promote sustainable development, including human rights, gender equality, the promotion of a culture of peace and non-violence, global citizenship, and an appreciation of cultural diversity and culture's contribution to sustainable development" (United Nations 2015, 21). As Target 4.7 shows, a transdisciplinary approach will be necessary to enable young people to develop the knowledge and competence (and the authors would also add the ethical values) needed for the socio-ecological transformation.

A global perspective in education has a long, nonlinear, record. Global Citizenship Education was developed primarily in the English-speaking world in the context of citizenship education and development education (Tarozzi and Torres 2016), but it also has roots in peace education (Reardon 1988) and concepts of cosmopolitan education (Seitz 2002). In education practice, a social-justice orientation implies a stronger emphasis on Global Citizenship Education as civic education for a world society, a continuation of the internationally established practice of Global Education. Both concepts address the issues of political, economic, social, and cultural interdependency and interconnectedness, and the understanding that global, national, and local well-being are strongly tied to one another, as well as emphasizing the necessity of addressing the historical origins of the current world order. While Global Education employs a more holistic approach, the educational goals and focus areas of Global Citizenship Education are more closely related to the traditional practice of civic education, with a focus on the political aspects of global citizenship (Böhme 2019, 165–69; Wintersteiner et al. 2015).

With the *Recommendation Concerning Education for International Understanding, Co-operation and Peace* (1974, currently under revision), UNESCO has presented an international milestone for the global perspective on education (UNESCO n.d.; Wintersteiner et al. 2015, 6). The declaration examines the connection between international understanding, human rights, cultural, and environmental education, and calls for the inclusion of the global perspective at all levels of education. Since 2013, UNESCO has fostered Global Citizenship Education as a multifaceted approach, to be implemented in national education systems.

The theoretical discussion of concepts and approaches to Global Citizenship Education, as well as of different perspectives on programmatic goals and perceptions, has intensified in recent years. This is manifested in the increased relevance of Global Citizenship Education to educational systems, and the often very differing concepts which are shaping both theory and practice (Davies et al. 2018) and constitute a highly controversial field of work and research (Akkari and Maleq 2020a; Torres and Bosio 2020).

Criticism and debate

Concepts of Global Citizenship Education vary mainly in three points, which are explained in more detail below – in particular with regard to the implications of the term *global*, the interpretation of (global) citizenship and the underlying educational philosophies. On the one hand, neoliberal or entrepreneurial approaches focus on the globalized economy and the need to prepare young people to participate in global markets. This reinforces largely depoliticized interpretations of global citizenship (Pashby et al. 2020; Shultz 2007; Stein 2015). In contrast, other perspectives of Global Citizenship Education emphasize global solidarity or a vision of global social justice, accentuating the need to address global challenges and crises and the underlying power relations and ideologies; they address the potential role of education in solutions, and for making a stronger distinction between local, national, and global dimensions of global developments, while also being sensitive to their interconnectedness. This orientation is referred to as liberal-humanist or critical Global Citizenship Education (Pashby et al. 2020; Stein 2015). Other differentiations are seen in the comparison of a humanitarian versus a political approach (Dobson 2005), or a soft versus critical approach (De Oliveira Andreotti 2014). The differences can be summed up as a focus either on forming individual values and attitudes of cosmopolitan and responsible global citizens or on power relations and global inequalities that need to be changed so that global citizenship may become a tangible option (Wintersteiner et al. 2015, 11–13).

The term *citizenship* adds an already controversial element to the concept of global education. The idea of global citizenship focuses on the idea of community and belonging beyond the nation state, without implying a legal status or a formal global state. Contemporary examples of how a world society could function are the UN and international human rights conventions (Wintersteiner et al. 2015, 13). The work of philosophers like Nussbaum (1996) and Appiah (2005) has influenced the ethical foundations of the concept and linked it to the cosmopolitan tradition of antiquity. In the educational context, the concept of global citizenship, even if it is only an idea or a utopia, can open new individual frameworks for thinking and taking action, as well as spaces for collective discourse.

Many research papers frame global citizenship in a narrow regional context, where specific focal points are elaborated, such as citizenship in multicultural societies or in conflict or post-conflict societies, or the connections between (global) citizenship and human rights or social inequalities (Akkari and Maleq 2020a; Isin and Nyers 2014). In societies with a high level of immigration, issues of political rights and political participation of migrants move to the foreground. High immigration rates may awaken resentment toward migrants and refugees, and lead to strong manifestations of nationalism. In this case, Global Citizenship Education can provide an opportunity "to value multiple identities within national citizenship, and move away from the view that membership of a nation-state is earned through cultural assimilation, rather than an acquired right" (Akkari and Maleq 2020b, 206).

With regard to the underlying educational philosophies, the concepts of Global Citizenship Education differ according to a more instrumental understanding of education and competence development or a more strongly emancipatory approach. Neoliberal approaches are more prevalent in state-controlled implementation of Global Citizenship Education, such as in South Korea (Kim 2019), or in the programs of international organizations, such as the OECD's Global Competence Framework (OECD 2018).

Postcolonial criticism and theory have also added an extremely important component to the current discourse on Global Citizenship Education. Education systems were essential to the enforcement of colonial regimes and colonial domination, and the construction of "white" or "Western" superiority as an effective means of colonial domination still influences contemporary thoughts and actions. The criticism of Global Citizenship Education mirrors this debate, especially regarding the referencing of traditions of cosmopolitanism in ancient Greece and in the writings of Immanuel Kant, which led to a Eurocentric worldview of our origins and the *Enlightenment* as a universal concept. International institutions such as Oxfam, UNESCO, OECD, and others have integrated these ideas into their conceptions of Global Citizenship Education and "now purport to apply [them] to 'the rest' of the world in a universalizing fashion" (Szakács-Behling 2020, 105).

Global Citizenship Education also operationalizes such powerful concepts and expressions as globalization, globality, sustainability, justice, and citizenship, often without questioning its own perspective: "Whose citizenship? Whose global?" (Szakács-Behling 2020, 104). In conjunction with criticism from a postcolonial perspective (Akkari and Maleq 2020a; De Oliveira Andreotti and Souza 2011), the call for historical contextualization, as well as demands for a consequent decentralization or decolonization of Global Citizenship Education (Abdi 2015), and the consideration of historically marginalized philosophies and systems of thought (e.g. Ubuntu and indigenous perspectives) have become more vehement (Szakács-Behling 2020, 105).

Current forms of implementation in higher education

Global Citizenship Education is a broad pedagogical field of research and practice with very diverse perspectives. Accordingly, programs and strategies for implementing the concept in the education system are numerous and multifaceted. The following examples provide insights into three key areas of higher education: UNESCO initiatives, especially master's programs to strengthen Global Citizenship Education, university research, and teacher training.

UNESCO has established chairs to promote higher education for international understanding and global citizenship towards a culture of peace at Universidad de San Andrés (Argentina), the University of the Region of Joinville (Brazil), Université du Québec en Outaouais (Canada), the University of Bologna (Italy), the National University Corporation Kyushu University (Japan), Universidad Veracruzana and the University of Guadalajara (Mexico), the University of Montenegro, the University of Klagenfurt (Austria), and the University of California (USA). In cooperation with the Government of the Republic of Korea, UNESCO has also founded the Asia-Pacific Centre of Education for International Understanding (APCEIU). These higher-level UNESCO institutions undertake research and policy development, capacity-building, development of teaching and learning materials, international teacher exchange, and information sharing and networking.

The UNESCO Chair at the University of Klagenfurt has established the Master's Program *Global Citizenship Education* (Grobbauer and Wintersteiner 2019) in cooperation with the nongovernmental organization KommEnt and the University College of Teacher Education Carinthia. The program is primarily aimed at teaching staff, teacher trainers, and other professionals in extra-curricular education, thus increasing the number of committed, well-trained professionals in the field. The curriculum is characterized by its interdisciplinarity: social sciences form a central part of the course, and the examination of the concept of (global) citizenship and the associated issues of transnational democracy are based on political science and political theory. The foundations of educational science and the knowledge and critical reflection of pedagogical concepts are fundamental. Moreover, the understanding of Global Citizenship Education is also based on the normative principle of global justice, whereby philosophy and ethics offer important theoretical points of reference. The critical orientation of the program questions concepts of Western hegemony, provides insights into European legacies of colonialism and neo-colonialism, and fosters the discussion of postcolonialism.

The field of research has expanded significantly in recent years, with an emphasis on teaching about ethical global issues. Based on Andreotti's critical education approach, Pashby and Sund (2019) have conducted research with teachers in England, Sweden, and Finland, with the goal of connecting decolonial theory with classroom experience in Europe. A particular focus was placed on identifying mainstream, as well as marginalized or underrepresented perspectives, and recognizing enabling factors and challenges in addressing barriers to teaching critical Global Citizenship Education. Their findings show that secondary and upper secondary schoolteachers welcome teaching guides that can be easily adapted to the classroom. The result of the ensuing participatory process is a resource which enables teachers to critically reflect upon their own teaching practice, guide students to understand the origins of different perspectives and worldviews, including their own, identify mainstream viewpoints and appreciate marginalized perspectives, recognize the complexity and root causes of global issues, and discuss possible activities or responses following the lessons.

Students' critical understanding of their roles as global citizens is a crucial element of international study programs at schools and universities, and has become an established focus of teacher training programs at the University of Alberta (Canada). Here, Schultz trains educators to examine theories and case studies focusing on the "constructions of global citizenship, and how citizenship is (has been) lived, denied, recreated and/or re-imagined" (University of Alberta 2023). Teachers learn to develop empathy for diverse opinions, analyze the interconnectedness of global events and problem-solving strategies, and understand the potential of the concept of global citizenship for their future students and the increasingly diverse social, economic, and political communities and contexts that they will deal with professionally.

The Center for Global Education (EPIZ) is involved in a number of transdisciplinary endeavors, including teacher training programs at the Berlin Department of Education and Master of Education programs at TU Berlin and the School of Vocational Education at the Münster University of Applied Sciences, and the project *Vocational Education Meets University* at the Otto-Suhr-Institute of the Freie Universität Berlin. Study tours and online conferences for teacher trainers in Berlin and at the University of KwaZulu-Natal in Durban, South Africa, enable an academic exchange on topics such as community-based approaches to climate education, and decolonial perspectives on structural inequality in South Africa's education system. The goal of a current lighthouse project with the Humboldt Universität zu Berlin is to identify core competencies crucial for the implementation of climate change prevention and mitigation in three vocational fields, and to design teaching strategies for developing these competencies within existing curriculum frameworks. These programs represent the vital contributions of current transdisciplinary collaborations between academic disciplines, civil society actors, and the field of education to the advancement of Global Citizenship Education in theory and practice.

References

Abdi, Ali A. 2015. Decolonizing global citizenship education. In *Decolonizing global citizenship education*, eds. Ali A. Abdi, Lynette Shultz, and Thasika Pillay, 11–26. Rotterdam: Sense.

Akkari, Abdeljalil and Kathrine Maleq. 2020a. Global citizenship education: Recognizing diversity in a global world. In *Global citizenship education*, eds. Abdeljalil Akkari and Kathrine Maleq, 3–13. Cham: Springer.

Akkari, Abdeljalil and Kathrine Maleq. 2020b. Rethinking global citizenship education: A critical perspective. In *Global citizenship education*, eds. Abdeljalil Akkari and Kathrine Maleq, 205–16. Cham: Springer.

Appiah, Kwame Anthony. 2005. *The ethics of identity*. Princeton: Princeton University Press.

Bellamy, Richard. 2015. Citizenship: Historical development of. In *International encyclopedia of social and behavioural sciences*, ed. James D. Wright, 643–49. 2nd edition. Oxford: Elsevier.

Böhme, Lars. 2019. *Politische Bildung für Schüler und Schülerinnen mit sonderpädagogischem Förderbedarf. Perspektiven Globalen Lernens an Förderzentren*. Frankfurt am Main: Wochenschau Verlag.

Bourn, Douglas. 2021. Pedagogy of hope: Global learning and the future of education. *International Journal of Development Education and Global Learning* 13 (2): 65–78.

Davies, Ian, Li-Ching Ho, Dina Kiwan, Carla L. Peck, Andrew Peterson, Edda Sant, and Yusef Waghid, eds. 2018. *The Palgrave handbook of global citizenship and education*. London: Palgrave Macmillan.

De Oliveira Andreotti, Vanessa. 2014. Soft versus critical global citizenship education. In *Development education in policy and practice*, ed. Stephen McCloskey, 21–31. London: Palgrave Macmillan.

De Oliveira Andreotti, Vanessa and Louis De Souza, eds. 2011. *Postcolonial perspectives on global citizenship education*. New York: Routledge.

Dobson, Andrew. 2005. Globalisation, cosmopolitanism and the environment. *International Relations* (19): 259–73.

Grobbauer, Heidi, and Werner Wintersteiner. 2019. *Global citizenship education: concepts, efforts, perspectives – an Austrian experience*. Salzburg: Klagenfurt.

Isin, Engin Fahri, and Peter Nyers, eds. 2014. *Routledge handbook of global citizenship studies*. London: Routledge.

Kim, Yeji. 2019. Global citizenship education in South Korea: Ideologies, inequalities, and teacher voices. In *Globalisation, Societies and Education* 17 (2): 177–93.

Nussbaum, Martha Craven. 1996. For love of country: *Debating the limits of patriotism*. Boston: Beacon Press.

OECD [Organisation for Economic Co-operation and Development]. 2018. *Preparing our youth for an inclusive and sustainable world. The OECD PISA Global Competence Framework*. Available from https://www.oecd.org/pisa/aboutpisa/global-competency-for-an-inclusive-world.pdf.

Osler, Audrey, and Hugh Starkey. 2005. *Changing citizenship: Democracy and inclusion in education*. Berkshire: Open University Press.

Pashby, Karen, Louise Sund, and Sin Corcoran. 2019. *Teaching for sustainable development through ethical global issues pedagogy: Participatory research with teachers*. Manchester: Metropolitan University.

Pashby, Karen, Marta da Costa, Sharon Stein, and Vanessa Andreotti. 2020. A meta-review of typologies of global citizenship education. *Comparative Education* 56 (2): 144–64.

Reardon, Betty A. 1988. *Comprehensive peace education: Educating for global responsibility.* New York: Teachers College Press.

Seitz, Klaus. 2002. *Bildung in der Weltgesellschaft: Gesellschaftstheoretische Grundlagen globalen Lernens.* Frankfurt am Main: Brandes & Apsel.

Shultz, Lynette. 2007. Educating for global citizenship: Conflicting agendas and understandings. *Alberta Journal of Educational Research* 53 (3): 248–58.

Stein, Sharon. 2015. Mapping global citizenship. *Journal of College and Character* 16 (4): 242–52.

Szakács-Behling, Simona, Jennifer Riggan, and Bassel Akar. 2020. Introduction: Rethinking global citizenship education from the ground up: Intentions, power, and accidents. *Tertium Comparationis* 26 (2): 100–15.

Tarozzi, Massimiliano, and Carlos Alberto Torres. 2016. *Global citizenship education and the crises of multiculturalism: Comparative perspectives.* London: Bloomsbury Academic.

Torres, Carlos Alberto, and Emiliano Bosio. 2020. Global citizenship education at the crossroads: Globalization, global commons, common good, and critical consciousness. *Prospects* 48 (3): 99–113.

Turner, Bryan S. 2016. We are all denizens now: On the erosion of citizenship. *Citizenship Studies* 20 (6-7): 679–92.

UNESCO. n.d. *Recommendation concerning education for international understanding, co-operation and peace and education relating to human rights and fundamental freedoms.* Available from https://www.unesco.org/en/education/1974recommendation.

UNESCO. 2014. *Global citizenship education: Preparing learners for the challenges of the 21st century.* Available from http://unesdoc.unesco.org/images/0022/002277/227729e.pdf.

United Nations. 2015. *Transforming our world: The 2030 Agenda for sustainable Development.* Available from https://sdgs.un.org/2030agenda.

University of Alberta. 2023. *Directory.* Available from https://apps.ualberta.ca/directory/person/lshultz.

Wintersteiner, Werner, Heidi Grobbauer, Gertraud Diendorfer, and Susanne Reitmair-Juárez. 2015. *Global citizenship education: Citizenship education for globalizing societies.* Available from https://www.peace-ed-campaign.org/global-citizenship-education-citizenship-education-for-globalizing-societies.

Hackathon

Sonia Massari, Sara Roversi, Steven Finn, Chhavi Jatwani, Alessandro Fusco, Erika Solimeo, Alessio Cavicchi, and Matteo Vignoli

Definition

The word hackathon results from the combination of two words: the first one, *hack* or *hacking*, is a term that derives from the Old English verb *haccian* ("cut in pieces") and recently, especially in the technological age, is used in the (investigative) programming field to overcome obstacles to accessing technological systems and information. The word element *thon* is related to *marathon* and refers to the ancient Greek city *Μαραθών* (Marathon) from where, according to legend, the messenger Pheidippides departed to arrive at the Acropolis in Athens to announce the victory of the Greeks over the Persians in 490 BC. Today, the term refers to the long-distance running competition.

Being a relatively recent word, as it was first used in the late 1960s in programming activities in the United States (Yarmohammadian et al. 2021, 1), hackathon combines the creative process outside the established rules (hack) and recalls the endurance and high degree of effort that is at the core of marathons (Briscoe and Mulligan 2014, 2). The nature of a hackathon is that of a collective challenge, in which participants – divided into teams and within a narrow time frame – have to overcome obstacles to come up with something new through collaborative dynamics. In detail, hackathons start from the definition of the problem to be solved, pass through the information needed to tackle it (including searching for available resources and existing solutions), and apply learnings to solve it, which also include imagining new ways to activate them. This structure is due to specific phases that characterize the hackathon and ensure its success: collective elaboration of the idea (Damen et al. 2019, 4), the realization of the project (Prieto et al. 2019), creation of the prototype (Karlsen and Løvlie 2017), verification using a test of the prototype (Rey 2017), presentation of the final product to an audience of people interested in the theme and especially in the proposed solutions (Gama et al. 2018). These phases are similar to those of a design thinking innovation process (Liedtka 2014). Considering their systemic approach (Nechkoska et al. 2023, 311), hackathons are increasingly gaining traction at the global and local scale.

Historically, hackathons originated in a specific sector: in information technology, where programmers and designers would gather for hours at a time to co-create innovative new software. Since its initial use, the hackathon has been applied to a variety of fields as an innovative transdisciplinary method of engaging as many stakeholders as possible to develop collaborative projects that respond to shared problems. Hackathons thereby contribute to the plurality of knowledge resources and enlarge the circle of knowledge producers.

Current hackathons are defined as interactive and cross-discipline events lasting from four to 36 hours, in which participants come together, discuss, and inform themselves to respond to a problem that they perceive in their community (work, housing, political, etc.). Participants, guided by experts and in thematic working groups, leverage their experiences and skills by devising solutions. At the conclusion of the hackathon, these solutions are presented to a panel of experts and stakeholders who will judge them on their feasibility and then identify a winning team. Hackathons, actively used nowadays in the world of education and training at different levels from elementary to higher education, have become a space to express oneself, collaborate, and be an active part of changing mindset and way of teaching and learning in a transdisciplinary way. Hackathons stimulate the enthusiasm of the different participants, allowing them to intensify relationships, and to join an educational community characterized by a high diversity of societal roles and educational biographies.

Background

Hackathons are the result of the capitalist market and were initially used as tools to foster digital innovation when digital technologies began to penetrate Western society (Uffreduzzi 2017; Yarmohammadian et al. 2021). After the rise of technologization exposed society to new challenges, hackathons began to be applied also in the business world: their approach, grounded in collaborative patterns, fostered the development of hackathons in research projects, design, and new solutions in the field of technology and digital innovation in the 1990s (Page et al. 2016, 246). During that period, when marketing strategies were focused exclusively on maximizing the needs of companies, hackathons boomed as tools to support companies in selling their products (Briscoe and Mulligan 2014).

In the early 2000s, when the markets in Europe and North America started to include the needs of the consumers, citizens, children, and families at the center of strategies (Richterich 2019), the hackathon started to play a role not only as a problem-solving tool but also for finding solutions closer to people's needs (human-value centered) through design thinking methods (Bell 2008). Hackathons were used as a fast problem-solving technique, drawing the attention of a transdisciplinary group of experts or novices to specific problems. The joint sleepless effort of this group created viable solutions for the given problem that could be implemented after a process of refinement. Scholars acknowledge that the hackathon approach has recently been hybridized with human-centered design to overcome the problem of obtaining technologically impressive solutions that lack a deep understanding of the problems from the users' point of view, always with an accelerated pace of implementation (Mincolelli et al. 2020). In the 2010s, several cases confirm that hackathons were also developed in non-Western countries, such as Asia and Africa (Chen 2018; Ghouila et al. 2018) as a tool to support farmers, to connect and empower citizens, and to expand scientific discoveries. More recently, civil society and leaders started to consider that the needs of humans pose the risk of ignoring systemic and complex problems such as climate change, biodiversity loss, or deforestation (Lodato and DiSalvo 2015). These challenges that global society has been facing have changed the scope and breadth of the problems that need to be hacked (Flores et al. 2018; Nechkoska et. al. 2023; Vignoli et al. 2021). New design methodologies were increasingly applied in hackathons to support the shift from ego-systems (characterized by a pyramidal approach where humans are at the peak) to eco-systems (Scharmer 2013), and embrace a more inclusive perspective.

Figure 1. Graphical representations of the Ego-system and Eco-system models. Source: Scharmer 2013.

One example is *prosperity thinking*, a design innovation methodology first developed in 2019 (Vignoli et al. 2021, 1801). Its aim is to reconnect and simultaneously consider both human and planetary needs in creative sessions. Prosperity thinking differs considerably from the prosperity thinking mindset as it emerged in the late 1990s, which is grounded in the attitude towards "abundance, confidence, and gain" (Napolitano 1999, 3), and goes beyond the concept of prosperity thinking intended to achieve a prosperous life as proposed by Gallamore (Gallamore and Gallamore 2011, 78).

The Hackathon model and approach, when implemented with prosperity thinking design innovation methodology, is a rapid problem-solving tool that considers people's needs within the limits of the planet, therefore breaking down complex problems. The participants work in solution hunting, building artifacts to make ideas tangible (Dorst 2011), and in community co-design and testing. Most hackathons end up in project presentations that shed light on new ways of solving complex problems at local and more specific levels (macro, meso, micro) while generating multiple solution alternatives (Lewis et al. 2015; Soligno et al. 2015). For example, the hackathon virtually launched by AccelerateEstonia, Garage48, and Guaana (Global Hack) in 2020 with the goal of hacking health and wellness issues triggered by the global pandemic was a perfect example of how hackathons can become tools in support of complex and multidisciplinary issues (Flood 2020). This 48-hour online hackathon virtually gathered startups and tech teams, global leaders and industry experts, policymakers, media, and investors to turn ideas into rapidly implementable solutions on well-being, mental well-being, and relations, medicine, work and productivity, human resources, leadership, and environment (including climate actions, sustainability, and clean energy). These aspects are increasingly favoring the use of hackathons in schools and universities applied not to maximize their commercial application but to generate social and cultural impacts.

A hackathon is a transdisciplinary method of teaching and learning as it applies participatory and cooperative techniques (Hope et al. 2019). It proposes and invites rigorous, inclusive, and creative work and thereby takes from a high plurality of available knowledge resources. Recognizing the multiplicity and interconnectedness of the various dimensions of the real world, transdisciplinarity has the ambition to cross boundaries that separate paradigms, rigid institutional norms, and disciplinary labels.

Various diversities come together to face challenges during a hackathon: participants create groups of three to five people, and form personal relationships with each other due to the intensity of the project; it is an opportunity for enrichment, and participants are engaged by a facilitator and a team of tutors and judged by a panel of experts. Multidimensionality, inclusiveness, and creativity are the terms that denote the transdisciplinary approach of the hackathon. Today hackathons can involve increasingly large groups and, through the use of the internet, even those located in different geographic areas. In addition, there are several web platforms available for free, which can be used to share results and make communication during hackathons easy and effective.

This type of teaching methodology has a threefold goal: to strengthen the community involved, to attract and welcome diversity, and to provide an opportunity for all participants to learn. In the case of training, teachers are an integral part of this co-learning process. The transdisciplinary hackathon method is certainly not easy to pursue, but it is also the current bet for particular and broad areas of research (such as those on co-creation processes).

Debate and criticism

The addition of hackathons into the educational system is an added value, as hackathons are tools that emphasize teamwork and collaboration, ensure bi-directional exchanges, stimulate dialogue, questions, and critical thinking in students, and can support in-presence lessons to shape the professional figures of the future (Holley 2009; Huerta and Romaní 2022). In this sense, hackathons can guide students toward a holistic and ecosystemic overview of problems (and solutions), enhancing interconnection and transdisciplinarity (Massari 2021, 320). Though hackathons were not initially conceived as educational, didactic tools, increasing literature is proving their potential and increasing applicability as a teaching and learning tool in class (Wallwey et al. 2022).

Similarities and connections can be identified between the hackathon's three moments (involvement of people, search for solutions, and action for the realization of the solutions) and the process followed by teachers in designing lessons. Lessons, just like hackathons, are aimed at helping the students acquire knowledge. A hackathon can therefore be compared to a learning unit and indeed can be accurately defined as challenge-based learning (Leijon et al. 2021; Malmqvist et al. 2015; Nichols and Cator 2008). Significant differences between the challenge-based model and the traditional model subsist. In a traditional model, students expect first to receive the information they need to understand the learning content, then memorize that information, and finally apply it to a problem. In challenge-based learning, on the other hand, the problem to be solved is posed first, and only afterward are the students asked to figure out where to acquire the necessary information and how to apply it to solve the problem (Gallagher and Savage 2020; Leijon et al. 2021). Whereas in the first case, learning is based on solving the problem according to a functional approach, the second approach defines learning as challenge-based and follows a systemic perspective (Sternad 2015, 252).

A similar approach is project-based learning, a method of learning based on the realization of projects (Krajcik and Blumenfeld 2005). In this case, the differences between the two approaches are less pronounced but still significant. Project-based learning can be carried out individually or in groups, in which case it is the teacher who defines a problem and identifies steps to solve it. Challenge-based learning, on the other hand, involves only group work. It is the students themselves who define the problem, identify the steps, and create the solution. Nonetheless, in both cases, real-world problems are tackled, students are at the center of the activity, and the teacher takes the roles of a guide and a learner at the same time. The ideal *forma mentis* of a teacher approaching such methodologies is openness to innovation.

It is precisely in the common area between these two approaches that the main characteristics of the hackathon are identified. The hackathon is configured as a strategy for inclusive and collaborative teaching (Flus and Hurst 2021) that has many entry points (to the problem) and many solutions, allowing it to focus on global problems and develop local solutions (Decker et al. 2015, 4). In addition, it allows connection with multiple disciplines (Yarmohammadian et al. 2021) and represents an opportunity to develop competencies and useful skills (Cwickel and Simhi 2021), including soft skills (Decker et al. 2015, 3) as well as targeted use of the internet and digital tools for organization, collaboration, and sharing (Wallwey et al. 2022). Applied to education, hackathons could therefore represent a key tool to offer students the chance to foster active learning (learning by doing rather than learning by listening), become responsible for their own learning (a process that can trigger a sense of agency as responsible citizens), and learn to face collective and transdisciplinary challenges.

Although a hackathon as a methodology is usually associated with the challenge-based approach, typically applied in entrepreneurial environments (Gregg 2015), challenge-based learning can be used in class settings and has a high impact on the learning process, fosters collaborative creativity and provides benefits to the class (Hope et al. 2019, Lodato and DiSalvo 2015). In addition, challenge-based learning has added to new ways of teaching and experiential learning in class and has influenced (both tangible and intangible) teaching methods and the role of the teacher, which can now be considered not just as coordinators of activities but as real facilitators of interactions among students and between students and societal actors (mentors, stakeholders, and other collaborators).

Current forms of implementation in higher education

The aforementioned potential of hackathons applied in education has also been supported by concrete examples. Applied at the local and global levels, hackathons are already being used as examples to teach and learn in virtual, hybrid, and in-presence formats while being applied to many different critical topics.

The role of hackathons in supporting and widening civic engagement in schools is exemplified by the hackathon on civic education, which was co-organized by the Association Amore per il Sapere (Apis) and the Future Food Institute in April 2021. This virtual hackathon was able to connect students of different ages and teachers around Italy. The challenges that students needed to hack were culture, wood, education, water, social innovation, food waste, companies, and territory. For each challenge, students could rely on the support of teachers, who were properly trained in advance to introduce the hackathon as an active teaching methodology, and of professionals working on the topic's relation to the challenge, to make their solutions viable.

The hackathon on community engagement, organized by the Paideia Campus in Pollica, Italy, in 2022, represents an example of a hackathon used for connecting formal educational environments to local community engagement. The main purpose was to find innovative solutions in the service of the cultural, social, and economic regeneration of the local territory, by gathering people from different sectors, ages, and experiences. Different community-based hackathons actively involved local citizens (outdoor lovers, community engagers, educators, and children) in designing new solutions to local problems. Local elementary schools and teachers were involved in all the intergenerational and transdisciplinary challenge-based dynamics.

Another application of a hackathon in higher education and geographically dislocated in different areas was the hackathon (called "Foodathon") organized both physically and virtually by the University of Wageningen (Netherlands) in 2018 grounded on the challenge of food security. As part of a conference on food and global hunger, the Foodathon was a special competition organized to engage students in achieving SDG 2 (Zero Hunger) and creating local solutions to the challenges of food systems. Geographically and culturally diverse participants had a 36-hour time frame to create intercultural and multidisciplinary teams and hack problems such as policy coherence for food production, food security and genetic diversity, sustainable dietary patterns, rural–urban linkages for food security, financing mechanisms to achieve SDG2, and circular food systems.

There have also been cases where hackathons have been applied by one single university but involved different departments and disciplines. The hackathon organized by the University of Pisa (Italy) and the Pisa Contamination Lab (CLab Pisa) in October 2022 started from the challenge of mobility and digital transformation fields and was aimed at fostering digital solutions in the agrifood and urban mobility sector for the creation of scalable and sustainable business models in compliance with the SDGs. Students, researchers, Ph.D. candidates, and professors from different departments of Pisa were asked to find a common language and a way to co-create digital and collaborative solutions for transparency and efficiency in food supply chains; city–country logistics and food and beverage delivery; food waste reduction and food recovery; healthier and more sustainable food regimes; and conscious, responsible and more sustainable better tourism in rural areas.

These examples are not exhaustive but are indicative of hackathons as a process and methodology that can promote and spread a culture of innovation and transdisciplinarity in schools at all levels, while also enabling participants to acquire tools and skills in planning, organization, communication, and soft skills. The examples aim to maximize the potential of hackathons as teaching tools and as a powerful example of a training scenario (for both students and teachers) for informal learning.

Finally, the cases demonstrate a hackathon as a method and approach able to engage an intergenerational, multidisciplinary, and multi-geographic group. The outcomes and impacts of a hackathon are both tangible (new and innovative final products, services, and systems) and intangible (new processes of learning, discussion, cross-cultural contaminations, new mindsets). On the one hand, a hackathon has a long-term impact as it supports teamwork and the value of diversity. The hackathon in education highlights how it is possible to co-create new physical and *phygital* (both digital and physical) spaces to reinvent the interaction between actors, schools, and local communities through experiences. As the cases presented here demonstrate, the most significant outcomes are often intangible and long-term: participants are "forced" to think with an ecosystemic perspective and propose solutions related to the entire value chain (production, distribution, marketing, consumption, post-consumption, and circularity to be related), embracing multi-level interaction to improve the ecosystem as a whole. The hackathon touches on diverse themes and issues as the cases present them: health and wellness, mental well-being and relationships, medicine, labor and productivity, human resources, leadership, environment (including climate action, sustainability, and clean energy), economics, banking, education, poverty, arts, and governance. The tangible short-term results of the hackathon are that it enables prototyping, testing, evaluating failures, and creating solutions that can generate change.

The aforementioned cases demonstrate not only the positive potential of the hackathon, but also the major difficulties: acquiring the methodologies of open innovation, working in teams, and facing multicultural and intergenerational difficulties. Through the hackathon, one can experience agility, learning to focus on the real needs to be met or the problems to be solved. A hackathon is therefore a strenuous but highly engaging and, at times, fun experience for all participants. It is no coincidence that global leaders, politicians, media, and investors today rely on the application of hackathon sessions to transform ideas into rapidly implementable solutions for positive impact on the community and for prototyping and implementation of solutions.

References

Bell, Steven. 2008. Design thinking. *American Libraries* 39 (1/2): 44–49.

Briscoe, Gerard, and Catherine Mulligan. 2014. *Digital innovation: The hackathon phenomenon*. Available from https://qmro.qmul.ac.uk/xmlui/bitstream/handle/123456789/11418/Briscoe%20Digital%20Innovation:%20The%20Hackathon%20Phenomenon%202014%20Published.pdf?sequence=2.

Chen, Liang-Chih. 2018. Developing technologies or learning institutions? Exploring the role of hackathons for developing innovation capability in emerging economies: The case of Taiwan. *Asian Journal of Technology Innovation* 26 (2): 202–21.

Cwikel, Julie, and Meital Simhi. 2021. Using the hackathon model in social work education. *International Journal of Social Work Education* 41(8): 1563-76.

Damen, Ida, Rens Brankaert, Mengru Xue, Xiaoyue Chen, Anne Grave, and Steven Vos. 2019. *Root: A multi-disciplinary approach to urban health challenges with HCI*. Available from https://dl.acm.org/doi/10.1145/3290607.3299051.

Decker, Adrienne, Kurt Eiselt, and Kimberly Voll. 2015. Understanding and improving the culture of hackathons: Think global hack local. *IEEE Frontiers in Education Conference* (FIE). Available from https://ieeexplore.ieee.org/document/7344211.

Dorst, Kees. 2011. The core of "design thinking" and its application. *Design Studies* 32 (6): 521–32.

Flood, Gary. 2020. *World's largest ever hackathon starts today to develop COVID-19 solutions*. Available from https://www.thinkdigitalpartners.com/news/2020/04/09/worlds-largest-ever-hackathon-starts-today-to-develop-covid-19-solutions/.

Flores, Myrna, Matic Golob, Doroteja Maklin, Martin Herrera, Christopher Tucci, Ahmed Al-Ashaab, Leon Williams, Adriana Encinas, Veronica Martinez, Mohamed Zaki, Lourdes Sosa, and Karina Flores Pineda. 2018. *How can hackathons accelerate corporate innovation?* Available from https://link.springer.com/content/pdf/10.1007/978-3-319-99704-9_21.pdf.

Flus, Meagan, and Ada Hurst. 2021. Design at hackathons: New opportunities for design research. *Design Science* 7: E4.

Gallagher, S. Elena, and Timothy Savage. 2020. Challenge-based learning in higher education: an exploratory literature review. *Teaching in Higher Education* 1–23. Avail able from https://www.tandfonline.com/doi/abs/10.1080/13562517.2020.1863354?journalCode=cthe20.

Gallamore, Larry E., and Jan Burk Gallamore. 2011. *Prosperity thinking: Recession-proof thinking*. Bloomington, IN: Balboa Press.

Gama, Kiev, Breno Alencar, Filipe Calegario, and Andre Neves. 2018. A hackathon methodology for undergraduate course projects. *IEEE Frontiers in Education Conference* 1–9. 2018 IEEE Frontiers in Education Conference (FIE)Oct 2018, 1-9.

Ghouila, Amel, Geoffrey Henry Siwo, Jean-Baka Domelevo Entfellner, Sumir Panji, Katrina A. Button-Simons, Sage Zenon Davis, and Faisal M. Fadlelmola, The DREAM of Malaria Hackathon Participants, Michael T. Ferdig, and Nicola Mulder. 2018. Hackathons as a means of accelerating scientific discoveries and knowledge transfer. *Genome Research* 28: 759–65.

Gregg, Melissa. 2015. FCJ–186 Hack for good: Speculative labour, app development and the burden of austerity. *Fibreculture Journal* 25: 183–201.

Holley, Karry A. 2009. Understanding interdisciplinary challenges and opportunities in higher education. *ASHE Higher Education Report* 35, 1–131.

Hope, Alexis, Catherine D'Ignazio, Josephine Hoy, Rebecca Michelson, Jennifer Roberts, Kate Krontiris, and Ethan Zuckerman. 2019. *Hackathons as participatory design: Iterating feminist utopias*. Available from https://dl.acm.org/doi/pdf/10.1145/3290605.3300291.

Huerta, H. Gardó, and Riera J. Romaní. 2022. Marco de competencias para el análisis de hackatones cívicas de transformación educativa. *Edutec. Revista Electrónica de Tecnología Educativa* 82: 9–28.

Karlsen, Joakim, and Anders Sundnes Løvlie. 2017. "You can dance your prototype if you like": Independent filmmakers adapting the hackathon. *Digital Creativity* 28 (3): 224–39.

Krajcik, Joseph, S., and Phyllis C. Blumenfeld. 2005. Project-based learning. In *The Cambridge handbook of learning sciences*, ed. R. Keith Sawyer, 317–34. Cambridge, MA: Cambridge University Press.

Leijon, Marie, Petri Gudmundsson, Patricia Staaf, and Cecilia Christersson. 2021. Challenge based learning in higher education – A systematic literature review. *Innovations in Education and Teaching International* 59(5): 609–18.

Lewis, Bridget A., James Parker, Lara W. S. Cheng, and Marc Resnick. 2015. UX day design challenge: Hackathon to apply rapid design ideation to a practical user experience challenge. *Proceedings of the Human Factors and Ergonomics Society Annual Meeting* 59(1): 304–6.

Liedtka, Jeanne. 2014. Perspective: Linking design thinking with innovation outcomes through cognitive bias reduction. *Journal of Product Innovation Management* 32 (6): 925–38.

Lodato, Thomas J., and Carl DiSalvo. 2015. Issue-oriented hackathons as material participation. *New Media and Society* 18 (4), 539–57.

Malmqvist, Johan, Kamilla Kohn Rådberg, and Ulrika Lundqvist. 2015. *Comparative analysis of challenge-based learning experiences*. Available from https://publications.lib.chalmers.se/records/fulltext/218615/local_218615.pdf

Massari, Sonia. 2021. Transforming research and innovation for sustainability: *Transdisciplinary design for future pathways in agri-food sector*. In Transdisciplinary case studies on design for food and sustainability, ed. Sonia Massari, 315–26. Cambridge: Woodhead.

Mincolelli, Giuseppe, Nicolò Cocchi, Clio Dosi, and Matteo Vignoli. 2020. "OPER. TEN" transform emergency now! Facing Covid-19 with open innovation and human centered design. *Strategic Design Research Journal* 13(3): 658–68.

Napolitano, John P. 1999. American Institute of Certified Public Accountants *Planner: Ideas from leading experts in financial planning*. Available from https://egrove.olemiss.edu/cgi/viewcontent.cgi?article=3971&context=aicpa_news.

Nechkoska, Renata P., Antonia C. Gonzalez, Alberto Bertello, Simona Grande, Marc Schmüser, Nataliia Rzhevska, Yulia Matskevich, Milen Baltov, Urska Jez, Eugenio Clavijo, Ekaterina Tsaranok, Montserrat D. Marín, Raouf Hajji, Rui Couto, Karolina Bolesta, Sara Abou Ibrahim, Elena Poughia, Man Yang, Paola De Bernardi, Geert Poels, Gordon Müller-Seitz, and Marcel Bogers (2023). Multi-vortex tornado blueprint for disruptive global co-creation (inspired by EUvsVirus). In *Facilitation in complexity: Contributions to management science*, eds. Petrevska R. Nechkoska, Gjorgji Manceski, and Geert Poels, 307–62. Cham: Springer.

Nichols, Mark H., and Karen Cator. 2008. *Challenge based learning white paper*. Cupertino, CA: Apple.

Page, Finlay, Sylvester Sweeney, Fraser Bruce, and Seaton Baxter. 2016. *The use of the "hackathon" in design education: An opportunistic exploration*. Available from https://www.designsociety.org/publication/39074/.

Prieto, Marcello, Krishnan Unnikrishnan, Colin Keenan, Kaochoy Danny Saetern, and Wendy Wei. 2019. *Designing for collaborative play in new realities: A values-aligned approach*. Available from https://ieeexplore.ieee.org/document/8811545.

Rey, Stéphanie. 2017. Museomix: *Lessons learned from an open creative hackathon in museums*. Available from https://ceur-ws.org/Vol-1861/paper7.pdf.

Richterich, Annika. 2019. Hacking events: Project development practices and technology use at hackathons. *Convergence: The International Journal of Research into New MediaTechnologies* 25 (5–6): 1000–26.

Scharmer, Otto. 2013. *From ego-system to eco-system economies*. Open Democracy. Available from https://www.opendemocracy.net/en/transformation/from-ego-system-to-eco-system-economies.

Soligno, Roberta, Francesco Scorza, Federico Amato, Giuseppe Las Casas, and Beniamino Murgante. 2015. Citizens participation in improving rural communities quality of life. In *Computational science and its applications – ICCSA 2015* Vol 9156, eds. Osvaldo Gervasi, Beniamino Murgante, Sanjay Misra, Marina L. Gavrilova, Ana Maria Alves Coutinho Rocha, Carmelo Torre, David Taniar, Bernady O. Apduhan, 731–46. Cham: Springer.

Sternad, Dietmar. 2016. A challenge-feedback learning approach to teaching international business. *Journal of Teaching in International Business* 26(4): 241–57.

Uffreduzzi, Massimo. 2017. *Hackathon as emerging innovation practice: Exploring opportunities and challenges through 8 in-depth case studies.* Milano: Politecnico di Milano.

Vignoli, Matteo, Sara Roversi, Chhavi Jatwani, and Margherita Tiriduzzi. 2021. Human and planet centered approach: Prosperity thinking in action. *Proceedings of the Design Society* 1: 1797-1806

Wallwey, Cassie, Meris M. Longmeier, Donnelley Hayde, Julia Armstrong, Rachel Kajfez, and Renee Pelan. 2022. Consider "HACKS" when designing hackathon challenges: Hook, action, collaborative knowledge sharing. *Frontiers in Education*, 7 (95-104).

Yarmohammadian, Mohammad H., Monsef Sanaz, Shaghayegh Haghjooy Javanmard, Youseph Yazdi, and Mostafa Amini-Rarani. 2021. The role of hackathons in education: Can hackathon improve health and medical education? *Journal of Educational Health Promotion* 10 (1): 334–39.

Indigenous Knowledge

Bem Le Hunte, Tyson Yunkaporta, Jacqueline Melvold, Monique Potts, Katie Ross, and Lucy Allen

Definition

Indigenous knowledges are knowledges that are developed by groups of people through "centuries of unbroken residence" in a place (Dei et al. 2022, 104). Importantly, they privilege the knowledge of ancient cultures whose practices have survived millennia, despite colonization. They have transdisciplinary value in the contemporary university environment – an environment that privileges mostly Western ways of knowing and being – as these ancient knowledges, or what some describe as spiritual knowledges, tend to offer a more unified, connected, relational field of knowledge as a starting point, which at closer examination may provide transdisciplinary insights that other disciplinary-bound knowledges may obscure (Lange et al. 2021). By providing a more unified understanding of knowledge as a connected whole, they balance Western ways of compartmentalizing knowledge in university institutional structures that are set up to create silos and fiefdoms that serve to separate us (Ross and Mitchell 2018).

Indigenous knowledges provide a more inclusive contextual framework that incorporate not just contemporary thought but ancient, spiritual, relational, Indigenous ways of thinking, knowing, and being. Infusing these into a transdisciplinary curriculum can be seen as an act of resistance – a way to detox compartmentalized thinking, whilst also transforming and expanding student worldviews and learning (Harvey and Russell-Mundine 2019).

Background

In his *Transdisciplinary Manifesto*, Nicolescu writes: "No single culture is privileged over any other culture. The transdisciplinary approach is inherently transcultural" (Nicolescu 2002, 150). Indigenous knowledges have found their way into the transdisciplinary curriculum as a way to counteract the somewhat "shrunken" notion of knowledge as being predominantly Western, single-disciplinary, and linear.

Indigenous knowledges have been categorized into three intersecting forms: (1) *traditional knowledge*, generally passed down by elders; (2) *empirical knowledge* that is knowledge sourced about and from nature, culture and society; and finally, (3) *revealed knowledge*, which privileges knowledge revealed through visions, dreams, and intuition (Castellano 2000).

Along with these forms are a set of qualities that Indigenous knowledges possess. For example, they are often personalized – in the sense that knowledge is inherently tied to the person who holds it or speaks its truth, and these received knowledges come with responsibilities. They are also orally transmitted and experiential – in other words, they depend on subjective experience and the inner workings of the self to generate social interpretations, meanings, and explanations. They are also holistic and relational. "They connect economic, cultural, political, spiritual, ecological and material forces and conditions. Indigenous epistemologies are grounded in an awareness and deep appreciation of the cosmos and how the self/selves, spiritual, known and unknown worlds are interconnected" (Dei 2000, 115). This idea of connectedness of knowledge – "of people to rocks to heavenly bodies to animals – is not a concept that comes from the Judeo-Christian world, but it is present in other worldviews", which can be challenging for Western thinkers (Le Hunte 2022b, 224). For example, it is simply too hard to translate the complexity of the word "Dreaming" in the worldview of Indigenous Australians – it might translate as a "supra-rational interdimensional ontology endogenous to custodial ritual complexes". But saying this every five minutes is a challenge, "so Dreaming it is" (Yunkaporta 2019, 22).

Academics have debated the responsibility of Western institutions to decolonize their curricula, acknowledging that knowledge is not exclusive to the powerful. There is also a growing trend to support students to understand the legacy of a white, patriarchal, heteronormative intellectual tradition that remains rooted in colonial and Western-centric worldviews (Keele University 2018).

> Colonialism has from the very beginning been a contest over the mind and the intellect. What will count as knowledge? And who will count as expert or innovator? Such questions have been central to the project of colonizing diverse cultures and their knowledge systems. Indigenous knowledges have been systematically usurped and then destroyed in their own cultures by the colonizing West [which has only served to keep them isolated from the Western curriculum]. (Shiva 2000, vii)

Debate and criticism

One of the conflicts around Indigenous knowledge is that it generally takes a very different stance on what knowledge is and where it comes from. Added to this, colonial influences in Western education systems have meant that the "intellectual heritage of non-Western societies was devalued" and "knowledge plurality [has] mutated into knowledge hierarchy, the horizontal ordering of diverse but equally valid systems [has been] converted into a vertical ordering of unequal systems, and the epistemological foundations of Western knowledge were imposed on non-Western knowledge systems with the result that the latter were invalidated" (Shiva 2002, vii).

According to Dei (2000), educators should recognize the coexistence of knowledges and their complementary nature (as well as their sometimes contradictory nature – yet this contradiction can be addressed if educators do not conceive the past and present as separate – "frozen in time and space" (Dei 2000, 120). In Indigenous cultures with ontologies embracing relatedness with all of life, learning is a process of inquiry involving coming to know those relations between knowledges, which include the innovations and discoveries of contemporary sciences (Frazer and Yunkaporta 2021).

The value of Indigenous knowledges is also that they challenge the notion that knowledges are separate – and force people to examine their notion of "discipline". Indeed, "understanding how we are 'disciplined' may be something of a precursor to transdisciplinary work" (Le Hunte 2022a, 1670). If people can undo the limited and limiting understanding of knowledge and where it resides, they might then be able to unpack the power dynamic that informs so much of their learning in Western institutions.

Whilst university structures have been responsible for curating how people learn, students have been learning for millennia in a systemic, non-linear, networked fashion – in a way that is fully embedded in the materiality and philosophy of a face-to-face culture that learns and adapts whilst we are learning. In a bid to move beyond the limitations of dominant paradigms, systems thinkers are creating words to describe this type of learning – words like *symmathesy,* from the Greek *syn* or *sym* ("together") and *mathesi* ("to learn") for "learning together". Nora Bateson created this word to describe a process of contextual mutual learning in an interconnected way outside of institutions (Bateson 2015a). For example, where learning to play a violin, learning may take place in the muscles, in the teacher, in the music, in the emotions and memory, in the culture and history of the instrument (Bateson 2015b).

Ancient and Indigenous cultures have long had notions of the connectedness and relatedness of all things in their philosophy – ideas that identify knowledges as non-separate, before Western thinking sought to exert power through the disciplines to "tame the wild profusion of existing things" (Foucault 2005, 16) – before knowledge was categorized into fields and professions – into faculties and divisions. Indigenous knowledge is categorized according to the sovereign context of knowledge keepers rather than discrete topical abstractions (e.g. by age, status, location, totemic relation, clan, and gender). Knowledge is distributed throughout the society, with each member and family responsible for specific content and the aggregate wisdom of the group only being accessible for governance through rituals of collective intellectual representation (Yunkaporta 2022).

By contrast, the intellectual ghettos that have evolved in contemporary global knowledge production rarely invite confluence between disciplines, and this has potentially detrimental consequences. If universities began with seven disciplines in the 13th century, this proliferation has resulted in more than 8,000 in 2012, meaning that "a great expert in a given discipline is totally ignorant in more than 7,999 disciplines" (Nicolescu and Ertas 2008, 17).

Legitimizing Indigenous knowledge sets students up for success, as these knowledges act as boundary objects to promote understanding between cultures (Aikenhead 1996). It also promotes the incorporation of core Indigenous values (Cajete 1999), as well as supporting contextual place- and problem-based learning (Newberry and Trujillo 2018). Indigenous knowledges also foster a sense of connectedness and relationality between disciplines and humans (as well as the beyond-human world), which might be a useful starting point in helping to understand a more expansive version of complex systems and the role of learning – for example, in the eco-social crisis that we now face (Ross 2020).

Current forms of implementation in higher education

The transdisciplinary curriculum globally is attempting to open up the boundaries of learning to other ways of knowing and being well beyond the scientific paradigms privileged in the West, to live the ambition of transdisciplinarity and advocate for the openness, inclusivity, and wisdom required to do transdisciplinary work that truly broadens students' worldviews. The diversity of Indigenous knowledges means that their incorporation into curricula around the world is highly contextualized and differentiated as it is often co-designed with Indigenous people.

The first case study offered is a transdisciplinary program at the University of Technology Sydney in the Bachelor of Creative Intelligence and Innovation. This degree combines with 26 other disciplines from every faculty within the university and opens up a new world for learners where different knowledges form confluence from across all fields, industries, and disciplines. In such a context, it would be unwise to privilege only Western or scientific ways of knowing. One subject, New Knowledge Making Lab, has been designed as a site for engaging with non-Western as well as Western knowledge from around the world and it covers the work of Australian Indigenous scholars through to Vedic philosophers and Chinese ways of knowing. Through this work, the notion of decolonization naturally arises. As Australia is a migrant nation, living on Indigenous land that was never ceded, it is easy to understand how some academics might feel they have a particular responsibility to decolonize their curriculum, acknowledging that knowledge is not exclusive to the powerful (Gothe and De Santolo 2022). Also, with students in this degree course coming from 26 disciplines, it is not within the remit of educators to teach any of these fields in depth. Instead, they privilege the Vedic (ancient Indian) idea that "there are many knowledges but only one knower" (Le Hunte 2020, 30). The approach to research within this program also follows the principles of the transformative research movement (Fazey et al. 2018) and the growing understanding that transformation in our systems involves transformation of self, because we are not disconnected from the systems we are attempting to change – a very Indigenous notion.

A case study from the United States introduces transdisciplinary education as a model of decolonizing curricula in climate change education (Newberry and Trujillo 2018). In this context, academics created a model to enable students to learn science from multicultural perspectives; the program integrates knowledge and understandings from the social sciences, water policy, traditional ecological knowledge, and climate change science to examine strategies for including Indigenous knowledge and cultural traditions into water policy and environmental decision-making. Students were provided background on the Tohono O'odham cultural perspectives and traditional practices, alongside knowledge on current and predicted climate change, and required to include Indigenous viewpoints on water as well as Tohono O'odham cultural values as part of their final projects. Newberry and Trujillio conclude by remarking on the importance of the incorporation of Indigenous knowledge in the curriculum for: (1) approaches to climate education in promoting resilience within Indigenous communities; (2) for the student's ability to navigate between different knowledges and methodologies in order to produce new knowledge, and (3) having well-trained Indigenous ecologists who also have a strong grounding in their own cultural knowledge can provide the scientific community with unique multiple contexts.

Another case study from the University of Fort Hare in South Africa centers around the Life Knowledge Action and Grounding Program that embraces African heritage in Western epistemology (Mahlangu and Garutsa 2019). The learning community is divided into three different levels of community – an individual home (or *amakhaya* in isiXhosa or Zulu), a community of homes (or *imizi* in isiXhosa or Zulu) and entire villages, where all relationships between stakeholders (think learners and community members) are reciprocal and all stakeholders are a part of an interconnected whole. Learners undertake a problem-solving project within their own community, and the whole program takes place in a particular language, a dialogue which is supported by visiting lecturers (community members). The use of vernacular terminology is important for its humanizing function. "This program allowed for resonance between students' real-life experiences, their histories and the learning process" (Keet and Porteus 2010).

Given the colonial heritage of many of our societies, there is ongoing debate around designing learning for the inclusion of radically different worldviews in a transdisciplinary curriculum, but educators are also cautioned to do so in a sensitive fashion, given that engaging with Indigenous ways of knowing as a non-Indigenous person requires a very active form of allyship and explicit intention to remove the structural racism, colonialism, and inequalities experienced by Indigenous people today (Williams 2018). And, where possible, it is argued that educators should attempt to invite Indigenous knowledge holders into the educational space to ensure that knowledge is not appropriated or simplified, but rather amplified, to create the much-needed insight into principles of unity – that which connects humans, disciplines, and fields – rather than focusing on that which separates people.

References

Aikenhead, Glen S. 1996. Science education: Border crossing into the subculture of science. *Studies in Science Education* 27 (1): 1–52.

Bateson, Nora. 2015a. *Symmathesy: A word in progress.* Available from https://norabateson.wordpress.com/2015/11/03/symmathesy-a-word-in-progress/.

Bateson, Nora. 2015b. *Beginnings?* Available from https://norabateson.wordpress.com/2015/09/22/beginnings.

Cajete, Gregory A. 1999. The Native American learner and bicultural science education. In *Next steps: Research and practice to advance Indian education*, ed. Clarice Baker, 135–60. Charleston: ERIC.

Castellano, Marlene Brant. 2000. Updating Aboriginal traditions of knowledge. In *Indigenous knowledges in global contexts: Multiple readings of our world*, eds. George J. Sefa Dei, Budd Hall, and Dorothy Goldin Rosenberg, 21–36. Toronto: University of Toronto Press.

Dei, George J. Sefa. 2000. Rethinking the role of Indigenous knowledges in the academy. *International Journal of Inclusive Education* 4 (2): 111–32.

Dei, George J. Sefa, Wambui Karanja, and Grace Erger. 2022. *Elders' cultural knowledges and the question of Black/African indigeneity in education.* Cham: Springer.

Fazey, Ioan, Peter Moug, Simon Allen, Kate Beckmann, David Blackwood, Mike Bonaventura, Kathryn Burnett, Mike Danson, Ruth Falconer, and Alexandre S

Gagnon. 2018. Transformation in a changing climate: A research agenda. *Climate and Development* 10 (3): 197–217.

Foucault, Michel. 2005. *The order of things*. London: Routledge.

Frazer, Baressa, and Tyson Yunkaporta. 2021. Wik pedagogies: Adapting oral culture processes for print-based learning contexts. *Australian Journal of Indigenous Education* 50 (1): 88–94.

Gothe, Jacqueline, and Jason De Santolo. 2022. Decolonising design practices and research in unceded Australia: Reframing design-led research methods. *Architecture_MPS* 21 (1): 1–13.

Harvey, Arlene, and Gabrielle Russell-Mundine. 2019. Decolonising the curriculum: Using graduate qualities to embed Indigenous knowledges at the academic cultural interface. *Teaching in Higher Education* 24 (6): 789–808.

Keele University. 2018. Keele manifesto for decolonising the curriculum. *Journal of Global Faultlines* 5 (1–2): 97–99.

Keet, Andre, and Kimberley Porteus. 2010. Life, knowledge, action: The grounding programme at the University of Fort Hare. Report on the Pilot July–December 2009. Available from https://www.academia.edu/8923662/Report_Life_Knowledge_Action_Programme.

Lange, Elizabeth A., Joy Kcenia Polanco O'Neil, and Katie E. Ross. 2021. Educating during the great transformation: Relationality and transformative sustainability education. *Andragoška spoznanja* 27 (1): 23–46.

Le Hunte, Bem. 2020. *A curriculum for being: Creativity for a complex world*. Available from https://opus.lib.uts.edu.au/rest/bitstreams/445b7a36-7419-4c03-a16e-62aa232335cf/retrieve.

Le Hunte, Bem. 2022a. Transdisciplinarity. In *The Palgrave encyclopedia of the possible*, ed. Vlad P. Glăveanu, 1669–76. Cham: Palgrave Macmillan.

Le Hunte, Bem. 2022b. The Anableps guide to serendipity: Intentional serendipity as creative encounter – a decolonised, literary perspective. In *The art of serendipity*, eds. Wendy Ross and Samantha Copeland, 221–38. Cham: Palgrave Macmillan.

Mahlangu, Petrus M., and Tendayi C. Garutsa. 2019. A transdisciplinary approach and Indigenous knowledge as transformative tools in pedagogical design: The case of the Centre for Transdisciplinary Studies, University of Fort Hare. *Africa Education Review* 16 (5): 60–69.

Newberry, Teresa, and Octaviana Trujillo. 2018. Decolonizing education through transdisciplinary approaches to climate change education. In *Indigenous and decolonizing studies in education*, eds. Eve Tuck, Linda Tuhiwai Smith, K. Wayne Yang, Eve Tuck, and K. Wayne Yang, 204–14. London: Routledge.

Nicolescu, Basarab. 2002. *Manifesto of transdisciplinarity*. Albany: State University of New York Press.

Nicolescu, Basarab, and Atila Ertas. 2008. *Transdisciplinary theory and practice.* New York: Hampton Press.

Ross, Katie E. 2020. *Transforming the ways we create change: Experiencing and cultivating transformative sustainability learning.* Available from https://opus.lib.uts.edu.au/bitstream/10453/149105/2/02Whole.pdf.

Ross, Katie E., and Cynthia Mitchell. 2018. Transforming transdisciplinarity: An expansion of strong transdisciplinarity and its centrality in enabling effective collaboration. In *Transdisciplinary theory, practice and education,* eds. Dena Fam, Linda Neuhauser, and Paul Gibbs, 39–56. Cham: Springer.

Shiva, Vandana. 2000. *Indigenous knowledges in global contexts: Multiple readings of our world.* Toronto: University of Toronto Press.

Williams, Lewis. 2018. Transformative sustainability education and empowerment practice on Indigenous lands: Part one. *Journal of Transformative Education* 16 (4): 344–64.

Yunkaporta, Tyson. 2019. *Sand talk: How Indigenous thinking can save the world.* Melbourne: Text Publishing.

Yunkaporta, Tyson. 2022. *Blackpilled: Conspirituality, backlash, and Indigenous online radicalisation.* Available from https://www.abc.net.au/religion/tyson-yunkaporta-indigenous-online-radicalisation/13877124.

Interdisciplinarity

Julie Thompson Klein and Thorsten Philipp (authors contributed equally)

Definition

Interdisciplinarity is both a programmatic term in higher education policy and a catch-all phrase across disciplines and fields. Hence, it labels many types of beneficial cooperation between forms of expertise, including equality among participants in teamwork. Further read through an etymological lens, the term implies a process conceived as operating *between* (lat. *inter*) and across *disciplines* (lat. *disciplinae*), even in some instances independently of them. Whether the process results in methodological connections between disciplines or even new communities of practice depends on the complexity and purpose of a given activity (Apostel 1972; NASEM 2005). Interdisciplinarity has also been presupposed in the past semantically and conceptually. Disciplines are individual bodies of knowledge, defined within their boundaries. Increased cross-fertilizations, however, have fostered and facilitated greater boundary crossing, ranging from assimilating approaches borrowed from other disciplines to formation of interdisciplinary fields.

Beyond this broad definition, however, a linguistic question arises from inconsistent terminology across domains and a political question marked by diverging research and education policy interests. The range of intentions and outcomes varies by context. Interdisciplinarity can refer to an act of translation between representatives of individual branches of knowledge, a methodical way of acquiring and generating new knowledge, a normative organizational and top-down objective, the answer to a complex question or solution to a complex problem, dialogue about preconditions and possibilities, limits for collaboration between disciplines, implications for teaching and research, and a transitional phase in the emergence of new disciplines or new interdisciplinary strands. Thus, interdisciplinarity can begin with exchanging ideas about complex problems or questions, continue formally integrating methodologies and epistemologies, be applied in exchanges of data, and ultimately even restructure research and teaching.

From a historical standpoint, interdisciplinarity has been predominantly construed as an academic endeavor that combines openness and contextual awareness, while still recognizing disciplinary boundaries. The combination influences willingness and ability to cooperate in collaborative work (Briggs and Michaud 1972, 192).

Background

Regardless of approaches or contexts, the most common motivations for transdisciplinary work are criticism of narrow approaches in single disciplines, rigid and inappropriate institutional structures, and excessive specialization and isolation of individual disciplines. Furthermore, indicating a gap between current needs and traditional classifications (Barthes 1987, 15), interdisciplinarity – similarly to transdisciplinarity – emerges in response to problems and questions too complex to be assigned to any one discipline or to be solved by any single branch of knowledge. Thus, interdisciplinarity is also an outgrowth of realization that traditional disciplinary patterns of thought and practice are inadequate in pressing global issues such as climate change, disease, urbanization, migration, food insecurity, and digital transformation. At the same time, interrogating and challenging established organizational systems of academic research and education fuels demand for transcending disciplinary boundaries and bridging the divide between society and science.

Here too, a historical perspective is illuminating. The term interdisciplinarity is conventionally dated to the 1920s in the context of social-scientific research on problems of the day and in alternative forms of general education and core curricula. During the 1930s and 1940s, the new field of area studies also arose, as well as problem-focused research such as the Manhattan Project to create an atomic bomb. Discussions around it, though, are much older and have influenced development of modern disciplines since their beginnings (Klein 1993, 19). By the 1960s and early 1970s the word appeared more widely as a level for educational experimentation and new fields such as environmental, urban, and culture-based topics that arose from sociopolitical movements outside the academy. From the 1980s onward, the term became more prevalent in industrialized nations in science-based fields and, concomitantly, philosophy of science and science policy.

Up to this historical point, subject differentiation and organization according to disciplines had been regarded as indispensable, while failure to adhere to disciplinary boundaries was deemed pejoratively as incompetence, "outsiderism", and dilettantism (see Hentig 1971, 866). Like any system, disciplines are decision- and experience-based constructs whose influence, stability, and boundaries result from socialization and institutionalization: power and resources, monopolization of knowledge, path dependencies, and hegemonies shape parameters of research and education. As a result, early interdisciplinary initiatives were regarded as anomalous or marginal. It has become apparent, not just since Foucault's habilitation thesis was rejected at Uppsala University in 1958 because it was deemed irreconcilable with the self-image of history as a discipline (Edelberg 2017, 286; Eribon 1989, 106–9), that disciplines can be used effectively as instruments of power to deny recognition and exclude the participation of alternative practices. Popper's dictum – "We are not students of some subject matter, but students of problems" (1963, 88) – captures the criticism of failure to respond to societal challenges.

Moreover, when interdisciplinarity is associated with innovation, and heightened in funding policies and grant applications, thinking within disciplinary boundaries is rather associated with amateurism, and narrow and decontextualized objectives (see Davis 2007; Nissani 1997; Palang 2003, 56). Nonetheless, the connecting threads across motivations and contexts is the responsibility for questions and problems that rigid disciplinarity cannot address. Interdisciplinarity is thus also a result of continuous accountability renegotiation. From a philosophical standpoint, since the world in its complexity can be apprehended neither encyclopedically nor categorically, interdisciplinarity does not represent overcoming, let alone abolition of disciplines. Rather, it explores their non-linear rhizome-like connections (see Deleuze and Guattari 1976): Individual disciplines remain the dominant structure of organization and classification. Interdisciplinarity is thus not just a program to reform university structures and educational systems, but should also be understood as a learning mode to recognize and deal with justice conflicts, path dependencies, hierarchies, control regimes, and techniques of marginalization.

At the same time, the increased number and size of scientific and professional teams has also resulted in increased awareness of collaborative learning dynamics and research. Moreover, heightened attention is being paid today to the involvement of stakeholders in government, industry, and communities, including the co-production of knowledge in transdisciplinary problem-oriented research. In fact, inter- and transdisciplinarity originate from the same malaise and offer different, complementary though interconnected methods of dealing with it. While interdisciplinarity originally encompassed purely scientific exchange, transdisciplinarity aims at the collaboration of diverse knowledge producers at the interface of science and society, theory and practice. A semantic, conceptual, and historical discernment, though, remains difficult. In contrast to earlier emphasis on epistemology, problem-solving today looms larger in interdisciplinarity discourse. This development is apparent on a global scale.

In the past, literature has been dominated by accounts from Europe and North America and in the English language. However, as examples indicate, boundary-crossing discourses are expanding awareness of inter- and trans-disciplinarity in the Global South. A growing body of reports from science-policy bodies and educational commission documents increased calls for support in both research and education. Gleed and Marchant's (2016) interdisciplinarity survey report includes examples from the Americas, Europe, the Asia-Pacific, the Middle East, and North Africa.

Debate and criticism

It is no surprise that the postulate of interdisciplinarity has spawned a plethora of literature, but a systematic definition of terminology and practices remains elusive. Attempts at systematization are numerous, misunderstandings emerge, and some argue "Babylonian confusion", despite clear patterns of consensus (Klein 2017, 21). In addition, appropriate evaluation criteria need to be used in assessing funding, publications, research performance, and program review. Disappointments are hardly surprising in light of the high expectations associated with interdisciplinarity. They include communication issues, empty phrases, political declarations of intent that dominate "interdisciplinary hype" (Jacobs 2009), and alignment with innovation and commercialization. The overarching concept has been misinterpreted as a panacea (Segal 2009). Some claim that it has supplanted disciplinarity as the primary *raison d'être* of research and education. The paradox of interdisciplinarity, however, refutes the latter claim. Science-policy bodies and educational commissions are increasingly endorsing it. Obstacles persist on all levels.

Key continuing hindrances include jargon and translation problems, discipline-based publication criteria and rigid discipline-based worldviews, concern about lack or loss of hierarchical status, insufficient incentive structures, and inadequate compensation structures accompanied by the need for increased financial support. In addition, cooperation across disciplines requires time to build trust in teams and wiliness to compromise, as well as joint goal setting, power sharing, and equitable work distribution. Sufficient opportunities for profile and career promotion are also needed. Since interdisciplinarity is aligned increasingly with complex and often global problems, long-term change in institutional structures is essential, too (Abbott 2007, 134). Mindful of the many obstacles and disincentives, Gleed and Marchant (2016, 7ff.) call for a robust "architecture" of programs in all countries, facilitated by physical and social spaces including centers, networks, and graduate education and research training. On the other hand, scholars, practitioners, and educators disagree on when students should be exposed to interdisciplinarity. The traditional hierarchy of expertise prioritizes mastery of a specialized body of knowledge first. However, pertinent skills are called for across contexts.

To illustrate, the Education Reimagined's *Partnership for 21st Century Learning* cited four significant competencies that are also aligned with interdisciplinarity: communication, collaboration, critical thinking, and creativity. In comparison, the World Economic Forum's *Future of Jobs Report 2020* (Zahidi et al. 2020) listed critical thinking, creativity, and coordination in problem-solving as among the top-ten skills students need. Spelt et al.'s (2009) systematic review of relevant literature in higher education spans multiple contexts. Some are not exclusive to interdisciplinarity, including disciplinary knowledge, ability to communicate, and critical thinking. Recently, though, there has been increasing interest in understanding the nature of interdisciplinarity, integration, and collaboration. Other abilities deemed crucial for dealing with complex questions and problems include curiosity and respect for other disciplines, empathy and emotional intelligence, and ethical concerns. Spelt et al. add the pedagogical goal of fostering collaboration in curriculum development and teaching, whereas Borrego and Newswander (2010) include capacity for teamwork, along with grounding in disciplines, integration and broad perspective, interdisciplinary communication, and critical awareness.

Current forms of implementation in higher education

Given the plurality of approaches, forms of implementation differ as well: from student-initiated research and living labs to large-scale inter-institutional programs and projects. They also include new subdisciplines and disciplines as well as integrated fields. Interdisciplinarity in research was initially organized in graduate colleges and project-based clusters of limited duration, but it has become well-established in dedicated units such as Berlin's Einstein Center Digital Future, Stanford University's Bio-X institute, and the Global Institute of Sustainability and Innovation at Arizona State University. Furthermore, it increasingly gained a place in curricula of many universities. In Europe, more study programs with interdisciplinary aspirations have emerged since the introduction of bachelor's degrees. For example, Philosophy, Politics, and Economics, developed at Oxford in 1920, then subsequently adopted in one form or another by several other universities. Attempts to establish the Anglo-American liberal arts tradition in other countries (e.g. since 2012 at University College Freiburg) follow a similar trajectory. All of these examples, however, do not conclusively prove that interdisciplinary aims are always achieved. In many cases, the term is merely a catchphrase. Proactive attention is required.

Universities and colleges need discursive spaces where experiences are shared and exchanged across boundaries, and integration and collaboration are explicitly cultivated in educational and training programs. Otherwise, thought patterns formed by preparation in individual sciences are not overcome, or at most are relativized and reorganized only to a modest degree. As a result, the goal of integrating interdisciplinarity into students' personal and professional identities diminish. Interdisciplinarity is often viewed solely as a phenomenon of application, while theory-building and reflection on epistemological, didactic, and methodological dimensions is short-changed (Philipp 2021, 169). Interdisciplinary learning further requires expansion of university-based counseling and guidance services and exposure to a range of forms and methods (Briggs and Michaud 1972, 228–29).

Two accounts in the authoritative *Oxford Handbook of Interdisciplinarity* furnish deeper understanding of current forms and strategies of implementation. In reporting on administrative structures, Holley (2017) sketches a typology of interdisciplinary programs organized into categories of students, faculty, curriculum, funding, and institutional location. Curriculum spans institutional- and student-designed programs and activities, theme-based learning communities, capstone or culminating classes within disciplinary majors, topic-based multidisciplinary course sequences, and prescribed coursework in recognized interdisciplinary fields or on emergent topics or interests. Holley concludes that no single model exists: forms may be autonomous, freestanding units or located within an established college or university, including new and renovated buildings that are centers for theme-based research with some educational and training opportunities. Institution-wide prioritizing of interdisciplinarity is rare. Related pedagogies are also implementing interdisciplinary learning.

DeZure (2017) reported that interdisciplinary teaching and learning do not claim any unique set of pedagogies. Following suit, teachers employ an array of instructional methods to support integrative learning outcomes. Dubbing them "productive pedagogies", DeZure aligns them with a broad-based shift from mastery of content to competencies, and the elevated importance of integrative and interdisciplinary learning outcomes. DeZure also reports proliferation of interdisciplinary curricula and programs both in disciplinary departments and beyond them, as well as pedagogies that promote active and discovery-based learning. Illustrating the constructivist philosophy of learning, teachers also engage students in team- and problem-based learning. The more the pedagogy engages students in experiences based in the complexities of the real world, DeZure adds, the more interdisciplinary approaches to problem solving and authentic assessment are advanced. "Inclusive pedagogies" also recognize multiple perspectives, to ensure all voices are heard.

Finally, Vienni Baptista and Klein (2022) illustrate the expanding scope of examples in a wide range of countries, spanning Africa, Europe and the United Kingdom, Russia and the South Caucasus, Latin and North America, Australia, and Japan. The overarching commonality is the need to address complex societal problems, including the global pandemic, climate change, and sociopolitical inequities. However, contexts differ in individual chapters. Political history, for example, was a decisive factor in countries where universities reflect top-down, centralized, and hierarchical relationships from the Soviet system. Mokiy and Lukyanova (2022) report a continuing authoritarian leadership style in Russia, though autonomous nonprofit organizations are advancing the potential for problem-focused interactions with civil society – even though transdisciplinary participation with stakeholders is a new concept in Russia, Armenia, and Georgia. Further south, in Ghana, Akua-Sakyiwah (2022) situates reform efforts against the backdrop of dependence on colonial masters and development partners and the dominance of Western forms of knowledge. Yet material realities differ from the North, including irregular financial support from the government.

Brazil also illustrates the impact of political history and, today, international momentum for solving complex societal issues. Since the 1980s, after 21 years of military dictatorship, reform has occurred against the backdrop of redemocratization. Litre, Lindoso, and Burstyn (2022) characterize several Brazilian universities as avant-garde social spaces. Interdisciplinary initiatives have grown in graduate programs, but they are subject to centralized government regulation of education. Innovative programs are also judged by traditional metrics, regarded as incubators rather than mature initiatives, and stigmatized as too general, shallow, and unevenly institutionalized. The Center for Sustainable Development at the University of Brasilia, though, illustrates potential in a geographical area rich in ecological and social diversity, while located strategically in the country's capital.

Accounting for China, Pearce (2022) cited precedents for interdisciplinarity and transdisciplinarity in the conception of knowledge as an integrated corpus, the common good, and holistic knowledge and education for character development and ability. However, dominance of the Soviet model of higher education between 1949 and 1966 prioritized a socialist agenda for economic development. Between 1966 and 1976, the Cultural Revolution shut down higher education, except for military institutions. In addition, China's state-driven technocratic approach does not foster holistic consideration of complex societal and cultural factors. At the national level, interdisciplinarity is aligned more with solving problems than a general concept.

As these case studies indicate, it is critical to recognize similarities and differences when comparing lessons from different countries and regions. Interdisciplinarity will have to prove itself less as a method and more as a fundamental academic and everyday attitude of graduates, particularly in post-secondary education. It is not the abundance of areas of application, but the ability to reappraise scientific methodology and to provide reflection spaces for interdisciplinarity-induced learning experiences which will determine whether the university remains the most crucial pillar of disciplinary *and* interdisciplinary knowledge structures.

References

Abbott, Andrew Delano. 2007. *Chaos of disciplines*. Chicago: University of Chicago Press.

Apostel, Léo, ed. 1972. *Interdisciplinarity: Problems of teaching and research in universities*. Paris: OECD.

Akua-Sakyiwah, Beatrice. 2022. A contextual approach to institutional change: Transdisciplinarity in Ghanaian higher education. In *Institutionalizating interdisciplinarity and transdisciplinarity: Collaboration across cultures and communities*, eds. Bianca Vienni Baptista and Julie Thompson Klein, 107–23. London: Routledge.

Barthes, Roland. 1987. *Image, music, text*. London: Fontana.

Borrego, Maura, and Lynita K. Newswander. 2010. Definitions of interdisciplinary research: Toward graduate-level interdisciplinary learning outcomes. *Review of Higher Education* 34 (1): 61–84.

Briggs, Asa, and Guy Michaud. 1972. Problems and solutions. In *Interdisciplinarity: Problems of teaching and research in universities*, ed. Léo Apostel, 181–251. Paris: OECD.

Davis, Lennard J. 2007. A grand unified theory of interdisciplinarity. *Chronicle of Higher Education* 53 (40): B9.

Deleuze, Gilles, and Félix Guattari. 1976. *Rhizome*. Paris: Les Éditions de Minuit.

DeZure, Deborah. 2017. Interdisciplinary pedagogies in higher education. In *The Oxford handbook of interdisciplinarity*, eds. Robert Frodeman, Julie Thompson Klein, and Roberto Pacheco, 558–72.Oxford: Oxford University Press.

Edelberg, Peter. 2017. Trans-Nordic neo-empiricism in a European setting – or, why did Foucault leave Uppsala? In *Making Nordic historiography. Connections, tensions and methodology* 1850–1970, eds. Pertti Haapala, Marja Jalava, and Simon Larsson, 286–310. New York: Berghahn.

Eribon, Didier. 1989. *Michel Foucault: 1926–1984*. Paris: Flammarion.

Gleed, Alasdair, and David Marchant. 2016. *Interdisciplinarity: Survey report for the Global Research Council 2016*. Stockholm: DJS Research. Available from https://www.djsresearch.co.uk/Free/published/DJS_GRCreport.pdf.

Hentig, Hartmut von. 1971. Interdisziplinarität, Wissenschaftsdidaktik, Wissenschaftspropädeutik. *Merkur* 25: 855–71.

Holley, Karri A. 2017. Interdisciplinary curriculum and learning in higher education. In *Oxford research encyclopedia of education*, ed. Karri A. Holley. Oxford: Oxford University Press.

Jacobs, Jerry A. 2009. Interdisciplinary hype. *The chronicle of higher education*. Available from https://www.chronicle.com/article/interdisciplinary-hype.

Klein, Julie Thompson. 1993. *Interdisciplinarity: History, theory, and practice*. 3rd edition. Detroit: Wayne State University Press.

Klein, Julie Thompson. 2017. Typologies of Interdisciplinarity: The boundary work of definition. In *The Oxford handbook of interdisciplinarity*, eds. Robert Frodeman, Julie Thompson Klein, and Roberto Pacheco, 21–34. 2nd edition. Oxford: Oxford University Press.

Litre, Gabriela, Diego Pereira Lindoso, and Marcel Bursztyn. 2022. A long and winding road toward institutionalizing interdisciplinarity: Lessons from environmental and sustainability science programs in Brazil. In *Institutionalizing interdisciplinarity and transdisciplinarity: Collaboration across cultures and communities*, eds. Bianca Vienni Baptista and Julie Thompson Klein, 57–71. London: Routledge.

Mokiy, Vladimir, and Tatiana Lukyanova. 2022. The development of interdisciplinarity and transdisciplinarity in modern Russian science and higher education. In *Institutionalizing interdisciplinarity and transdisciplinarity: Collaboration across cultures and communities*, eds. Bianca Vienni Baptista and Julie Thompson Klein, 124–38. London: Routledge.

NASEM [National Academies of Sciences, Engineering and Medicine], ed. 2005. *Facilitating interdisciplinary research*. Washington: National Academies Press.

Nissani, Moti. 1997. Ten cheers for interdisciplinarity: The case for interdisciplinary knowledge and research. *Social Science Journal* 34 (2): 201–16.

Palang, Hannes. 2003. How does an elephant look like? Some experiences and some more fears about interdisciplinary landscape research. In *Interdisciplinary and transdisciplinary landscape studies: Potential and limitations*, eds. Bärbel Tress, Gunther Tress, Arnold van der Falk, and Gary Fry, 55–58. Wageningen: Alterra.

Pearce, BinBin J. 2022. "Leaping over" disciplines: Historical context and future potential for interdisciplinarity and transdisciplinarity in Chinese higher education. In *Institutionalizing interdisciplinarity and transdisciplinarity: Collaboration across cultures and communities*, eds. Bianca Vienni Baptista and Julie Thompson Klein, 139–51. London: Routledge.

Philipp, Thorsten. 2021. Interdisziplinarität. In *Handbuch Transdisziplinäre Didaktik*, eds. Tobias Schmohl and Thorsten Philipp, 163–73. Bielefeld: transcript.

Popper, Karl R. 1963. *Conjectures and refutations*. London: Routledge and Kegan Paul.

Segal, Robert A. 2009. Crossing borders can lead to gold – But so can digging deep. *Times Higher Education*. Available from https://www.timeshighereducation.com/407028.article.

Spelt, Elisabeth J., Harm J. Biemans, Hilde Tobi, Pieternel A. Luning, and Martin Mulder. 2009. Teaching and learning in interdisciplinary higher education: A systematic review. *Educational Psychology Review* 21 (4): 365–78.

Vienni Baptista, Bianca, and Julie Thompson Klein, eds. 2022. *Institutionalizing interdisciplinarity and transdisciplinarity: Collaboration across cultures and communities*. London: Routledge.

Zahidi, Saadia, Vesselina Ratcheva, Guillaume Hingel, and Sophie Brown, eds. 2020. *The future of jobs report 2020*. Geneva: World Economic Forum.

Internship

Ewald Terhart and Ulrike Weyland

Definition

Internships as part of higher education provide practical experience in the field of study and are considered an integral part of the curriculum. They are different from voluntary internships or compulsory practical training after graduation. For maximum benefit, internships should be linked to critical reflection and research-based learning, with a focus on transdisciplinarity and continuous learning. Internships can range from several weeks to several months and are considered "macro-forms of practical relevance" (Hedtke 2000, own translation) in comparison to shorter, situational experiences. Internships fulfill the demand for practical relevance in high-quality academic programs.

In addition to theoretical and knowledge-based studies (lectures, seminars, exercises), the vast majority of higher education programs include learning opportunities for practical experience that often, if not always, take place in the later professional field. In higher education, these learning opportunities are commonly called internships, "field-based experience", or "practicum" (Barron 2020; Kosinar et al. 2019; Ryan et al. 1996).

In its broadest sense, every time-limited change of location or social context can be considered an "internship". According to this broad understanding, "internships may be located anywhere in the world" (Jones 2006, 17). In a narrower sense, internships are elements of university courses and integral components of higher education. Voluntary internships undertaken with potential future employers after graduation or additional compulsory practical training required for the final licensing of state professions (doctors, lawyers, teachers, etc.) are not part of this narrower definition.

From an etymological point of view, the term *practicum* relates to practical activities, in the sense of doing or wanting to do something (from ancient Greek *πράσσειν, prássein* "to do, to accomplish"), while the term *internship* traces back to the Latin *internus* ("inward, internal"). In higher education, "practice" or "the practical" is associated with the idea of real experiences rather than imaginary or theoretical ones. Thus, a practicum or internship is often considered an additional or complementary element to theoretical studies. During the internship, students gain practical experience by testing things out and getting involved in the problems and realities of their practical field. Yet this simplified view reduces the epistemic and didactic value of internships to mere practical activity.

If, however, internships are regarded and realized as integral parts of the curriculum in higher education and if they purposefully develop the students' academic, research-based attitudes, the practical experience gained in internships is always linked to critical reflection and scientific, research-based learning – not just acquaintance with the realities of the field. Internships as complex, holistic learning opportunities implement the idea of transdisciplinarity as an acknowledgment of plural knowledge resources. Internships also include learning how to keep an analytical, critical distance out in the field and aim to lay the foundations for continuous learning throughout further phases of professional development.

Internships in academic study programs may run for several weeks or several months and can therefore also be characterized as "macro-forms of practical relevance" (Hedtke 2000, 5, own translation). Such forms must be distinguished from practice-related "micro-forms" (Hedtke 2000, 5), such as situationally and temporally limited practical experiences or practical elements within conventional study structures (e.g. creating an action plan, simulations of practice situations, project seminars with practice contacts, work in teaching-learning laboratories with practitioners, etc.). As macro-forms, internships fulfill the continuous and urgent demand for practical relevance and practical orientation as a feature of high-quality academic study programs.

Background

University study programs aim to provide students with academic disciplinary, interdisciplinary, or transdisciplinary knowledge and reflective skills that will enable them to work competently and responsibly while remaining open to new ideas based on new research and a (self-)reflective attitude throughout their professional life. For the few students and graduates who remain in the academic system (initially as junior researchers), their professional practice is research, it is theory. However, the vast majority of students go on to work in professional fields outside the university. Nevertheless, research is the basis of their professional work.

Accordingly, the question of the relationship between *theoretical* knowledge and forms of reflection acquired through study and the *practical* ability to act and make appropriate decisions in the professional setting is key: How does a course of study instill a science-based professionalism characterized by an understanding that this foundation is also a mandate for lifelong science-based and research-based learning in the profession? This question outlines a central theme in both *theoretically* oriented discourses of self-understanding on the relationship between (academic) disciplines and (practical) profession and – quite *practically* – in the context of the structure and specific curricular and didactic design of academic study programs (Hessler et al. 2013): How can practical learning experiences be meaningfully designed and integrated into academic studies? What kind of curricular formats lead to didactic and educational support for the initiation of transformative learning? Obviously, there are several possible answers to this question as the academic systems, traditions, cultures, and pathways from graduation to employment, etc. are different in all parts of the world.

Internships have long had a place in study programs but to varying degrees. In Europe, their systematic inclusion marks a movement in education and higher education policy associated with the Bologna process, i.e. the introduction of European higher education degrees that are comparable. With the introduction of the bachelor's degree, intended as the first qualification for a profession, the concepts of professional and practical relevance advanced to become guiding formulas for study program design (Schubarth et al. 2012). This not only brought the debate about internships and their significance into the spotlight in higher education policy, it also brought it to the attention of employers. Since then, study programs have included several internships, with various purposes, throughout the course of study. General, introductory internships occur early in the bachelor's degree, while more in-depth internships, possibly abroad (for example in foreign-language studies), occur as the bachelor's degree progresses, and, finally, research-oriented, specialized internships are undertaken during the master's degree.

Teacher education programs around the world typically include (several) internships and the "practicum" in teacher education is probably the most researched internship (Cohen et al. 2013; Degeling et al. 2019; Hodges and Baum 2019; Lawson et al. 2015). In the context of academic internships, teacher education plays a special role because it is not linked to one single academic subject; the study program for teachers is essentially multidisciplinary in terms of at least one subject in addition to educational studies.

Depending on the study program, internships are linked to sometimes overlapping, sometimes competing objectives. They test the stability of the participant's chosen field of study or profession, provide an opportunity to apply acquired theoretical knowledge or research methods "in the field", and see whether the professional field is a good fit for the participant. In some cases, internships also open doors for future employment. In the context of university studies, an internship can also prompt further study, career counseling, or – last but not least – inspire a topic for a final thesis.

However, internships are not simply meant to open up additional learning opportunities for students on a *personal* level. On a general, *societal* level they are regarded as an open space for exploration that connects the academic world of university study, the academic world of acquiring, producing, and reflecting new knowledge, with the broader contexts outside university: society and culture, the system of professions, the world of industry, commerce, and administration, etc. This contextualizes internships within discussions about the Third Mission of universities: the transfer of knowledge and expertise between academic and non-academic worlds and its transformative role in society.

The idea of *transference* can, in principle, be understood through different motives: The idea of "transfer" is linked with the outward-looking, socio-critical, and innovative tasks of universities (the Third Mission in addition to research and teaching). Innovative and *critical science* should also carry its results and methods out – not least through internships – into society, institutions, professions, etc., thus putting the idea of science into practice in social responsibility. This understanding guided the university reforms and student movements of the 1960s and 1970s and inspired the current concepts of "service learning" and "citizen science" (Angelique 2001; Böhm and Weissköppel 2022; Rieckmann 2015).

During Europe's Bologna Reform and the conversion of degree programs to the bachelor–master system, and in general in the course of a worldwide adaption of higher education to neoliberal ideas and practices, a different understanding of "transfer" dominated. This understanding focused on a stronger alignment between university and student qualification processes and the requirements of the professional world – workplaces of both industry and administration (employability). In this *functional* understanding, "transfer" means preparing graduates to meet the requirements of the target occupational field and rapidly changing labor market (Kapareliotis et al. 2019). This approach must be viewed critically, especially when students only need to meet the demands of their future employers. As long as internships are integrated into academic study, students must be able to study independently of the necessities and specific demands of later workplaces and employers.

However, the relationship between the critical and functional understanding of internships is characterized by a fundamental ambivalence (Weyland and Terhart 2021). When connecting with the two places of learning, namely the university and the internship, students are often confronted with conflicting demands and expectations from different actors. As a result, if clear and explicit goals are not established for the participants, this can lead to role conflicts and also to unintended and counterproductive learning effects for some students. In extreme cases, a fundamental problem can arise: Students, in their role as interns, may not be seen as learners but as a cheap labor force. Therefore, it is of utmost importance that university supervisors prepare their students, keep in contact, and support them by discussing and evaluating their students' experiences during and after the internship (Myers Kiser 2011; Schweizer and King 2018).

This raises the question of who is responsible for planning and organizing internships in the respective degree programs. If the degree programs are *mono-disciplinary*, then each discipline, each subject is responsible for their internships. However, the majority of study programs are not mono-disciplinary. Indeed, the number of multi-, inter- and transdisciplinary study programs is growing: Who organizes, supervises, and evaluates their internships? How can the organization of transdisciplinary internships be secured within the discipline-based institutional structure of universities and colleges? What inter- and transdisciplinary models exist for, for example, environmental science, health science, educational science, sustainability science, and gender studies degree programs?

And finally: How can students successfully combine, integrate, and transcend knowledge from different disciplines and create new transdisciplinary knowledge if the discipline representatives themselves have difficulty with or are skeptical about transdisciplinarity? (Bain et al. 2019; Gibbs 2015; Pohl et al. 2018; Rieckmann 2015; Yeoung 2015). Transdisciplinarity is challenging for academics and researchers, more so than for their students. Ultimately, universities must transform and find solutions for these theoretical and epistemic problems on organizational and curricular levels. Some universities establish new working units or institutions (e.g. interdisciplinary or transdisciplinary centers) to institutionalize interdisciplinarity and transdisciplinarity (for an overview about international developments see Vienni Baptista and Thompson Klein 2022). These new units are explicitly independent from the traditional disciplines and faculties, which is both an opportunity and a challenge. The centers have to convince the disciplines and groups involved (teaching personnel, mentors, administrators, the representatives of relevant fields of practice, etc.) about the advantages and specific academic and professional value of inter- or transdisciplinary internships to keep all actors "on board". They must also attend to the practical side of the preparation, monitoring, and evaluation of their internships to make sure that "everything works". It is obvious that the *rhetoric* of transdisciplinarity and transformative learning is flourishing in academic discourses about the future of higher education – but it is doubtful that the *reality* of the various internship programs in higher education is being developed in the same way. On this ultimately decisive level, there is still a lot of work to do.

Debate and criticism

There is a growing body of empirical research that accompanies the various theories, programs, and concepts on internships in higher education (Gibbs 2015; Kosinar et al. 2019; Merz et al. 2014; Ryan et al. 1996). This research focuses on various questions: What expectations do students and hosting institutions have of internships? To what extent do internships achieve their goals? How do different actors evaluate the success of internships? What central learning effects do internships have on further academic studies and later professional biographies? Some of the central research questions and findings related to internships are presented below.

Students generally have very high expectations of internships. "Going intern" embodies hopes for specific, practical experience – in contrast to what they gain from their theoretical courses – and students associate these new practical experiences with personal developmental goals.

Student satisfaction with internships is high overall. According to a survey from the mid-2000s, only 10–13 percent reported being "rather dissatisfied" or "not at all satisfied" (Krawietz et al. 2006) and recent studies have returned similar results. Medical students in particular rate the supervision of their internships very highly (BMBF 2012, survey period 2010; Piedmont and Robra 2015).

The extent to which internships help students transition from the academic to the professional world cannot be assessed reliably as many factors contribute to a successful career transition and not all graduates launch their careers straight after graduation. Nevertheless, research has demonstrated that internships undertaken while studying – especially longer internships in the final phases of the degree – combined with good final grades and timely completion do contribute to a quick and successful start to a career. Studies have also shown that some fundamental personality traits positively impact the transition from study to work (Sarcletti 2009; see also Silva et al. 2016) and that internships have a positive impact on salary (Margaryan et al. 2019).

If the curricular, organizational, and personnel conditions are ensured, internships offer productive learning opportunities to develop and reflect upon academic knowledge, while developing professional competence related to the perception, assessment, and practical management of professional situations.

Current forms of implementation in higher education

Changing knowledge and systems within the (academic) professions are reflected in the growing importance of new forms of internship and, more broadly, gaining practical and field experience during academic studies. Currently, there is a tendency towards immense heterogeneity among the developmental statuses and trajectories of the new kinds of internship. Three main factors have led to these new forms.

(1) *New order of disciplines*: In addition to continuing the more traditional mono-disciplinary or discipline-adding study programs and their internships, there are increasing numbers of interdisciplinary and transdisciplinary forms. These forms offer broader access to problem areas of society and the professional world that are not typically "sorted" into the classical disciplines (e.g. nutrition, environmental, cultural, sustainability, or midwifery sciences). In these contexts, intern-

ships have to be organized primarily as communities of practice, as locally integrated bottom-up forms of cooperation between academic and practical worlds (Suh and Jensen 2020).

(2) *New biographical pathways*: Traditional structures for academic, educational, and professional biographies (school or secondary school followed by vocational training, college, or university, graduation, career entry, followed by various phases of professional life, and finally retirement) still persist, primarily in the so-called state professions. However, with increasing modernization and globalization, more flexible biographical patterns and pathways are being sought and realized, alternating between phases of education, renewed education, professional work, study, family, further education, career change, and moving up and down the career ladder. Living, learning, and working are becoming indistinguishable. To meet these growing needs for continuing and further academic education and learning, more and more universities are offering "dual forms" of continuous learning that are placed in both the academic and vocational worlds (Duncan et al. 2017). As another element, universities encourage non-standard students with alternative qualifications to take part in continuing training and education.

(3) *International internships in the context of post-colonial movements*: The conventional practice of conducting international internships to promote the geographical and cultural mobility of students (see Di Pietro 2022) is challenged by debates about a new balance in the relationship between Global North and South. Opportunities to participate in international placements are unequally distributed worldwide. The support for students from the "Global South" needs to be expanded and must be organized in a non-patriarchal way. The development of (inter) cultural competences can only succeed if a participatory approach is pursued (Lambert Snodgrass et al. 2021; Fortune et al. 2019).

In light of these complex developments, the classic internship, a specific period in the course of studies that students experience as "something special", is being phased out. Meanwhile, constant changes between different forms of learning, places of learning, and learning rhythms are increasingly observable in processes of education and training. The modern media and knowledge professions are stimulating and accelerating the still traditionally determined academic and non-academic professional and educational biographies. This is likely to affect the didactic design of and access to internships further.

References

Angelique, Holly L. 2001. Linking the academy to the community through internships: A model of service learning, student empowerment, and transformative education. *Sociological Practice* 3 (1): 37–53.

Bain, Bernice, Keely Griffith, and Jennifer Varney. 2019. Transdisciplinarity practice in higher education. In *Handbook of research on transdisciplinarity knowledge generation*, ed. Victor X Wang, 115–31. Hershey: IGI Global.

Barrón, Nancy G. 2020. Internship models: Acknowledging social and academic expectations. *International Journal of Educational Management* 34 (6): 1049–61.

BMBF. 2012. Forschung und Praxis im Studium. Befunde aus Studierendensurvey und Studienqualitätsmonitor. https://kops.uni-konstanz.de/bitstream/handle/123456789/22246/Multrus_222461.pdf?isAllowed=y&sequence=2.

Böhm, Urte, and Angela Weissköppel. 2022. Hacking Hochschuldidaktik. Ein Plädoyer für transdisziplinäre, transformative und kritische Denkbewegungen. In *Hochschullehre als Gemeinschafsaufgabe. Akteur:innen und Fachkulturen in lernenden Organisationen*, eds. Nora Leeben, Katja Reinecke, and Ulrike Sonntag, 79–88. Bielefeld: wbv.

Cohen, Ester, Ron Hoz, and Haya Kaplan. 2013. The practicum in preservice teacher education: A review of empirical studies. *Teaching Education* 24 (4): 345–80.

Gibbs, Paul, ed. 2015. *Transdisciplinary professional learning and practice*. Heidelberg: Springer.

Degeling, Maria, Nadine Franken, Stefan Freund, Silvia Greiten, Daniela Neuhaus, and Judith Schellenbach-Zell. eds. 2019. *Herausforderung Kohärenz: Praxisphasen in der universitären Lehrerbildung*. Bad Heilbrunn: Klinkhardt.

Di Pietro, Giorgio. 2022. International internships and skill development: A systematic review. *Review of Education* 10: 1–25.

Duncan, Ken, Downing, Jillian, D'Oliveira Singo, Brigida, Papier, Joy, Hartmann, Martin D., Ogwo, Benjamin A., Ezekoye, Benadeth, and Gessler, Michael. 2017. Session 1 – Vocational education and training: Basics for teaching and research in vocational education and training at universities. In *Vocational education and training in sub-Saharan Africa: Current situation and development*, eds. Gesine Haseloff, Friedhelm Eicker, and Bernd Lennartz, 39–133. Bielefeld: Bertelsmann.

Fortune, Tracy, Shinead Borkovic, Anoo Bhopti, Renee Somoza, Ha Chan Nhan, and Shabnam Rangwala. 2019. Transformative learning through international project-based learning in the Global South: Applying a students-as-partners lens to a "high-impact" capstone. *Journal of Studies in International Education* 23 (1): 49–65.

Hedtke, Reinhold. 2000. Das unstillbare Verlangen nach Praxisbezug. Zum Theorie-Praxis-Problem der Lehrerbildung am Exempel Schulpraktischer Studien. *sowi online journal* 0: 1–17. Available from https://www.sowi-online.de/sites/default/files/hedtke.pdf.

Hessler, Gudrun, Mechthild Oechsle, and Ingrid Scharlau, eds. 2013. Studium und Beruf: *Studienstrategien – Praxiskonzepte – Professionsverständnis: Perspektiven von Studierenden und Lehrenden nach der Bologna-Reform*. Bielefeld: transcript.

Hodges, Thomas, and Angela Baum. eds. 2019. *Handbook of research on field-based teacher education*. Hershey: IGI Global.
Jones, Elka. 2006. Internships: Previewing a profession. *Occupational Outlook Quarterly* 50, Summer: 16–18.
Kosinar, Julia, Alexander Gröschner, and Ulrike Weyland, eds. 2019. *Langzeitpraktika als Lernräume. Historische Bezüge, Konzeptionen und Forschungsbefunde*. Münster: Waxmann.
Kapareliotis, Ilias, Katerina Voutsina, and Athanasios Patsiotis. 2019. Internship and employability prospects: Assessing student's work readiness. *Higher Education, Skills and Work-based Learning* 9 (4): 538–49.
Krawietz, Marian, Peter Müßig-Tapp, and Janka Willige. 2006. Praktika im Studium. Available from http://www.hisbus.de/results/pdf/2006_HIS_Praktika_im_Studium.pdf.
Lambert Snodgrass, Lisa, Margaret Hass, and Mehdi Ghahremani. 2021. Developing cultural intelligence: Experiential interactions in an international internship program. *Journal of Global Education and Research* 5 (2): 165–74.
Lawson, Tony, Melek Cakmak, Müge Gündüz, and Hugh Busher. 2015. Research on teaching practicum – a systematic review. *European Journal of Teacher Education* 38 (3): 392–407.
Margaryan, Shushanik, Nils Saniter, Mathias Schumann, and Thomas Siedler. 2019. *Do internships pay off? The effects of student interships on earnings. Discussion Paper Series*. Bonn: IZA Institute of Labor Economics.
Merz, Carl P., Philipp A. Stoeberl, and Jill Marks. 2014. Building successful internships: Lessons from the research for interns, schools, and employers. *Career Development International* 19 (1): 123–42.
Myers Kiser, Pamela. 2011. *The human services internship: Getting the most from your experience*. Boston, MA: Cengage.
Piedmont, Silke, and Bernt-Peter Robra. 2015. Praxis und Wissenschaft im Studium – Erwartungen und erlebte Kompetenzförderung von Studierenden der Humanmedizin im Vergleich mit Studierenden anderer Fächer *GMS Zeitschrift für Medizinische Ausbildung* 32: 16–30.
Pohl, Christian, Pius Krütli, and Michael Stauffacher. 2018. Teaching transdisciplinarity appropriately for students' education level. GAIA 27: 250–52.
Rieckmann, Marco. 2015. Transdisziplinäre Forschung und Lehre als Brücke zwischen Zivilgesellschaft und Hochschule. *Zeitschrift für internationale Bildungsforschung und Entwicklungspädagogik* 38: 4–10.
Ryan, Greg, Susan Toohey, and Chris Hughes. 1996. The purpose, the value and structure of the practicum in higher education: A literature review. *Higher Education* 31: 355–77.
Sarcletti, Andreas. 2009. *Die Bedeutung von Praktika und studentischer Erwerbstätigkeiten für den Berufseinstieg*. München: IHF. Available from https://www.

ihf.bayern.de/fileadmin/news_import/ihf_studien_hochschulforschung-77.pdf.

Schubarth, Wilfried, Karsten Speck, Andreas Seidel, Corinna Gottmann, Caroline Kamm, and Maud Kroh, eds. 2012. *Studium nach Bologna: Praxisbezüge stärken?! Praktika als Brücke zwischen Hochschule und Arbeitsmarkt*. Wiesbaden: Springer.

Schweizer, H. Frederik, and Mary M. King. 2018. *The successful internship: Personal, professional, and civic development in experiential learning*. 5th edition. Boston: Cengage.

Silva, Patricia, Betina Lopes, Marco Costa, Ana Melo, Goncalo P. Dias, Elisabeth Brito, and Dina Seabra. 2016. The million-dollar question: Can internships boost employment? *Studies in Higher Education* 72 (1): 2–21.

Suh, Emily, and Darin Jensen. 2020. Examining communities of practice: Transdisciplinarity, resilience, and professional identity. *Journal of Basic Writing* 39 (2): 33–59.

Vienni Baptista, Bianca, and Julie Thompson Klein, eds. 2022. *Institutionalizing interdisciplinarity and transdisciplinarity: Collaboration across cultures and communities*. New York: Routledge.

Weyland, Ulrike, and Ewald Terhart. 2021. Praktikum. In *Handbuch Transdisziplinäre Didaktik*, eds. Tobias Schmohl and Thorsten Philipp, 231–42. Bielefeld: transcript.

Yeung, Raymond. 2015. Transdisciplinary learning in professional practice. In *Transdisciplinary professional learning and practice*, ed. Paul Gibbs, 89–96. Heidelberg: Springer.

Knowledge Transfer

Alhassan Yakubu Alhassan and Alexander Ruser

Definition

Despite its ubiquity, the term *knowledge transfer* lacks a clear and agreed-upon definition. Broadly speaking, it refers to the diffusion of knowledge as a "process of communicating research, innovations and knowledge to individuals, groups or organizations" (Thompson et al. 2006, 91). The etymological roots of the term are both Greek and Latin. *Knowledge* refers to the ancient Greek γνῶσις (gnōsis), which meant knowing through observation or experience. Transfer consists of the two Latin words trans, which means "across" or "beyond", and the verb *ferre*, which stands for "to bear" or "to carry, to bring" (Lewis and Short 2020, 428 and 1097). The literal translation refers to an activity of carrying knowledge between and beyond the places where it is produced.

The term *knowledge transfer* evades a clear-cut definition as neither the boundaries that knowledge has to cross nor the pathways along which knowledge can be "moved" are clearly defined. Naïve depictions of knowledge portray it as a kind of weightless entity that – in contrast to goods – can transgress physical and ideological boundaries and thus improve people's lives. Such romantic conceptions fail to acknowledge power imbalances and hierarchies within science and between science and societies, which pose considerable obstacles to transferring knowledge (Ruser 2021a).

Knowledge transfer can occur within scientific communities as the exchange of research findings, methodological innovation and theoretical debate, or the transmission of knowledge between researchers and distinct political, social, or economic environments. In the first case, knowledge transfer is an essential aspect of inter- or transdisciplinary research. In the latter case, it is addressed as the cornerstone of the universities' *Third Mission*, which is the "generation, use, application and exploitation of knowledge and other university capabilities outside academic environments" (Molas-Gallart et al. 2002, iii).

Analytically, functional and normative approaches to knowledge transfer can be distinguished (Ruser 2021b). Functional approaches tend to emphasize the importance of rules for transmitting knowledge, the significance of curricula and the need to agree upon standards. Functional approaches regard the transfer of knowledge within and beyond academia as an indispensable prerequisite for maintaining, controlling, and advancing knowledge. Accordingly, the transfer of knowledge needs to be organized in ways that allow for the dissemination of knowledge between places and across generations to guarantee the best application of available knowledge and the further advancement of that knowledge.

Normative approaches emphasize the importance of sharing and advancing knowledge as a basic human need. While scholarly thought tended to root this human need in ideas of European enlightenment, frequently invoking Immanuel Kant's call to emerge from "self-imposed immaturity" (Kant 2009, 1), more recently, the acknowledgment of the significance of non-Western and indigenous knowledge has gained importance (Adeyeye 2019; Al-Roubaie 2010).

In modern "knowledge societies" (Stehr 1994), knowledge is increasingly depicted as essential for economic growth and social and political inclusion. Consequentially, training and research facilities such as schools and universities become crucial determinants for a society's ability to remain competitive and develop knowledge-based and knowledge-driven economies. Likewise, more recent research has identified knowledge as an essential driving force of development while pointing out that low-income countries in the Global South, particularly in sub-Saharan Africa, struggle to put their domestic universities in a position to contribute to national development (Kruss et al. 2012). This apparent failure of transferring knowledge from the laboratories, academic hallways, and lecture theatres to broader society cannot be reduced to a lack of resources but often indicates a more general mismatch between academia and society (Kruss et al. 2012, 523–24), a lack of interaction or understanding between scientific and non-scientific communities.

Background

How can and how should knowledge be transferred? These questions have been a primary concern for centuries and has led to the development of specialized institutions and distinct social rules for how and to whom knowledge should be transferred. Moreover, the transfer of knowledge poses technical challenges and is inherently political. Current debates about the role of scientific and technical knowledge in transforming economies and societies, for instance, are implicitly or explicitly rooted in a Western understanding of scientific research. Likewise, the *politics* of enabling knowledge transfer – for instance, the "modernization" of curricula, the strengthening of relations between academic research or teaching and the wider society, and the encouragement of private–public partnerships to foster the co-creation of knowledge and life-long learning – are more often than not modeled after European or North American practices. The dominance of hegemonic, Western "specific forms of knowledge and knowledge production" (Okolie 2003, 235) and their impact on higher education and science politics in developing countries have increasingly drawn criticism from scholars from the Global South. Moreover, knowledge transfer and the "scholars, intellectuals, experts, and researchers implicated in the universalization of the dominant Euro-American knowledge" (Okolie 2003, 236) have been identified as important sources for the perpetuation of global power imbalances (Connell et al. 2017; Noda 2020).

Debate and criticism

Current debates about and criticism of the dominant understanding of knowledge transfer begin to touch upon the inherent power imbalances and criticize the dominance of specific cultural understanding of both critical terms, *knowledge* and *transfer*.

It is fruitful to make an analytical distinction between the three critical dimensions around which the debates evolve in order to better understand what drives current debate and to disentangle political and structural aspects of it. Moreover, such a systematization allows differentiation between three distinct thresholds knowledge must cross to transfer. This theoretical distinction enables a differentiated debate about how technical and cultural factors relate and how they contribute to or hinder distinct types of knowledge transfer.

As displayed in Figure 1, the analytical model distinguishes between three different conceptualizations of knowledge transfers. First, it can be understood as an instrumental and technical problem. Second, it can pose an epistemological and empirical challenge or, thirdly, it can be approached as mainly a reflexive and hierarchical task.

Figure 1: Categories of knowledge transfer. Source: Ruser 2021b, 410.

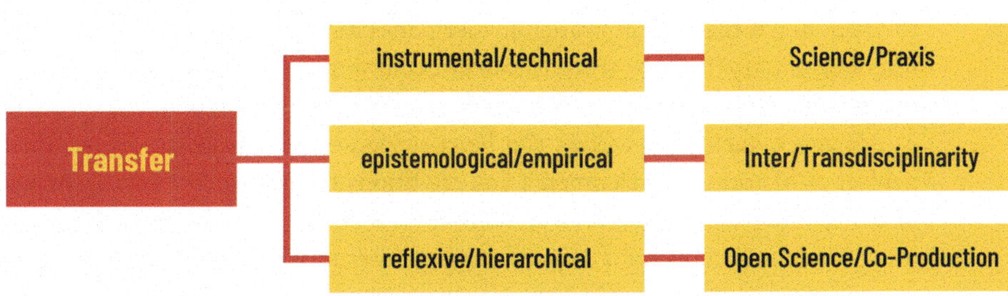

The instrumental and technical perspective focuses on transfer problems between "science" and "practice". It thus directly follows the demarcation of scientific knowledge production from other areas of society, such as commercial research and development activities or analyses conducted by governments or NGOs. Moreover, the underlying understanding of science implicitly reflects Robert Merton's understanding of a scientific ethos made up of four overarching norms: universalism, communality, disinterestedness, and organized skepticism (Merton 1973, 270). The normative structure of science, however, reflects the historical developments and value judgments that led to the emergence of Western science.

Consequentially, transfer issues have been located in and explained by discrepancies between these core values of the scientific ethos and normative concepts predominant in the surrounding societies. The organized skepticism of scientists, for instance, was (and is) at odds with religious convictions which are based on absolute, eternal truths. Accordingly, transferring scientific knowledge (for instance, about the origin and evolution of species) runs into obstacles when the receivers of the transfer hold contradictory beliefs (for instance, Christian ideas of God's creation).

Subsequently, the *impact* of transferred knowledge is conceptualized as a linear relation in which science transforms established practices. The in-principle compatibility between scientific insights (new knowledge) and social practices (sedimented knowledge) is assumed. Accordingly, prime examples of this perspective of knowledge transfer problems include questions of how to improve social organization, implement scientific findings into practice, to professionalize science communication further, or to establish executive education courses to shorten the time lag between "discovery" and "application".

The second epistemological and empirical perspective focuses on transfer problems *within* academia and scientific communities. The primary transfer arena is between scientific discipline and research fields, with "translation problems" posing the most important practical questions. In this interdisciplinary understanding, the main challenge is to guarantee or sustain the mutual ability of contributors from different disciplines and backgrounds to connect and collaborate. Solutions to these – often considerable – challenges are believed to depend on reforms in the academic context of knowledge production. Institutional reform to break down the silos of academic disciplines and the encouragement of new interdisciplinary thinking within academic communities are often seen as crucial components in overcoming translation problems (Townsend et al. 2015).

However, such perspectives can be criticized on two grounds – first, their neglect of historical developments and traditions in science. The establishment of disciplinary boundaries was an achievement in the development of modern universities and reflected a certain degree of definitional sovereignty of the respective field of research. Moreover, it also expresses the claim to autonomy concerning the transmission of knowledge. The canonization of classical theories, accepted methods, and central questions was key for the formation of disciplinary scientific communities since it allowed for establishing rules and agreed-upon regulations about what newcomers to the discipline had to learn, which knowledge they had to command, and what exams they had to pass in order to become an accepted member of the respective community (Stichweh 1992).

The second criticism focuses on the lack of awareness of the cultural distinctiveness of the premises of interdisciplinary knowledge transfer. The depiction of the emergence of academic disciplines as an indicator of increasingly differentiated scientific autonomy is plausible for the educational systems in Central European and North American countries. In many developing countries, educational systems were modeled after the examples of former colonial overlords, thus reflecting the academic understanding, rules, and agreed-upon regulations of others.

For universities in the Global South, current calls for more interdisciplinary research (Townsend et al. 2015) and learning as a form of more efficient knowledge sharing within academic contexts thus represents a double-edged sword. Universities in developing countries are still striving to establish binding norms of academic freedom and disciplinary autonomy that correspond with local rather than colonial ideals of organizing higher education and research. International calls for rearranging disciplinary context to improve knowledge transfer are thus not seen as "some neutral, apolitical technical" (Tabulawa 2017, 13) invitation to reform domestic higher education systems but as an expression of a more comprehensive neoliberal agenda to globalize a narrow, instrumental understanding of knowledge and its transfer (Tabulawa 2017).

From this perspective, knowledge transfer is rooted in the idea that disciplinary boundaries can be overcome, and new forms of scientific research and innovative new academic practices can be established (Mittelstraß 1987, 2005, 19). A crucial aspect of breaking down disciplinary barriers and establishing new practices is the emphasis on the *reflexive* character of knowledge transfer. Transdisciplinary thinking cannot be limited to scientific communities. As the boundaries between scientific disciplines fade, the distinction between academic and non-academic scientific knowledge production becomes more blurred, as does the differentiation between knowledge production and knowledge transfer itself.

This last depiction might be the most promising for catching the intricacies and particularities of knowledge transfer challenges in non-Western and postcolonial contexts: Drawing on debates about "post-normal science" (Ravetz 1999, 647), scientific knowledge is conceived as "deeply enmeshed" in social debates (Ravetz 1999, 647) and thus a reflection of distinct cultural, political, and historical circumstances. Accordingly, and in sharp contrast to the first reading of the transfer problem, transferring scientific knowledge cannot be compared to shipping a fixed and finished product of a research process. Knowledge transfer is instead depicted as a process of mutual, collaborative exchange, the fabrication of agreed-upon ways of conceptualizing problems and sense-making. Accordingly, the demarcation between scientific and non-scientific agents of knowledge production, producers and receivers of knowledge, and the distinction between scientific research and lived practices loses importance.

Current forms of implementation in higher education

Transferring knowledge takes place in specific contexts and, as outlined above, requires access to target groups and a way of breaking down knowledge to make it accessible. Respective projects in the Global South face peculiar, additional hurdles since educational systems often retain colonial heteronomy characteristics. Intended to serve the interests of colonial masters, education systems in large parts of Africa were designed to train office assistants who would help in the operations of the colonial administration. However, in a limited sense, knowledge transfer was at the center of such other-directed educational institutions, as educated locals were trained to act as translators between the administration and indigenous groups or to serve as clerks for bookkeeping and general upkeep of colonial offices.

These historical roots create severe problems for knowledge transfer today. First, the narrow orientation and lack of independent development of disciplinary self-images create a mismatch between curricula – which are still borrowed from former colonial powers and the current domestic economic and social needs. For example, in Ghana, despite a large number of university graduates, many companies recruit foreign labor to fill certain positions where there is a lack of local expertise. Likewise, the peculiar relationship between academic research, higher education, and praxis and the lack of independent, emancipatory disciplinary development (Nukunya 2003) continue to hamper knowledge transfer. To overcome these obstacles to transferring academic knowledge into practice, important research initiatives have started to explore the impact of social networks on knowledge transmission and how graduates find jobs and inject their knowledge into local communities (Dwumah et al. 2018). Moreover, Yang (2018) points out that higher education in East Asia was not only based on Western knowledge for one and a half centuries but is still analyzed and understood through the lens of Western concepts, thus creating a disconnect from local, traditional schools of thought, which in turn limits the exchange of knowledge and knowledge-based practices.

In addition, the instrumental understanding of academic knowledge as a critical tool to solve specific domestic problems increasingly shapes the expectations towards social science research. Thus, scholars orient their research agendas to address concrete political problems such as the "chieftaincy crisis in Northern Ghana" (Anamzoya and Tonah 2016, 255) and offer practical solutions. A similar bias for practical problem solving in shaping national research and educational priorities in the social sciences can be found in India, China, and Brazil (Gupta et al. 2009).

Despite many difficulties, initiatives to transfer knowledge from universities to the praxis, such as "Third Trimester Field Practice", established in 1993 at the University for Development Studies in Ghana (Abonyi 2016), facilitate interdisciplinary research on the transfer of academic knowledge into local communities. Sharing the same instrumental and technical understanding of knowledge transfer as Third Trimester Field Practice, universities in the Global South adopt "executive education" schemes. Modeled after American and European business schools, these programs aim to professionalize management and business leadership by providing uniform knowledge packages and standards developed in the Global North and distributing them to domestic contexts (Amdam 2019). Moreover, recent work points to the potential of rediscovering research and knowledge transfer traditions in Latin America which are based on the eradication of the distinction between researchers and researched and can thus widen our understanding of reflexive knowledge transfers (Lomeli et al. 2018).

Attempts to overcome epistemological and empirical divisions include the facilitation of inter- and transdisciplinary learning. Unlike many Western universities, some higher education institutions in the Global South, such as the Kwame Nkrumah University of Science and Technology, have adopted a system of interdepartmental collaboration (Simpson et al. 2008) in the form of so-called "service courses" where different departments introduce students to the basics of their respective fields. The aim is to bridge disciplinary boundaries and find ways to combine different subjects to make university education more relevant to the local context. However, because of the instrumental character of cross-departmental collaboration and since disciplines lack the opportunity to develop strong national identities and a domestic canon of social science knowledge, these attempts run the risk of falling short of achieving transdisciplinary knowledge production and transfer.

Examples of open science and co-production approaches to knowledge transfer in the global south include the implementation of scenario workshops to identify future challenges (Sagasti 2004) and the implementation of living labs, in South Africa (Coetzee et al. 2012) and Indonesia (Supangkat et al. 2020) for instance, to facilitate development and transition to smart cities. However, research on participatory models for knowledge transfer reveals that Western models cannot simply be transferred to different national or cultural contexts. Living labs, as concrete didactical methods for bringing people together, e.g. in South Africa not only have "characteristics unique to the context in which they operate" (Coetzee et al. 2012, 23) but at times violate cultural practices for selecting stakeholders, contradict established norms of knowledge sharing, and thus require a more thorough investigation of the very meaning of knowledge transfer or co-production in non-Western environments (Coetzee et al. 2012, 25).

The application and the spread of modes and models for transferring knowledge within disciplinary contexts and between science and society tell us very little about the actual content, contexts, and impact of the knowledge transfer. Therefore, an international perspective in knowledge transfer requires not considering contextual factors – such as economic and institutional limitations of educational organizations in the Global South – but taking different, diverging, and potentially conflicting cultural understandings of knowledge and transfer seriously.

References

Abonyi, Usman Kojo. 2016. Universities' role in regional development: A case study of University for Development Studies, Ghana. *Journal of Education and Practice* 7 (26): 11–20.

Adeyeye, Biliamin Adekunle. 2019. African indigenous knowledge and practices and the 2030 Sustainable Development Goals: Exploring its uniqueness for quality knowledge sharing. *International Journal of Humanities and Education Development* 1 (4): 147–52.

Al-Roubaie, Amer. 2010. Building indigenous knowledge capacity for development. *World Journal of Science, Technology and Sustainable Development* 7 (2): 113–29.

Amdam, Rolv Petter. 2019. The internationalization of executive education. In *The Routledge companion to the makers of global business*, eds. Teresa da Silva Lopes, Christina Lubinski, and Heidi J. S. Tworek, 125–37. London: Routledge.

Anamzoya, Alhassan Sulemana, and Steve Tonah. 2016. *Managing chieftaincy and ethnic conflicts in Ghana*. Accra: Woeli Publishing Services.

Coetzee, Hendri, Ina-Mari Du Toit, and Marlien Herselman. 2012. Living Labs in South Africa: An analysis based on five case studies. eJOV: *The Electronic Journal for Virtual Organizations and Networks* 14: 29.

Connell, Raewyn, Fran Collyer, João Maia, and Robert Morrell. 2017. Toward a global sociology of knowledge: Post-colonial realities and intellectual practices. *International Sociology* 32 (1): 21–37.

Dwumah, Peter, Kofi Osei Akuoko, and Eric Henry Yeboah. 2018. Family networks' support to employment paths of rural youth in a Ghanaian community. *International Journal of Social Science Studies* 6 (2): 32–46.

Gupta, Brij Mohan, Surinder Dhawan, and Ugrasen Sing. 2009. Social science research in India, China and Brazil – A comparative study. *DESIDOC Journal of Library & Information Technology* 29 (2): 15–23.

Kant, Immanuel. [1779] 2009. *Answer to the question: What is enlightenment?* London: Penguin.

Kruss, Glenda, John Adeoti, and Dani Nabudere. 2012. Universities and knowledge-based development in sub-Saharan Africa: Comparing university–firm interaction in Nigeria, Uganda and South Africa. *The Journal of Development Studies* 48 (4): 516-30.

Lewis, Charlton T., and Charles Short. 2020. *A Latin dictionary: Founded on Andrews' edition of Freund's Latin dictionary*. Chapel-en-le-Frith: Nigel Gourlay.

Lomeli, Robles, Jafte Dilean, and Joanne Rappaport. 2018. Imagining Latin American social science from the Global South: Orlando Fals Borda and participatory action research. *Latin American Research Review* 53 (3): 597–12.

Merton, Robert King. [1942] 1973. The normative structure of science. In *The sociology of science*, ed. Norman W. Storer, 267–78. Chicago: University of Chicago Press.

Mittelstraß, Jürgen. 1987. Die Stunde der Interdisziplinarität? Interdisziplinarität. In *Interdisziplinarität: Praxis – Herausforderung – Ideologie*, ed. Jürgen Kocka, 152–58. Frankfurt am Main: Suhrkamp.

Mittelstraß, Jürgen. 2005. Methodische Transdisziplinarität. *Technikfolgenabschätzung – Theorie und Praxis* 2 (14): 18–23.

Molas-Gallart, Jordi, Ammon Salter, Pari Patel, Alister Scott, and Xavier Duran. 2002. *Measuring third stream activities*. Available from https://www.researchgate.net/profile/Jordi-Molas/publication/246796517.

Noda, Orion. 2020. Epistemic hegemony: The Western straitjacket and post-colonial scares in academic publishing. *Revista Brasileira de Politica International* 63 (1): 1–23.

Nukunya, Gordwin Kwaku. 2003. *Tradition and change in Ghana: An introduction to sociology*. Accra: Ghana Universities Press.

Okolie, Andrew C. 2003. Producing knowledge for sustainable development in Africa: Implications for higher education. *Higher Education* 46 (2): 235–60.

Ravetz, Jerome R. 1999. What is a post-normal science. *Futures: The Journal of Forecasting Planning and Policy* 31 (7): 647–53.

Ruser, Alexander. 2021a. Knowledge as world capital: Global knowledge. In *Knowledge for the Anthropocene*, eds. Francisco J. Carillo and Günter Koch, 240–48. Cheltenham: Edward Elgar.

Ruser, Alexander. 2021b. Wissenstransfer. In *Handbuch Transdisziplinäre Didaktik*, eds. Tobias Schmohl and Thorsten Philipp, 407–15. Bielefeld: transcript.

Sagasti, Francisco. 2004. Thinking about the future: Trends and scenarios in Latin America. *Development* 47 (4): 15–25.

Simpson, Tim, Russel Barton, and Dave Celento. 2008. Interdisciplinary by design. *ASME. Mechanical Engineering* 130 (9): 30–33.

Stehr, Nico. 1994. *Knowledge Societies*. London: Sage.

stichweh, rudolf. 1992. the sociology of scientific disciplines: on the genesis and stability of the disciplinary structure of modern science. *Science in Context* 5 (1), 3–15.

Supangkat, Suhono Harso, Arry Akhmad Arman, Yuti Ariani Fatimah, Ryan Adhitya Nugraha, and Hendra Sandhi Firmansyah. 2020. The role of living labs in developing smart cities in Indonesia. In *Data-driven multivalence in the built environment*, ed. Nimish Biloria, 223–41. Cham: Springer.

Tabulawa, Richard. 2017. Interdisciplinarity, neoliberalism and academic Identities: Reflections on recent developments at the University of Botswana. *Journal of Education* 69: 11–42.

Thompson, Genevieve N., Carole A. Estabrooks, and Lesley F. Degner. 2006. Clarifying the concepts in knowledge transfer: A literature review. *Journal of Advanced Nursing* 53 (6): 691–701.

Townsend, Tony, John Pisapia, and Jamila Razzaq. 2015. Fostering interdisciplinary research in universities: A case study of leadership, alignment and support. *Studies in Higher Education* 40 (4): 658–75.

Yang, Rui. 2018. Foil to the West? Interrogating perspectives for observing East Asian higher education. In *Researching higher education in Asia: History, developments and future*, eds. Jisun Jung, Hugo Horta, and Akiyoshi Yonezawa, 37–50. Cham: Springer.

Learning in Transformation

Isa Jahnke and Johannes Wildt

Definition

The noun *transformation* was first mentioned as verb *to transform* in the mid-14th century and means "change the form of" (transitive), from Old French transformer (14th century), from Latin *transformare* "change in shape, metamorphose", from trans "across, beyond" and *formare* "to form" that relates to the meaning of "undergoing a change of form" (Harper 2023).

Learning in transformation (in short: *transformative learning*) appears as a novel form of teaching and learning in higher education that is transdisciplinary and provides conditions and opportunities for investigating and shaping learning through the participation of the learners in *transformation* processes in work and society. From this viewpoint, learning is embedded in processes of societal transformation that are also situations in which meaning-making (learning) occurs. As such, the theory of meaningful learning (Howland et al. 2012), including digital technologies, seems appropriate.

The meaning of the term transformative learning is derived from scientific linguistic usage. For example, in natural science or technology, the term *transformer* (English) is used to describe an apparatus which converts kinetic energy into electrical energy. Here, transformation means change as a transition into a new form (of energy). In medicine, transformation refers to the conversion of a healthy cell into a malignant cancer cell. Transformation processes in the behavioral or social sciences are characterized by transitions of individual or social actors in which the actors change themselves in the transformations. For example, the invention of the printing press and the introduction of money led to behavioral changes in society.

In higher education, transformative learning is a learning process in which students use scientific knowledge to solve problems outside of school while working with people outside the university. While doing so, they use the scientific knowledge they learned in the context of higher education, then decontextualize that knowledge (from the setting in which they learned it) and recontextualize it (to apply and develop in the setting in which they are now working). This learning experience is a process of transformation. In addition, transformative learning is not just a learning design approach. Instead, it becomes part of a transformation process itself; thus, learning itself becomes the object of transformation processes.

A difference between what is presented in this chapter and Mezirow's (1991) original understanding of transformative learning is that while Mezirow looks at the individual process of learning, *learning in transformation* is based, theoretically, on social learning. This approach views all forms of learning as social; learning only occurs if it is linked to a social dimension, such as a community or society in which learners experience themselves as participants, either physically or online. The synthesis of psychological and societal aspects makes *learning in transformation* an interdisciplinary task. As this task is oriented towards practical problem solving and, therefore, receives its scientific structure, it detaches itself from the disciplinary order and becomes the concept of *transdisciplinary learning*.

Background

The concept of *transformative learning* cannot be understood without understanding the historical roots of both (a) transformative learning by Mezirow (1991) and (b) transformation processes in societies. Both concepts were developed independently of one another. From a behavioral or social science perspective, the path to understand transformative learning leads back to Polanyi (1944). For Polanyi, the collapse of the global economic and social system at the end of World War II was followed by the *Great Transformation*, a complete reorganization of the global economic and social system.

After the collapse of the Soviet Union (and in later comparable upheavals in Latin America, Asia, and Africa), political and social scientists described the transformation of post-Soviet states into democratic societies with capitalist economic systems as societies in transition (Merritt 1980). To capture the specifics of this systemic change, Kollmorgen et al. (2015) distinguish transformation from other forms of social change. They recognize evolutionary adaptations of social systems and changes in their environment as forms of revolutions (which are usually disruptive but not necessarily violent), which refer to planned and controlled changes within social systems that serve the system's functionality. Transformations, however, involve fundamental changes in the systems themselves.

Currently, the understanding of the term *transformation* includes the view of socio-ecological transformation, such as the worldwide movement for sustainable development (UNESCO 2017). A widespread agreement is that such ambitious goals cannot be achieved without the participation of the actors involved or affected. Interaction of groups in socio-ecological transformation projects are needed (in education, politics, business, churches, civil society, and science). Higher education is one relevant actor in this process of transformation, where teaching and learning must leave the wall of the ivory tower and enter professional and social practices ("transformative turn", Wildt 2022, 201–2). While science (and in this context research-based learning) is still at the center of all teaching and learning activities driving transformative learning, working on real-world problems creates opportunities that go beyond the possibilities of traditional learning (e.g. Rein and Wildt 2022). These opportunities lie in students learning through participation in planned collaboration with involved or affected groups of diverse actors in professional or societal practice.

Hence, transformative learning involves a shift in learning that goes beyond the traditional classroom setting, and involves engaging with real-world problems, perspectives, actions, and interests outside of the university. This type of change is facilitated by actors beyond the university who challenge students to move away from passive learning and instead immerse themselves in authentic experiences that foster deeper understanding and growth. In other words, transformative learning is an approach that takes students from dryland swimming and immerses them in the dynamic, multifaceted world of applied knowledge. This type of learning will encourage students to learn to decontextualize scientific knowledge acquired at academic institutions and then recontextualize it in outside fields.

Furthermore, transformative learning started in the field of adult education (Mezirow 1991). Until then, education had been predominantly directed toward the continuation of determined learning paths based on earlier phases of life. Instead, Mezirow paved the way toward lifelong learning that made room for deep shifts in career biographies and lifestyles. For higher education, this understanding of transformative learning is particularly relevant because it creates possibilities and opportunities for connecting conventional concepts of learning with newer concepts such as education through science. In an international comparison of 100 member universities of the European University Association, study programs were compared to show how the focus on transformative learning can be integrated into the development of study programs as a learning goal related to employability (professional relevance) and citizenship (social relevance). Recent reports (Jankowki 2022; Wagenaar 2022) show that these development tasks are also on the agenda today. However, this requires teaching and learning concepts that support a kind of student autonomy in the sense of a "shift from teaching to learning" (Barr and Tagg 1995). Meanwhile, in the United States, universities have agreed on standards for recognizing relevant achievements such as service learning also known as community-related learning (Jankowski 2022).

Finally, transformative learning has its roots in active, meaningful, and situated learning (Fry et al. 2003), which has been further developed by Howland et al. (2012). Active, meaningful learning is an umbrella term referring to a group of pedagogical strategies that the instructor applies to help students engage and learn. Its premise is that learners do not learn because the instructor performs an activity, but learners learn through their own activity and reflections through facilitating the learner's interaction with the course material and with peers. Active learning contrasts traditional methods where students are rather passive, tending to listen, read, or watch something (e.g. lectures). Research has shown that active learning increases learning outcomes and improves learning performance, grades, and higher order competencies (Deslauriers et al. 2019; Freeman et al. 2014). Transformative learning is a type of active learning that takes places in a collaborative learning environment where practical contexts (situated learning) facilitate a community of practice (Lave and Wenger 1991).

Debate and criticism

The development of a transformative learning approach is still in its infancy. However, it can be argued that it is becoming apparent that the approach faces issues.

Epistemological issue: With the crossing of borders from university into practice, students run the risk of getting into normative conflicts between options for action and interests in truth, and disregarding the different logics between scientific and practical contexts. However, transformative learning has the potential to help people develop better solutions to problems, and thus keep people from holding onto inadequate solutions. In the long run, no one is served by (false) legitimation and, for students, the (possible) experience of a conflict between scientific truth and practical interests is a unique chance to reflect on epistemological differences between scientific theories and methods, or tensions between fundamental and applied research.

Self-regulation issue: Transformative learning raises questions about the extent of self-regulation of student groups. It does not reduce the responsibility of teaching or facilitating learning but requires new skills and competencies of teachers (e.g. coaching, supporting structures, providing feedback). Furthermore, transformative learning also reveals the *education paradox*: The learner needs autonomy for doing such projects, but regular courses with clear instructions often do not offer any student autonomy. The paradox is that in the learning process a space for learner autonomy must be created, which is not yet present, but which can only be achieved through the learning process.

VUCA (volatility, uncertainty, complexity, and ambiguity) conditions: One advantage of higher education institutions is the creation of stable learning conditions (e.g. rooms, resources, teaching staff, curricula, examination requirements). All this can be difficult to achieve in the field outside the university. These uncertain conditions create the need for improvisation. For instance, the university should provide support by assuming the costs of teaching materials, room rentals, travel expenses, teaching staff, and appropriate exam formats.

Communication and interaction issues: Experiences and practices of knowledge sharing are needed in transformative learning. However, students, who go into the field and encounter non-scientific practitioners, are typically not used to communicating their scientific knowledge. Therefore, they must be provided with opportunities for reflection in scientific writing and oral communication.

Digitalization issues: Digital or technology-enhanced environments in educational institutions provide various methods of digital communication which may not be fit for transformative learning. For example, students are challenged to engage in a variety of digitally enhanced cultural practices and thus have to learn to communicate in these spaces. Usually, the university provides the framework, but outside the university, different stakeholders have different tools or practices. Thus, the challenge for students is to create a communicative environment for all participants.

Current forms of implementation in higher education

In transformative learning, concepts of traditional teaching and learning practices grow together in a new design under continuously changing contexts. In *(1) project-based learning,* students conduct research-based projects. Project-based learning can be seen as a direct precursor to transformative learning concepts due to the integration of practical and research-based learning. Project-based learning can become transformative learning in higher education only when the projects are based on scientific knowledge and are being conducted together with stakeholders outside the university to solve a real-world issue (Wildt 2021).

Close to the concept of project-based learning is *(2) problem-based learning*. In this type of learning, students solve predefined problems set by the instructor. Like project-based learning, it has developed in practical contexts, solving problems that arise there (e.g. medicine, where the ability to solve problems is expected). It differs from project-based learning in the narrower definition of the tasks, distinctive structuring of learning, and the standardization of problem-solving procedures (Gibson 2005). However, the choice toward one or the other depends on the intended learning outcomes. Problem-based learning becomes transformative learning in higher education only when the problems are open-ended and the answers are not known, when it is related to real-world issues, and when learners are solving the problems along with stakeholders outside the university.

(3) Research-based or inquiry-based learning integrates learning activities into a research process that ideally ranges from the development of the research question to the theoretical and methodological elaboration of the research process

to the presentation and communication of the results (Jenkins and Healey 2011). Its potential toward transformative learning only unfolds when educational settings and students are connected with actors outside the university (Schneider and Wildt 2002). Research-based learning only becomes transformative learning when the research projects being conducted are linked with real-world problems and students work together with stakeholders outside the university.

(4) *Service learning* includes sociocultural engagement of students, especially in regional or communal initiatives, and becomes transformative learning in higher education only in connection with research and science.

(5) *Living labs* are open or public spaces designated to test innovative practice. They can also function as learning spaces suitable for transformative learning.

(6) *Simulation*: Transformation-relevant topics can also be offered in lectures, seminars, and tutorials, and thus can prepare students for non-academic contexts. One option is simulation, which allows students to learn in a protected space without any real-life consequences. In medicine, for example, simulations are applied in the use of actors as patients, virtual doctor's offices, or hospital wards in the medical clinic. Other examples include case studies at the Harvard Business School, business games in economics, and simulated court cases. Furthermore, virtual spaces open unexploited possibilities for a transformative learning design (e.g. online or remote labs in engineering education to teach sustainability in product design, augmented or virtual reality simulation with immersive learning experiences).

In summary, such learning formats can support transformative learning by slightly revising those learning designs, adding real-world problems for which the answers are not known, and providing opportunities for collaboration with stakeholders outside the university. Transformative learning in higher education has expanded considerably through digital options, while teaching and learning in person limits the potential of transformative learning. For example, the design of transformative learning experience with digital technologies, with consideration of user experience research, opens wide-ranging possibilities for communication and interaction in transformative learning projects.

Learning experience design is built on the concept of active learning through technological support. It is one possible way to design for transformative learning experiences in digital environments and evaluate the quality of such designs (Schmidt et al. 2020). Learning experience design emerged from the field of instructional design and educational technologies. It focuses on the idea that traditional instructional design or learning design lacks the design for enjoyable or memorable experiences (Jahnke 2023). When adding the viewpoint of *experiences*, a design also ensures students experience something special, something they will not forget, or something that leaves an impression on them. This is *memorable experience*, which is typically connected with a positive emotion (Pekrun 2014). To enable such positive learning experiences, methods from user experience research can be applied to digital transformative learning. However, contemporary user experience methods do not sufficiently tackle the issues of the learner's interaction with the pedagogical design, the sociocultural dimension, and the diversity of learners (Jahnke 2023). Therefore, the design and development of digital transformative learning focuses on three dimensions of experiences: technological, pedagogical, and social (Jahnke et al. 2020). These three dimensions combine in what is called sociotechnical-pedagogical experience, which embraces social presence, shaping roles and interactions in social relationships so that learners recognize themselves as a community of learners in a digital context. The element of roles alerts the learning designer that teachers should actively design a role shift away from the I-present-myself-and-my-knowledge role to the what-can-I-do-for-you role, so that learning is experienced positively and individual learners are supported individually. The challenge of transformative learning is to include such role structures.

In conclusion, transformative learning, defined as learning in contexts of collaboration with professional and societal actors and supported via digital environments, needs specific conditions to be successful. It offers students an opportunity to solve open-ended, real-world problems that are complex and for which the answer is not yet known (Fischer et al. 2022). That includes considering how to support students in developing competencies so they can learn how to decontextualize and recontextualize scientific knowledge in different contexts both within and outside of higher education. Furthermore, digitally supported environments for transformative learning should allow for a variety of conditions, tools, and apps that may be embedded in a digital learning ecology, which suits both the demands of higher education institutions and the involved groups outside higher education. Educators need to be trained to apply transformative learning. It requires a mindset shift away from traditional teaching to the new roles of the designer of learning processes, including coaching, consulting, and learning companion.

Additionally, students also need training to become active agents (Deslauriers et al. 2019). Therefore, higher education institutions may offer workshops covering project management or solving conflict in teams, for example. The kind of skills needed during transformative learning need to be explicitly communicated. Transformative learning requires certain conditions to establish and promote cultures of participation, to build a learning community where all students feel a sense of belonging, to include diverse group of learners, and to support equity. It also must include the chance to develop skills needed in professional and societal contexts.

By examining the development of transformative learning over time and by observing the success it has produced, the need to further explore and implement this kind of learning is evident. Universities, foundations, and governmental and political institutions have the duty to lead with more research and development programs, along with necessary funding, to explore transformative learning. More evidence is needed, specifically empirical data on the effectiveness and efficiency of transformative learning designs in use. Such evidence can provide necessary recommendations on how to apply transformative learning and to ask what works, and, more specifically, how it works. In such a vein, we argue for an iterative *research-to-improve* program (see details in Honebein and Reigeluth 2021) for further development of evidence-informed transformative learning.

References

Barr, Robert B., and John Tagg. 1995. From teaching to learning – A new paradigm for undergraduate education. *Change Management* 27 (1): 13–15.

Deslauriers, Louis, Logan S. McCarty, Kelly Miller, Kristina Callaghan, and Greg Kestin. 2019. Measuring actual learning versus feeling of learning in response to being actively engaged in the classroom. *Proceedings of the National Academy of Sciences* 116 (39): 19251–57.

Fischer, Gerhard, Johan Lundin, and Ola J. Lindberg. 2022. The challenge for the digital age: Making learning a part of life. *International Journal of Information and Learning Technology* 40 (1): 1–16.

Freeman, Scott, Sarah L. Eddy, Miles McDonough, Michelle K. Smith, Nnadozie Okoroafor, Hannah Jordt, and Mary Pat Wenderoth. 2014. Active learning increases student performance in science, engineering, and mathematics. *Proceedings of the National Academy of Sciences* 111 (23): 8410–15.

Fry, Heather, Steve Ketteridge, and Stephanie Marshall. 2003. Understanding student learning. In *A handbook for teaching & learning in higher education: Enhancing academic practice*, eds. Heather Fry, Steve Ketteridge, and Stephanie Marshall, 9–25. London: Routledge.

Gibson, Ivan. 2005. Designing projects for learning. In *Handbook of enquiry and problem-based learning*, eds. Terry Barrett, Iain Mac Labhrainn, and Helen Fallon, 27–36. Galway: AISHE and CELT.

Harper, Douglas, ed. 2023. *Transform (v.)*. Available from https://www.etymonline.com/word/transform.

Honebein, Peter C., and Charles M. Reigeluth. 2021. To prove or improve, that is the question: The resurgence of comparative, confounded research between 2010 and 2019. *Educational Technology Research and Development* 69 (2): 465–96.

Howland, Jane L., David Jonassen, and Rose M. Marra. 2012. *Meaningful learning with technology*. Boston, MA: Pearson.

Jahnke, Isa. 2023. Quality of digital learning experiences - Effective, efficient, and appealing designs? *International Journal of Information and Learning Technology* 40 (1): 17–30.

Jahnke, Isa, Matthew Schmidt, Minh Pham, and Kanupriya Singh. 2020. Sociotechnical-pedagogical usability for designing and evaluating learner experience in technology-enhanced environments. In *Learner and user experience research*, eds. Matthew Schmidt, Andrew A. Tawfik, Isa Jahnke, and Yvonne Earnshaw. Available from https://edtechbooks.org/ux/sociotechnical_pedagogical_usability.

Jankowski, Natasha. 2022. Participatory learning system paradigm in the US: Prospective potentials, options, and challenges. In *Professional-scientific education*, eds. Volker Rein and Johannes Wildt, 273–94. Opladen: Budrich.

Jenkins, Alan, and Mick Healey. 2011. Research based learning – A collection of case studies in different disciplines. In *Fachbezogene und fachübergreifende Hochschuldidaktik. Blickpunkt Hochschuldidaktik*, eds. Johannes Wildt and Isa Jahnke, 37–46. Bielefeld: wbv.

Kollmorgen, Raj, Wolfgang Merkel, and Hans-Jürgen Wagener. 2015. *Handbuch Transformationsforschung*. Wiesbaden: Springer.

Lave, Jean, and Etienne Wenger. 1991. *Situated learning: Legitimate peripherical participation*. Cambridge: Cambridge University Press.

Merritt, Richard L. 1980. On the transformation of systems. *International Political Review* 1 (1): 13–22.

Mezirow, Jack. 1991. *Transformative dimensions of adult learning*. San Francisco: Jossey-Bass.

Pekrun, Reinhard. 2014. *Emotions and learning*. Educational Practices Series – 24. Available from http://www.iaoed.org/downloads/edu-practices_24_eng.pdf.

Polanyi, Karl. 1944. *The great transformation: The political and economic origins of our time*. Boston: Beacon Press.

Rein, Volker, and Johannes Wildt. 2022. *Professional-scientific education: Discourses, Perspectives, implications, and options for science and practice*. Opladen: Budrich.

Schmidt, Matthew, Andrew Tawfik, Yvonne Earnshaw, and Isa Jahnke. 2020. *Learner and user experience research: An introduction for the field of learning design & technology*. Available from https://edtechbooks.org/ux.

Schneider, Ralf, and Johannes Wildt. 2002. Forschendes Lernen in Praxisstudien – Das Beispiel des Berufspraktischen Halbjahres in der Lehrerausbildung. In *Neues Handbuch Hochschullehre*, eds. Brigitte Berendt, Hans-Peter Voss, and Johannes Wildt, G 3.1. Berlin: DUZ.

UNESCO. 2017. *Education for sustainable development goals: Learning objectives*. Available from https://www.unesco.de/sites/default/files/2018-08/unesco_education_for_sustainable_development_goals.pdf.

Wagenaar, Robert. 2022. Competence-based development of learning outcome through reference tools: Potentials and limits of cross-educational qualifications frameworks. In *Professional-scientific education*, eds. Volker Rein and Johannes Wildt, 235–44. Opladen: Budrich.

Wildt, Johannes. 2021. Projektstudium. In *Handbuch Transdisziplinäre Didaktik*, eds. Tobias Schmohl and Thorsten Philipp, 243–54. Bielefeld: transcript.

Wildt, Johannes. 2022. Entwicklungslinien der Hochschuldidaktik: Ein Blick nach vorn auf einen "transformative turn" in der Hochschulbildung. In *Hochschuldidaktik gestern – heute – morgen*, eds. Gisela Schutti-Pfeil, Antonia Darilion, and Barbara Ehrensdorfer, 196–207. Linz: FH Öberösterreich

Living Lab

Julia Backhaus, Stefan Böschen, Stefan John, Andrea Altepost, Frederik Cloppenburg, Frances Fahy, Julia Gäckle, Thomas Gries, Christoph Heckwolf, Kaisa Matschoss, Joost Meyer, Daniel Münderlein, Marco Schmitt, Alexander Sonntag, Axel Timpe, and Gabriele Gramelsberger

Definition

The Living Lab(oratory) inspires as a notion, a setting, and a methodology encouraging participatory approaches to the co-production of knowledge for innovation and transformation. Etymologically, the term refers to a place or space for work and exertion (lat. *lăbor*) where things are made, prepared, or strived for (lat. *lăbōro*), and that is lively (lat. *vivus*) or alive (Lewis and Short 2020, 594, 1146). While alchemist Thomas Knight has been accredited with coining the term *Living Laboratory* in 1749 as a metaphor for the human body, the term eventually became used for research and open innovation processes in real-life or lifelike settings, initially in the context of psychological research on viewers' reactions to television commercials (Leminen and Westerlund 2019, 254). Nowadays, the diversity in projects, programs, and institutions called Living Lab indicates that "the term 'living lab' is at risk of becoming a buzzword in the innovation domain because it lacks a consistent or commonly accepted definition" (Leminen 2015, 29). While the range and amount of projects, programs, and institutions called Living Lab continue to increase, a number of characteristics commonly shared by Living Labs has been identified: (1) a transdisciplinary approach to research and knowledge creation; (2) an iterative, experimental design committed to learning and reflexivity; (3) a long-term orientation towards societal transformation and an accompanying interest in transferability or scalability; (4) a focus on a real-life environment (Compagnucci et al. 2021).

In 2015, the German term *Reallabor* (Real-world Laboratory) was coined to demarcate spaces for transdisciplinary real-life experimentation towards sustainability from the mostly technology-driven living labs (Schäpke et al. 2015). By now, both terms relate to a broad range of real-life experimentation in transdisciplinary settings but a rough distinction can be made between Living Labs focusing on the optimization and application or implementation of innovative technologies and Living Labs engaging non-academic actors in participatory processes for sustainability transformations. Other commonly used terms for the later types of initiatives include Urban (Living) Lab or Transition Lab.

The focus on learning and iterative experimentation in all types of Living Labs suggests a recognition of the complexity characterizing the challenges that societies are currently facing. This complexity also poses a challenge to the educational system that is institutionalized and oftentimes remains organized by scientific disciplines. The remainder of this chapter examines opportunities for transdisciplinary learning garnered through the use of Living Labs for educational purposes.

Background

Historically, Living Labs have always been concerned with interaction – between individual actors, (potentially viewed as representatives of) relevant stakeholder groups, and certain stimuli, tests, experiments, or interventions in a particular setting. Since transformation has emerged as the order of the day and digitalization and sustainability are pursued with increased urgency, spaces for collaborative experimentation, learning, and development appear out of sheer necessity. In this context, Living Labs can function as an infrastructure (Schneidewind et al. 2018) for transdisciplinary experimentation towards societal transformation.

In the 1990s, Living Labs emerged in human–computer interaction research as physical places. Around the same time, Living Labs appeared as innovation spaces in the form of multi-stakeholder processes tackling complex problems in local settings (Leminen and Westerlund 2019, 254). This dual application and development of the concept continues until today. For instance, transformation researchers from diverse disciplinary backgrounds are co-designing Living Labs as catalysts for sustainable living, urban design or regional development in collaboration with local, regional, national, or international stakeholders (Hahne 2019; Matschoss et al. 2021). Likewise, IT specialists and engineers in academia, business, and industry, often supported by professional associations and policymakers, collaborate in Living Labs to improve human–computer interaction (Alavi et al. 2020) and other issues related to technological innovation. This type of Living Lab is primarily concerned with application, implementation, and marketization. Occasionally, for example on open days, it may be used for the purpose of science communication. In other words, the two most easily differentiable types of Living Lab are those addressing challenges pertaining to technological innovation to pave the way for wide-scale use, and those addressing societal challenges pertaining to sustainable development. Both types seek to explore possible future developmental pathways.

A significant part of the state-of-the-art literature on Living Labs is concerned with institutional, procedural, or methodological success factors (Bergmann et al. 2021) or with categorizing Living Labs based on literature reviews (Alavi et al. 2020; Compagnucci et al. 2021; Hossain et al. 2019) or empirical research. Attempts at classifying empirical examples of Living Labs focus, for example, on research aims, varying degrees of transdisciplinarity and resulting stakeholder constellations (Backhaus et al. 2022) or on objectives such as "commercialisation (1), research (2), teaching and qualification (3), ideation (4), integration of societal actors (5) and sustainable development (6)" (Held et al. 2022, ii). Typically, Living Labs serve more than one objective.

Debate and criticism

Heralded as platforms for open and participatory innovation, Living Labs represent sites of power struggles between hegemonic incumbents and advocates of alternative strategies for technological innovation and social transformation (Stirling 2008, 264). Current debates revolve around the questions of how justice and fairness may be served and how democracy may be preserved or even strengthened through transdisciplinary research and learning. Criticism regarding insufficient participation and reflexivity recurringly erupts at familiar fault lines, many of which also run between scientific disciplines. Living Labs thus present opportunities for the transdisciplinary exploration of problems and participatory experimentation with solutions. Or they may provide the backdrop for practicing "solutionism" (Morozov 2014). Where a particular Living Lab falls on the spectrum between transdisciplinary learning, on the one end, and the pursuit of pathways predefined by select actors or groups, on the other, depends on procedural aspects (such as who is involved in decision-making, when and how) and on the social construction of boundaries around the stakeholders involved, problems addressed, and solutions considered. While some consider this an issue of innovation and expectation management, others view it as a fundamental issue in technology R&D and research policy (Grunwald 2019, 36–42).

Regarding different modes or levels of experimentation and participation, three important "tensions" have been observed in Living Labs, between "controlled experimentation vs. open co-creation; learning from failure vs. public demonstration of success; [and] local embedding vs. scalability" (Engels et al. 2019, 1). Notably, stories of prospective transferability and scalability that are frequently spun around Living Labs blend in well with a "politics of technology" that is preoccupied with "solutionism, experimentalism and future-oriented valuation" (Pfotenhauer et al. 2021). In other words, the significant increase in Living Lab activities and publications since 2015 (Hossain et al. 2019) may at least partially emerge from mission-oriented innovation governance.

Considering that Living Labs are viewed and established as experimental spaces of our collective futures, the dominant focus on technological fixes paired with an ignorance of questions related to justice, plurality, and equality requires reflection and action. It has been argued that more participatory and pluralist approaches to understanding and addressing sustainability-related challenges would aid in delivering more rigorous and robust scientific findings and ameliorating democratic deficits (Böschen et al. 2021, 294–95).

Current forms of implementation in higher education

Increasingly, the Living Lab is also recognized as an infrastructure or institution for integrated inter- and transdisciplinary education. Aside from fostering students' personal and professional development with respect to conceptual and methodological learning objectives, Living Labs provide a setting for hands-on learning experiences, enabling and empowering students to acquire and advance "21st century skills" (World Economic Forum 2015, 3) and to lead on transformative change. Living Lab approaches – whether simulated or implemented – stimulate learning about system, target, and transformation knowledge (Pohl and Hirsch Hadorn 2021, 36). By encountering successes and setbacks in multi- or interdisciplinary teams, students profit from hands-on learning, receive more immediate feedback, and experience self-efficacy. Moreover, Living Labs can help deliver on the third-mission requirement of higher education institutions by providing a platform for encounters and exchanges with (local) societal stakeholders, including companies, public actors, and civil society . Since Living Labs emerge from particular goals, settings and actor constellations, there is no one-size-fits-all approach or single formula for success. However, countless examples from across the globe, albeit so far concentrated in the Global North, can serve as examples and provide some guidance and inspiration for setting up and operating educational Living Labs.

Since Living Labs first emerged in Europe and North America, it is not surprising that the concept has also spread furthest in these regions. To highlight the earlier noted richness of the concept and the various possibilities for application, institutionalization, and use, the first examples of current forms of implementation are all based at the same institution, RWTH Aachen University, the largest technical university in Germany. Recognizing the importance of transdisciplinary research and teaching, RWTH Aachen University and the Karlsruhe Institute of Technology both integrated Living Labs in their Excellence Strategies, which are funded by the German federal and state governments. At RWTH Aachen University, Living Labs that are exclusively or also used for teaching can be found across faculties and disciplines. Mirroring the dual meaning of the notion, a broad distinction can be made between those Living Labs addressing societal challenges related to sustainable development and those concerned with advancing digitalization and automation in business, industry, and society. The former include the

Büchel:Lab, which provided master's students of architecture and urban planning with the opportunity to develop concepts for temporary usage and participatory urban development for the redesign of an old town quarter (Digi-Sandbox. NRW 2023); the student-led Living Lab *nAChhaltig angezogen* (Sustainably dressed), which started as a graduate project seminar in sociology on the topic of sustainable fashion and has turned into a continuing initiative; the *Waldlabor Köln* (Forest Lab Cologne; Palm et al. 2023), which was set up by the City of Cologne to study the forest of the future, to enable students to test and evaluate design options in forest management, and more recently also to experiment with 3D printing technologies using wood mass, and the *project module "Green Blue Streets"* (lala.ruhr 2021), in which master's students in architecture and urban planning (and recently also students in transport engineering) developed design proposals for the water-sensitive transformation of an urban regeneration area in the city of Gelsenkirchen and presented the proposals to policymakers and other stakeholders. Two noteworthy examples of educational Living Labs addressing puzzles and problems pertaining to Industry 4.0 are the *Learning Factory "Textil vernetzt"* (Textile networked) operated by the Institute of Textile Technology as a real-life learning environment for students and professionals-in-training to address the challenges of the Internet of Things in manufacturing, and the *WIRKsam (Efficacious) Competence Centre*, which provides a collaborative space and develops a comprehensive set of training on the integration of Artificial Intelligence (AI) into industrial processes for industry professionals as well as graduate and undergraduate students from various disciplinary backgrounds such as computer science, sociology, engineering, and psychology (ifaa 2023). In addition, a simulation game which is offered as part of the master's programs in Sociology and Governance of Technology and Innovation challenges students to devise a Living Lab strategy for the Rhenish mining region, which, like many former mining areas, is undergoing large-scale, long-term structural changes. Through this game, students get a glimpse of the complexity of the issues and of what is at stake for the different actors involved.

Since Finland spearheaded the promotion of the notion in Europe, not least by initiating the founding of the European Network of Living Labs when holding the presidency of the European Council in 2006, it is also worth exploring current forms of implementation in higher education in the Finnish context. Laurea University of Applied Sciences, for example, has embraced the concept, running several Living Labs addressing various topics. The most recently set up *Laurea Circular Economy Living Lab* combines education, R&D, and regional development by providing educational modules and a networking space with regional partners for undergraduate students in Hospitality Management and Service Design to devise circular economy solutions (Laurea 2023). The *TAMK Catering Studio Living Lab*, concerned with sustainable food transformation, was created by Tampere University of Applied Sciences (TAMK 2023) as part of the EU project "Fostering the Urban Food System transformation through Innovative Living Labs Implementation" (FUSILLI) in the new urban area of Hiedanranta, home to numerous sustainability Living Labs, allowing students to participate in grassroots city development and experiment with various ideas, including a business model for regenerative urban micro-farming. Still under construction but promising to provide an institutional and infrastructural home for multidisciplinary research, teaching, and learning, the *Hyytiälä Forestry Field Station Living Lab* comprises a sustainable, wooden building complex to study sustainability, the climate, and well-being in the built environment (University of Helsinki 2023).

Upon gaining a strong foothold in Europe's northwest, the success story of the Living Lab currently continues globally, suggesting that collaborative experimentation with innovative technologies and social innovations yields valuable insights and experiences – also for students. Cases in point are the *Virtual FabLab* (vFabLab 2022) at King Abdullah University of Science and Technology in Saudi Arabia, which provides a web-based state-of-the-art 3D gamified interactive virtual fabrication lab to train students and enthusiasts around the world in nanofabrication techniques including sputtering, atomic layer deposition, and more, in a safe-to-fail virtual environment; the *BELgrade urban living LAB* (Centar za eksperimente i urbane studije 2023), which was set-up as the first Urban Living Lab in Serbia and the Western Balkans to co-create solutions with citizens, the public sector, planning experts, and private companies; and the *Rijeka iLivingLab*, which comprises four labs (a Maritime Navigation, Safety and Security lab, a Logistics lab, an E-learning lab, and an E-government lab), focuses on the entire coastal region of the Republic of Croatia, and has trained over 2,500 students on a range of issues such as artificial intelligence, agriculture and agri-food, culture and creativity, energy, innovation, social inclusion, and (health and well-being in a maritime environment. ENoLL 2023a).

Some Living Labs are specifically set up to increase international and intercultural collaboration. Two examples are the *Living Lab field course*, which forms part of the ICP Connect master's program Sustainable Development at KU Leuven (2023) in Belgium, North-West University in South-Afrika, Vietnam National University and Pontifical Catholic University of Peru during which small teams of students from the Global North and Global South engage in intensive field research, ideally involving key stakeholders, to devise a strategy to address a sustainability challenge in a Global South context; and *LivingLab SHANGHAI*, based at the Sino-Finnish Centre at Tongji University in Shanghai, China, which collaborates with Aalto University in Helsinki, Finland to involve key stakeholders in the development, prototyping, and testing of technological solutions to complex sustainability challenges in megacities surrounded by low-resource surroundings (ENoLL 2023b).

An interesting subset of educational Living Labs engages in large-scale experimentation using the entire campus or – in collaboration with other stakeholders

– additional urban areas for Living Lab experimentation and education. Examples are the *Living Lab Tomsk* (LEVS 2021) in Russia, a network of Living Labs involving seven universities, ten academic institutes, the Tomsk Scientific Center, innovative companies and architectural bureaus, regional and city administrations, and European partners, allowing students to experiment at seven locations, each with a specific focus (public space design, smart greening, water management, smart management, dialog of generations, healthy lifestyle, multicultural environments, street art and creativity), and the *Learn–Live–Lead approach to sustainability of University of Galway*, which uses the campus buildings and estate as a Living Lab to promote sustainability scholarship, environmental stewardship, and global citizenship. Initial successes include the city council working with the university to form a sustainable energy community and, in 2021, *Decarbonization Zones* (areas with a goal of 51% reduction in GHG emissions by 2030) have been designated in the city and on campus (University of Galway 2023).

The collection of examples suggests that educational Living Labs can fulfill the triple role of enabling research and education while at the same time advancing the sustainability transformation of higher education institutions and their local or regional surroundings. As the long and varied history of the notion suggests, Living Labs can function as open spaces for collaborative experimentation beyond disciplinary and social boundaries and offer learning opportunities for every participating individual and (stakeholder) group. Crucially, students' ideas and perspectives also enrich and diversify research and experimentation in Living Labs in novel ways. As an important transformative impulse, the experience, expertise, and in some cases entrepreneurial mindset acquired by students through the use of Living Labs for educational purposes helps transfuse transdisciplinary experimental research and practices into society.

References

Alavi, Hamed S., Denis Lalanne, and Yvonne Rogers. 2020. The five strands of Living Lab: A literature study of the evolution of Living Lab concepts in HCI. *ACM Transactions on Computer–Human Interaction* 27 (2): 1–26.

Backhaus, Julia, Stefan John, Stefan Böschen, Ana de la Varga, and Gabriele Gramelsberger. 2022. Reallabore um die RWTH Aachen: Rückblicke, Einblicke, Lichtblicke. *pnd - planung neu denken* 1: 104–23.

Bergmann, Matthias, Niko Schäpke, Oskar Marg, Franziska Stelzer, Daniel J. Lang, Michael Bossert, Marius Gantert, Elke Häußler, Editha Marquardt, Felix M. Piontek, Thomas Potthast, Regina Rhodius, Matthias Rudolph, Michael Ruddat, Andreas Seebacher, and Nico Sußmann. 2021. Transdisciplinary sustainability research in real-world labs: Success factors and methods for change. *Sustainability Science* (16): 541–64.

Böschen, Stefan, Julia Backhaus, Ana de la Varga, Stefan John, and Gabriele Gramelsberger. 2021. Reallabore: Simulierte Experimente – Simulierte Demokratie? In *In digitaler Gesellschaft: Neukonfigurationen zwischen Robotern, Algorithmen und Usern*, eds. Kathrin Braun and Cordula Kropp, 275–302. Bielefeld: transcript.

Centar za eksperimente i urbane studije, ed. 2023. *BELgrade Urban Living LAB*. Available from http://bellab.rs.

Compagnucci, Lorenzo, Francesca Spigarelli, José Coelho, and Carlos Duarte. 2021. Living Labs and user engagement for innovation and sustainability. *Journal of Cleaner Production* 289: 125721–38.

Digi-Sandbox.NRW, ed. 2023. *Büchel:Lab*. Available from https://www.digi-sandbox.nrw/projekte/buechel_Lab.

Engels, Franziska, Alexander Wentland, and Sebastian M. Pfotenhauer. 2019. Testing future societies? Developing a framework for test beds and living labs as instruments of innovation governance. *Research Policy* 48 (9): 103826.

ENoLL, ed. 2023a. *Rijeka iLivingLab*. Available from https://enoll.org/network/living-labs/?livinglab=rijeka-ilivinglab#description.

ENoLL, ed. 2023b. *LivingLab SHANGHAI*. Available from https://enoll.org/network/living-labs/?livinglab=livinglab-shanghai#description.

Grunwald, Armin. 2019. Das Akzeptanzproblem als Folge nicht adäquater Systemgrenzen in der technischen Entwicklung und Planung. In *Akzeptanz und politische Partizipation in der Energietransformation: Gesellschaftliche Herausforderungen jenseits von Technik und Ressourcenausstattung*, eds. Cornelia Fraune, Michèle Knodt, Sebastian Gölz, and Katharina Langer, 29–43. Wiesbaden: Springer.

Hahne, Ulf. 2019. Reallabore: Katalysator für die nachhaltige Stadt- und Regionalentwicklung? *RaumPlanung* 200 (1): 53–58.

Heckwolf, Christoph. 2023. *Wir sind nAChhaltig angezogen*. Available from https://nachhaltig-angezogen.de/wir/.

Held, Tobias, Sophie Kaiser, Felix Schneider, and Alexandra Hausstein. 2022. *"From Lab to Tab" – eine empirisch gestützte Typologie von Innovation Labs in Deutschland*. TRANSFORM Diskussionspapier. Karlsruhe: Karlsruher Institut für Technologie. Available from: https://publikationen.bibliothek.kit.edu/1000148809

Hossain, Mokter, Seppo Leminen, and Mika Westerlund. 2019. A systematic review of Living Lab literature. *Journal of Cleaner Production* 213: 976–88.

ifaa, ed. 2023. *Kompetenzzentrum WIRKsam – Arbeit mit KI gestalten*. Available from https://www.arbeitswissenschaft.net/wirksamweb.

KU Leuven, ed. 2023. *Sustainable Development Living Lab*. Available from https://susdev.eu/living-lab/.

lala.ruhr, ed. 2021. *Green-Blue Streets: Die studentischen Entwürfe sind ausgewertet*. Available from https://www.lala.ruhr/2021/11/01/green-blue-streets-die-studentischen-entwuerfe-sind-ausgewertet/.

Laurea, ed. 2023. *Service business and circular economy*. Available from https://www.laurea.fi/en/research/service-business-and-circular-economy/.

Leminen, Seppo. 2015. Q&A. What are Living Labs? *Technology Innovation Management Review* 5 (9): 29–35.

Leminen, Seppo, and Mika Westerlund. 2019. Living Labs: From scattered initiatives to a global movement. *Creativity and Innovation Management* 28 (2): 250–64.

Lewis, Charlton T., and Charles Short. 2020. *A Latin dictionary: Founded on Andrews' edition of Freund's Latin dictionary*. Chapel-en-le-Frith: Nigel Gourlay.

LEVS architecten, ed. 2021. *LEVS supports Living Lab Tomsk*. Available from https://www.levs.nl/nieuws/levs-supports-living-lab-tomsk.

Matschoss, Kaisa, Frances Fahy, Henrike Rau, Julia Backhaus, Gary Goggins, Eoin Grealis, Eva Heiskanen, Tuija Kajoskoski, Senja Laakso, Eeva-Lotta Apajalahti, Audley Genus, Laurence Godin, Marfuga Iskandarova, Annika-Kathrin Musch, Marlyne Sahakian, Christian Scholl, Edina Vadovics, and Veronique Vasseur. 2021. Challenging practices: Experiences from community and individual Living Lab approaches. *Sustainability: Science, Practice and Policy* 17 (1): 135–51.

Morozov, Evgeny. 2014. *To save everything, click here: The folly of technological solutionism*. London: PublicAffairs.

Palm, Stefan, Christoph Preuß, and Sandra Tibor, eds. 2023. *Waldlabor Köln*. Available from https://www.koeln-waldlabor.de.

Pfotenhauer, Sebastian M., Brice Laurent, Kyriaki Papageorgiou, and Jack Stilgoe. 2021. The politics of scaling. *Social Studies of Science* 52 (1): 3–34.

Pohl, Christian, and Gertrude Hirsch Hadorn. 2021. *Principles for designing transdisciplinary research: Proposed by the Swiss Academies of Arts and Sciences*. Munich: Oekom.

Schäpke, Niko, Mandy Singer-Brodowski, Franziska Stelzer, Matthias Bergmann, and Daniel J. Lang. 2015. Creating space for change: Sustainability transformations: The case of Baden-Württemberg. *GAIA – Ecological Perspectives for Science and Society* 24 (4): 281–83.

Schneidewind, Uwe, Karoline Augenstein, Franziska Stelzer, and Matthias Wanner. 2018. Structure matters: Real-world laboratories as a new type of large-scale research infrastructure: A framework inspired by Giddens' structuration theory. *GAIA – Ecological Perspectives for Science and Society* 27 (1): 12–17.

Stirling, Andy. 2008. "Opening up" and "closing down": Power, participation, and pluralism in the social appraisal of technology. *Science, Technology, & Human Values* 33 (2): 262–94.

TAMK [Tampere University of Applied Sciences], ed. 2023. *TAMK Catering Studio Living Lab of Food & Sustainability.* Available from https://sites.tuni.fi/catering-studio.

University of Galway, ed. 2023. *Learn–Live–Lead model.* Available from https://www.universityofgalway.ie/sustainability/learn-live-lead-model.

University of Helsinki, ed. 2023. *Hyytiälä Forestry Field Station.* Available from https://www.helsinki.fi/en/research-stations/hyytiala-forestry-field-station.

vFabLab, ed. 2022. *The first virtual cleanroom environment for semiconductor manufacturing technology.* Available from https://vfablab.org.

World Economic Forum. 2015. *New vision for education: Unlocking the potential of technology.* Geneva: World Economic Forum.

Mode 2

Ines Langemeyer and Eike Zimpelmann

Definition

The Latin word *modus* is often translated simply as "manner". It originally meant "measure, weight", but is also etymologically related to the diminutive *modulus*, meaning "scale, small measure" and "model" (Müller 2009, 638). There is little evidence in the research literature on Mode 2 that the word describes a model of science. Such an interpretation is obvious, because the differences between the two modes of science, which are the starting point of the discussion, can be understood primarily in terms of ideal types (Nowotny 1993, 70; Schauz 2014, 49): Mode 2, for example, stands for forms of knowledge creation that take place under the influence of industrial technology developments, and public- and private-sector organizations and state institutions involved in them, and bear the character of application-oriented, practice-integrated, and cross-disciplinary research. Consequently, the scientific aspiration to overcome contradictions in knowledge recedes into the background while practical solutions come to the fore. Mode 2 stops when a practical solution is found and implemented.

Mode 2 indicates the historical development of knowledge societies, whose knowledge – according to Max Weber (1934) – is distinguished from impartial, value-, and contradiction-free knowledge and rather finds recognition through its robustness and functionality in transformation processes. Mode 1, on the other hand, is regarded as the product of basic research, which can also refrain from practical solutions and decisions within the protected framework of universities and research institutions.

The two modes thus become ideal-types, distinguishable as models of science. Regardless of whether ideals (such as that of unbiased and contradiction-free knowledge) have actually been fulfilled, they become for Mode 1 one of its essential characteristics. Furthermore, it is said that Mode 1 and 2 do indeed coexist and interact in real terms (Gibbons et al. 1994, 9). What belongs to this or that mode is to be distinguished analytically.

Mode 2 does not follow rigorously the ideal of science to generate universal knowledge. It accepts value-based judgments and contradictions; heterogeneity, utility orientation, commercialization, dialogicity and reflexivity, transdisciplinarity, and fluctuating forms of collaboration are seen as aspects of the Mode 2 knowledge production (Gibbons et al. 1994, 3–8). For higher education that addresses, for example, problem-based learning in real-world laboratories, through service learning or citizen science, the decrease in rigorousness raises fundamental questions of transdisciplinary learning.

Background

The distinction of two modes of knowledge creation sparked an international discussion, particularly in the fields of science and technology studies, technology assessment, philosophy of science, and management (Bartunek 2011; Etzkowitz and Leydesdorff 2000; Nowotny 1999; Nowotny et al. 2001, 2003). It also inspired analyses of the transdisciplinarity of research practices, such as those found in the marketplace of digital information goods and services and other areas of high complexity (Holtgrewe 2012; Klein et al. 2001; Weingart 1997). Furthermore, forms of participation in practice-oriented research and their importance for gaining knowledge have been discussed (Jahn et al. 2012).

Several assumptions and views of the Mode-2 approach were already widespread in sociology and science studies (Bender 2001, 9; Jasanoff 2003; Nowotny 1993) when the approach was launched. Long before the 1990s, the "unity of the sciences" (*Einheitswissenschaft* as Neurath, Carnap and others termed it in the 1930s) was already doubted. At the beginning of the 20th century, representatives of the neo-positivist Vienna Circle such as Carnap (1931, 465) still defended the basic idea that the conflict between disciplines could be overcome and that true knowledge could be unified. In the 1950s, van Orman Quine's (1951) thesis of "two dogmas" and Snow's (1959) thesis of "two cultures" sought to describe the fact that certain subjects were sufficiently akin, so that scientists from similar disciplines would be able to communicate, while an unbridgeable gap had formed in relation to other disciplines. In this way, they primarily underpinned the difference between the humanities and the natural sciences. Research on "knowledge (science) cultures" (Arnold and Fischer 2004; Nerland 2012), sometimes with reference to Fleck's (1980) "thought collectives", sometimes to Kuhn's (1962) "paradigms". Also Polanyi's (1962) theory on the tacit dimension, among others, assumed an increasing differentiation of the sciences. The argument against a positivist notion of a unified science was that all knowledge is socially constructed (Knorr Cetina 1991, 2002). Disciplines create "life", e.g. in laboratories incorporating various "phenomenotechniques" (Bachelard 1998) and socially establish orders which form the rational background of scientific practices (Latour and Woolgar 1979). This epistemic argument did not express a new unification thesis (*science as one*) against the diversification of disciplines. Rather, the blaming of every kind of social influence that would contaminate scientific research was contested. Latour and Woolgar (1979, 23–24) concluded that the influence of the social on the scientific endeavor is not only visible in errors: "Scientific achievements held to be correct should be just as amenable to sociological analysis as those thought to be wrong". Knorr Cetina (1999, 4) similarly refuted the idea of a unified science or "one enterprise" of science in favor of analyzing a "whole landscape – or market – of independent epistemic monopolies producing vastly different products". From an epistemological rather than a sociological point of view, Rheinberger (2005, 316) resumed Bachelard's conclusion that "with the ever tighter interplay between ever more specific forms of knowledge and the phenomenal world, the sciences necessarily become fragmented into different epistemological regions" and "their conceptual dynamics finally became inseparable from the phenomena in which and through which they expressed themselves".

Since the 1970s, the concept of the transdisciplinarity of research (Bernstein 2015; Scholz 2020) has replaced a possible integration of all bodies of knowledge into one science and increasingly referred to a spillover of scientific efforts to many areas of society (Jahn et al. 2012). At the same time, while examining the socio-cultural dependency of thinking and knowledge creation, a practice-philosophical turn was also proclaimed – the *practice turn* (Schatzki et al. 2000), which borrowed from Marx the overcoming of the subject–object dualism, without understanding the turn itself as a Marxist one. Marx's first Feuerbach thesis – written in 1845, intended only for his eyes, not for the public – phrased the essential point in a nutshell: "The chief defect of all hitherto existing materialism – that of Feuerbach included – is that the thing, reality, sensuousness is conceived only in the form of the *object* or *of contemplation*, but not *as sensuous human activity, practice*, not subjectively" (Marx 1995; italics in the original). This thesis thus already contained the radical demand to place the epistemic subject back into the practical context – where it sometimes does not even approximately possess an all-encompassing and disinterested gaze, but must immanently reflect on its particular relationship to the object of knowledge (including the method and means of research), i.e. how the things became objects of knowledge, how they adopted the *form* of an object (Marx 1996). Thus, objects are seen as something changeable, something that can adopt different forms and therefore need to be interpreted in the practical contexts. The unity of recognition and change thus became the key to a new epistemology.

The practice-philosophical paradigm shift took place in various currents in the 20th century: for example, with feminist philosophy of science (Haraway 1988; Harding 1986), laboratory studies in the context of science and technology research (Knorr Cetina 1991; Rip 1997), historical epistemology (Rheinberger 1994), and sociological approaches (Bourdieu 1979). It involved understanding science and theory not as the opposite of practice, but as *doing science* and uncovering its power relations and hidden mechanisms. Further approaches emphasized, similar to Berger and Luckmann's (1966) social constructivist or Alfred Schütz's (1971) phenomenological conception of "everyday knowledge", that knowledge becomes effective in local contexts of practice. In the 1980s, studies emerged which explored how knowledge is distributed – and at least partly shared – among people as members of a community and thus must be understood as situated knowledge (Lave and Wenger 1991; Suchman 1987).

Since science creates communities with different cultures and contexts of research, scientific knowledge too is then regarded as situated knowledge. Thus, these currents were also concerned with the insight that science should not be considered as an institution somehow outside of society and completely independent of it. The postulate of impartiality was rejected as an insufficiently reflected idealization of scientific knowledge (Haug 2004). Scientific knowledge always articulates a standpoint and a perspective (Haug 2004).

The Mode-2 approach added to the practice-philosophical paradigm shift in the social sciences that there was a historically new way of producing knowledge in all areas of society through a stronger interconnection of segments such as industry, politics, and research, which would no longer be institutionalized in the conventional pattern of science in universities and similar research institutions (Gibbons et al. 1994, 10). With respect to technological development, the theses of the Mode-2 approach partly overlap with those of Machlup (1962), Drucker (1969), or Bell (1973), who emphasized the increasing influence of science on (industrial) production and political society (Hack 2001, 25), thus also addressing the transformation of the "scientification of society" into a "politicization of science" (Weingart 1983, 235) as a problem. Interpreted as a loss of power or as a phenomenon of dissolution, there is also a warning of the danger of the "de-professionalization" (Weingart 1983, 235) of scientific expertise.

The views that Gibbons et al. put forward touch on and flank theses from the currents that were outlined in the previous sections. The affirmed tendencies of a politicization and an economization of science are subsumed under the term 'Mode 2'. The authors' specific concern is to present a heuristic assumption that can be used to elaborate historical changes in the social role of science (Gibbons et al. 1994, 1).

However, since the boundary between modes runs between ideal-typical opposites, the diagnosis is not developed empirically, not in a historical-critical way. Its soundness has therefore become the subject of a debate (Birrer 2001; Gläser 2001). This is particularly evident in claims that Mode 1 still harbors an "epistemic core of the sciences", while science in Mode 2 exists ubiquitously, rhizome-like, without center and goal and without inner and outer boundaries (Nowotny 1999, 30–31, 118), alluding to a metaphor used by Deleuze and Guattari (1987). Consequently, whether a Mode 2 has actually emerged depends on the question of whether a Mode 1 existed at all.

From the perspective of Gibbons et al. (1994), however, it is helpful not to take Mode 1 as the only mode of science for breaking down narrowed notions. A central argument here is that the transdisciplinary and more participatory forms of research practice create a different pattern and thus different rules of institutionalization. Processes of institutionalization are important for the creation of scientific knowledge to be societally relevant (Langemeyer 2021). Mode 1 can be understood not only as a way of being, but at the same time as a "necessary myth" (Nowotny 1999, 81), a "symbolic resource" (44), that ensures the public's lack of trust in scientific reason. By relying on context-independent knowledge and – as the Vienna Circle historically did – on the unity of the sciences, Mode 1 could sufficiently legitimize the generated knowledge of science and give it social authority and power (Drori et al. 2006; Nowotny 1999, 22). According to another assessment – Nowotny (1999, 115) also sees this – the changes of the institution of science reveals the deeper problem of "democratic participation" in and through science. Since scientification runs deep into the fabric of Western societies (into its culture, see Drori et al. 2006) with their ideal of democracy and scientific progress, a caricature of science emerges: Mode 1 serves only as the foreshadowing of knowledge with social authority, whereas the critical reflection of knowledge shifts to ideas vaguely connected to Mode 2 – but the rigorousness of Mode 1 is gone. The construction of opposite modes ends in a deadlock.

Debate and criticism

The sociologically understood distinction between two modes of knowledge production ignited science-theoretical debates of the late 20th century. The debates revolve around a historical transformation of science. The mere obsolescence of ideas, theories, methods, and paradigms is not meant here, but rather the specific pressure for change, which affects science and brings it into closer interdependence with economic, political, and other societal actors who profit from research or scientific standards and norms and thus establish power relations.

Bora (2005, 755–56) highlights that Nowotny et al. grasp the emergence of "open systems" of knowledge production – which raises the question whether the notion of a "closed system" could ever apply to science. According to Frederichs (2001, 73), the Mode-2-approach touches on the setting: Knowledge is created where problems arise and are to be solved. This in turn provokes the question whether all problems are scientific problems, and whether solutions naturally reveal the scientific explanation of why something works.

From an epistemological point of view, science is not accomplished by registering and stating what "there is" just because something seems evident. It is not realized by merely solving problems. Scientists rather need to critically reflect on how they interpret a particular phenomenon, how they distinguish its components or the different states of its development, how they construct and identify the underlying causal or systemic relations, and how they find empirical evidence for this. They need to develop an argument for why that data is relevant, valid, and usable for the research object, and why this object is adequately scrutinized by a certain research method. In practice, when people solve problems, for example, there are limits and obstacles for ensuring that these reflections together with the coordination between theoretical and empirical steps are undertaken with rigor. Therefore, Hack (2001, 55) also criticizes that adherents of the Mode-2 diagnosis erroneously assume that "traditional" scientific knowledge (which they term Mode 1) would emerge without this kind of reflection. He alleges that in their point of view, Mode 1 would already be deprived of everything that constitutes the distinctive features of scientific knowledge: as a form of reflection and as way of retracting and restarting trains of thought that have been shown to be flawed.

Carayannis and Campbell (2012, 4) also see a deficit of reflection in Mode 2. However, without defining the necessities of reflection, they advocate a model of "knowledge production systems" – "Mode 3". "Mode 3" indicates higher-order learning processes and thus means a somewhat higher reflexivity and reflectiveness in change and innovation processes. In doing so, they ignore the fact that science is always based on reflexivity – even in Mode 1. An imagined increase in reflexivity from Mode 1 to Mode 3 creates the fiction of a success story. Hence, Hack's critique of a truncated understanding of science is also applicable to Carayannis and Campbell.

And there is another issue: With Mode 3, as with other innovation paradigms, the argument was made that "models of science" could be deliberately chosen to foster, for example, sustainable development (Liyanage and Netswera 2022, 1128). Regardless of what goals should be achieved (sustainability, innovation, etc.), reflexivity is not a quantitative, but a qualitative matter. It does not make sense to speak of an increased reflexivity without identifying why reflections in the concrete research processes are needed and in what ways.

In contradistinction, if models of science were only a matter of choice (like choosing an instrument or a technique), the distinction between the different paradigms *emerging from* research in different disciplines and the overall concept of science is blurred. Consequently, it seems possible to merely create and change definitions of science deliberately beforehand *without reflecting* on developments of disciplines from within. Ironically, this would be a loss of reflexivity.

Furthermore, the practices of knowledge described by Mode 2 raise the questions of which power relations emerge with them and how existing ones are changed by the fact that other institutions and actors are involved in knowledge production than the traditional ones (such as universities and research institutions which are legally protected to sustain their independence). Similarly, the different forms of knowledge, which arise partly from scientific disciplines and partly from professional, political, and economic expertise, can be addressed as a problem. How they come together in a transdisciplinary way so that a new form of scientific knowledge (expertise) and not just a kaleidoscopic assemblage of different elements emerges has not been clarified. A few works deal with the problem of how this might be accomplished through intellectual cooperation (Jahn et al. 2012: 5; Langemeyer 2015). Nevertheless, it can be assumed that the less planned and more unstructured forms of collaboration depend on new ways of working and expertise becoming institutionalized at some point. The efficacy gained from this is particularly relevant for long-term project goals such as the energy transition or climate protection. How this institutionalization could meaningfully take place may be a central task of research on transdisciplinary didactics in the future, in the context of which dealing with societal complexity and unsolved problems is examined as a form of learning.

Current forms of implementation in higher education

So far, only partial implementations of knowledge creation in Mode 2 are known in the field of innovations in higher education didactics, for example in approaches of service learning, citizen science, or in forms of research-based or problem-based learning. However, a systematic introduction of such Mode 2 elements into an entire study program would hardly be conceivable without a balancing act between traditional and novel ideals and orientations (Balsiger 2015). However, if universities and colleges, in their function as knowledge institutions that develop and pass on academic knowledge, were to act more as "change agents" (Scholz 2020) in the future, there would have to be a fundamental agreement on the following aspects of transdisciplinary learning.

Nine aspects should be taken into consideration: (1) If knowledge is no longer measured by the ideal of an impartial or value-free form of science without contradictions, how can *values relevant to practice* be meaningfully addressed in concrete content and qualification goals of a degree program? What normative viewpoints should transdisciplinary teaching be about, and how can students critically engage with competing values of practice?

What transformational knowledge can be anchored and taught in (2) *degree programs* in a planned and lasting way? To what extent can the curriculum address real-world problems that spring from different conjunctures? At what point does it become arbitrary and lose its effectiveness?

Can real-world problems, their respective context, and their (possible) solution processes be taken up in the framework of (3) a *curriculum* in such a way that teaching and learning, and the necessary reflections of students and teachers on the subject matter, take place meaningfully? How often can and must learning processes go through a cyclical structure of (4) *iterating actions and reflections* to achieve a meaningful learning outcome or acquisition of skills? How do students acquire the ability to overcome barriers and boundaries created by different (5) *professional languages* and cultures? To what extent can students protect themselves from capture by (6) *stakeholder interests* in the field? How can they be protected? Is academic (7) *socialization* made more difficult by the fact that students are involved in practice-oriented research projects? How can consciousness be raised about (8) *professional cultures* and their limits? How can the distance between the academic world and everyday contexts be established in such a way that learning does not come under increasing pressure to make premature or even (9) *erroneous conclusions*?

These questions point to a profound process of change in academic conceptions of education, insofar as transdisciplinarity becomes more prominent for learning processes. To be sure, some voices will continue to warn of disintegration and threats to academia – and their arguments will be cogent, like those of critics of the Mode 2 diagnosis. Nevertheless, the opportunities that become attainable with the transformations of learning should also become apparent. For scientific research is, at its best, transformative learning, whose participants not only find solutions, but also learn to consciously shape research and knowledge processes, which ultimately contribute to the democratization of the sciences.

References

Arnold, Markus, and Roland Fischer, eds. 2004. *Disziplinierungen. Kulturen der Wissenschaft im Vergleich.* Wien: Turia & Kant.

Bachelard, Gaston. [1949] 1998. *Le rationalisme appliqué.* 3rd edition. Paris: Presses Universitaires de France.

Balsiger, Jörg. 2015. Transdisciplinarity in the class room? Simulating the co-production of sustainability knowledge. *Futures* 65: 185–94.

Bartunek, Jean M. (2011). What has happened to Mode 2? *British Journal of Management* 22 (3): 555–58.

Bell, Daniel. 1973. *The coming of post-industrial society.* New York: Basic Books.

Bender, Gerd, ed. 2001. *Neue Formen der Wissenserzeugung.* Frankfurt am Main: Campus.

Berger, Peter, and Thomas Luckmann. 1966. *The social construction of reality.* London: Penguin.

Bernstein, Jay H. 2015. Transdisciplinarity: A review of its origins, development, and current issues. *Journal of Research Practice* 11 (1): 1–20.

Birrer, Frans. 2001. Combination, hybridisation and fusion of knowledge modes. In *Neue Formen der Wissenserzeugung*, ed. Gerd Bender, 57–68. Frankfurt am Main: Campus.

Bora, Alfons. 2005. Rezension: Helga Nowotny, Peter Scott und Michael Gibbons: Wissenschaft neu denken. Wissen und Öffentlichkeit in einem Zeitalter der Ungewissheit. *KZfSS Kölner Zeitschrift für Soziologie und Sozialpsychologie* 57 (4): 755–57.

Bourdieu, Pierre. 1979. *Entwurf einer Theorie der Praxis auf der ethnologischen Grundlage der kabylischen Gesellschaft* (Corula Pialoux and Bernd Schwibs, trans.). Frankfurt am Main: Suhrkamp.

Carayannis, Elias G., and David. F. Campbell. 2012. *Mode 3 knowledge production in quadruple helix innovation systems.* New York: Springer.

Carnap, Rudolf. 1931. Die physikalische Sprache als Universalsprache der Wissenschaft. *Erkenntnis* 2: 432–65.

Deleuze, Gilles, and Félix Guattari. 1987. *A thousand plateaus: Capitalism and schizophrenia.* Minneapolis: University of Minnesota Press.

Drori, Gili. S., John W. Meyer, and Hwang, Hokyu, eds. 2006. *Globalization and organization: World society and organizational change.* Oxford: Oxford University Press.

Drucker, Peter F. 1969. *The age of discontinuity.* New York: Harper & Row.

Etzkowitz, Henry, and Loet Leydesdorff. 2000. The dynamics of innovation: From national systems and "Mode 2" to a triple helix of university–industry–government relations. *Research Policy* 29 (2): 109–23.

Fleck, Ludwik. [1935] 1980. *Entstehung und Entwicklung einer wissenschaftlichen Tatsache.* Frankfurt am Main: Suhrkamp.

Frederichs, Günther. 2001. Mode 2 und Erkenntnis. *Neue Formen der Wissenserzeugung,* ed. Gerd Bender, 69–82. Frankfurt am Main: Campus.

Gibbons, Michael, Camille Limoges, Helga Nowotny, Simon Schwartzman, and Peter Scott. 1994. *The new production of knowledge: The dynamics of science and research in contemporary societies.* London: Sage.

Gläser, Jochen. 2001. Modus 2a und Modus 2b. In *Neue Formen der Wissenserzeugung,* ed. Gerd Bender, 83–99. Frankfurt am Main: Campus.

Hack, Lothar. 2001. "Ich habe da eine Theorie" oder: Neue Fokussierung von Kontext/en und Kompetenz/en. In *Neue Formen der Wissenserzeugung,* ed. Gerd Bender, 23–56. Frankfurt am Main: Campus.

Haraway, Donna. 1988. Situated knowledges: The science question in feminism and the privilege of partial perspective. *Feminist Studies* 14 (3): 575–99.

Harding, Sandra. 1986. *The science question in feminism.* Ithaca: Cornell University Press.

Haug, Wolfgang Fritz. 2004. Parteilichkeit und Objektivität. *Das Argument* 255: 207–26.

Holtgrewe, Ursula. 2012. Gibt es die public domain? Institutionen und ihre Grenzen in der Wissensgesellschaft. In *Soziale Ungleichheit, kulturelle Unterschiede: Verhandlungen des 32. Kongresses der Deutschen Gesellschaft für Soziologie in München,* ed. Karl-Siegbert Rehberg, 213–28. Frankfurt am Main: Campus

Jahn, Thomas, Matthias Bergmann, and Florian Keil. 2012. Transdisciplinarity: Between mainstreaming and marginalization. *Ecological Economics* 79: 1–10.

Jasanoff, Sheila. 2003. Breaking the waves in science studies. *Social Studies of Science* 33 (3): 389–400.

Klein, Julie Thompson, Walter Grossenbacher-Mansuy, Rudolf Häberli, Alain Bill, Roland W. Scholz, and Myrtha Welti, eds. 2001. *Transdisciplinarity: Joint problem solving among science, technology, and society. An effective way for managing complexity.* Basel: Birkhäuser.

Knorr-Cetina, Karin. 1999. *Epistemic cultures: How the sciences make knowledge.* Cambridge, MA: Harvard University Press.

Knorr-Cetina, Karin. 2013. *The manufacture of knowledge: An essay on the constructivist and contextual nature of science.* London: Pergamon.

Kuhn, Thomas S. 1962. *The structure of scientific revolutions.* Chicago: University of Chicago Press.

Langemeyer, Ines. 2015. *Das Wissen der Achtsamkeit. Kooperative Kompetenz in komplexen Arbeitsprozessen.* Münster: Waxmann.

Langemeyer, Ines. 2021. Optimierung von Arbeits-, Lern-und Vergesellschaftungsprozessen mittels KI–Anmerkungen aus psychologischer und pädagogischer Sicht. In *Optimierung: Anschlüsse an den 27. Kongress der Deutschen Gesellschaft für Erziehungswissenschaft*, eds. Sandra Hofhues, Elke Kleinau, and Henrike Terhart, 231–48. Opladen: Barbara Budrich.

Latour, Bruno, and Steve Woolgar. 1979. *Laboratory life: The construction of scientific facts.* Princeton: Princeton University Press.

Lave, Jean, and Etienne Wenger. 1991. *Situated learning: Legitimate peripheral participation.* Cambridge: Cambridge University Press.

Liyanage, Shantha I. H., and Fulu G. Netswera. 2022. Greening universities with Mode 3 and quintuple helix model of innovation–production of knowledge and innovation in knowledge-based economy, Botswana. *Journal of the Knowledge Economy* 13 (2): 1126–56.

Machlup, Fritz. 1962. *The production and distribution of knowledge.* Princeton: Princeton University Press.

Marx, Karl. [1845] 1995. *Theses ad Feuerbach.* Marx/Engels Internet Archive. Available from https://www.marxists.org/archive/marx/works/1845/theses/theses.pdf.

Marx, Karl. [1867] 1996. *Marx & Engels collected works.* Vol. 35: *Karl Marx Capital.* Chadwell Heath: Lawrence & Wishart.

Müller, Roland. 2009. The notion of a model: A historical overview. *Philosophy of technology and engineering sciences*, eds. Anthonie Meijers and Dov Gabbay, 637–64. Amsterdam: Elsevier.

Langemeyer, Ines. 2021. Modus 2. In *Handbuch Transdisziplinäre Didaktik*, eds. Tobias Schmohl and Thorsten Philipp, 185–94. Bielefeld: transcript.

Nerland, Monika. B. 2012. Professions as knowledge cultures. In *Professional learning in the knowledge society*, eds. Karen Jensen, Leif Christian Lahn, and Monika Nerland, 27–48. Amsterdam: Brill.

Nowotny, Helga. 1993. A new branch of Science, Inc. In *Science, politics, and morality: Scientific uncertainty and decision making*, ed. René Von Schomberg, 63–84. Berlin: Springer.

Nowotny, Helga. 1999. *Es ist so. Es könnte auch anders sein.* Frankfurt am Main: Suhrkamp.

Nowotny, Helga, Peter Scott, and Michael Gibbons. 2001. *Re-thinking science: Knowledge and the public in an age of uncertainty.* Cambridge: Polity.

Nowotny, Helga, Peter Scott, and Michael Gibbons. 2003. Introduction: "Mode 2" revisited: The new production of knowledge. *Minerva* 41 (3): 179–94.

Polanyi, Michael. 1962. *Personal knowledge: Towards a post-critical philosophy.* London: Routledge.

Rheinberger, Hans-Jörg. 1994. Experimentalsysteme, Epistemische Dinge, Experimentalkulturen. Zu einer Epistemologie des Experiments. *Deutsche Zeitschrift für Philosophie* 42 (3): 405.

Rheinberger, Hans-Jörg. 2005. Gaston Bachelard and the notion of "phenomenotechnique". *Perspectives on Science* 13 (3): 313–28.

Rip, Arie. 1997. A cognitive approach to relevance of science. *Social Science Information* 36: 615–40.

Schatzki, Theodor R., Karin Knorr-Cetina, and Eike von Savigny, eds. 2000. *The practice turn in contemporary theory.* London: Routledge.

Schauz, Desiree. 2014. Wissenschaftspolitische Sprache als Gegenstand von Forschung und disziplinärer Selbstreflexion. Das Programm des Forschungsnetzwerks CASTI. *Forum Interdisziplinäre Begriffsgeschichte* 3 (2): 49–61.

Scholz, Roland W. 2020. Transdisciplinarity: Science for and with society in light of the university's roles and functions. *Sustainability Science* 15: 1033–49.

Schütz, Alfred. 1971. *Gesammelte Aufsätze I. Das Problem der sozialen Wirklichkeit.* Den Haag: Martinus Nijhoff.

Snow, Charles. P. [1959] 1990. Two cultures. *Leonardo* 23 (2–3): 169–73.

Suchman, Lucy. 1987. *Plans and situated actions.* New York: Cambridge University Press.

Van Orman Quine, Willard. 1951. Two dogmas of empiricism. *The Philosophical Review* 60: 20–43.

Weber, Max. 1934. *Wissenschaftslehre.* Tübingen: Mohr.

Weingart, Peter. 1983. Verwissenschaftlichung der Gesellschaft – Politisierung der Wissenschaft. *Zeitschrift für Soziologie* 12 (3): 225–41.

Weingart, Peter. 1997. From "finalization" to "Mode 2": Old wine in new bottles? *Social Science Information* 36 (4): 591–613.

Participatory Action Research

Loni Hensler, Gerardo Alatorre Frenk, and Juliana Merçon

Definition

The three basic concepts that constitute participatory action research refer to the direct and active interaction (participation) required for the construction of knowledge (research) and collective practices (action). The adjective *participatory* comes from Latin *participationem* and means "sharing, having a share or part in common with others". The noun *action* derives from Latin *actionem*: "a putting in motion; a performing, a doing; lawsuit, legal action", from the past-participle stem of *agere* "to do". And *research* comes from French *recerche* (1530s), meaning "act of searching closely" for a specific person or thing (Harper 2023). As a research approach, participatory action research focuses on the co-construction of knowledge and action by integrating multiple perspectives to strengthen the social transformation of unjust structures. It is a political-epistemic research paradigm that originated in Latin America, based on the work of Orlando Fals Borda (1970) and Paulo Freire (1970). Participatory action research can be characterized by three main principles: (1) it is geared towards transforming injustices; (2) as they are directly involved in sociopolitical processes, researchers become activists, intertwining a constant dynamic of reflection and action; and (3) social actors become co-researchers, identifying needs and potential problems of study, information-gathering, analysis, and decision-making.

The methodology of participatory action research is cyclical, recursive, transdisciplinary, and transformative. It assumes critical reflection and dialogue in practice as the basis for research and action. This implies that people are not considered objects to be studied, but subjects, co-actors, and co-constructors of knowledge and actions from their diverse perspectives. This also includes co-constructing the different elements of the process (analysis of the problem, definition of objectives, agenda, planning, organization, etc.) that open spaces for collective analysis and interpretation as the research develops. Participatory action research offers powerful methods for collaborative research with a focus on power relations and transforming unjust and unsustainable circumstances.

Background

Kurt Tsadek Lewin (1946) was the first person to use the term "action research" to refer to a type of research that seeks both to understand the conditions and effects of various forms of social action, and to lead to transformative collective action. Lewin's epistemological proposal already contained some of the principles of what would later become participatory action research, such as the direct involvement of researchers in the processes that are studied and the use of multiple methods to promote democratic interactions.

Almost unrelated to Lewin's approach, in the early 1970s, several experiences around the world linked academic work and social action aimed at transforming oppressive and unjust circumstances (Rahman and Fals Borda 1992). Such was the case of the *Bhoomi Sena* (English: Earth Army) in Maharashtra, India, where social scientist Kaluram Dhodade, who formulated the principles of participatory action research (Rahman 1984), was involved in peaceful land seizures. In the village of Bunju, Tanzania, anthropologist Marja-Liisa Swantz's (1975) participatory immersion became a referent for alternative research in Africa and beyond. In Latin America, research for social transformation had important educational and cultural movements as allies. Paulo Freire's (1970) popular education, based on horizontal and emancipatory forms of learning, was key to the political-epistemological turn of the emerging participatory action research. Similarly, the work of Guillermo Bonfil Batalla (1987) and Pablo González Casanova (2004) reoriented the directions of social sciences by critiquing academia's colonialist practices (Fals Borda 1999).

In order to go beyond Lewin's social psychology and liberal theories of participation, Colombian sociologist Orlando Fals Borda (1970, 1990) focused on the development of participatory action research as a process capable of bringing together, on the one hand, the systematicity of science and, on the other, the knowledge and actions of marginalized populations. To this end, Fals Borda (1999) mentions three key features of participatory action research.

1. *Relations between science and ethics*: Participatory action research is based on the convergence of popular, traditional, and scientific thinking to support just causes. For ethical reasons, priority is given to working with the most vulnerable and marginalized groups. Fals Borda (1999, 77–78) warns that the claim of neutrality and absolute objectivity coincides, often unintentionally, with a position that supports the status quo or hegemonic social order.
2. *Dialectics between theory and practice*: One of the basic principles of participatory action research is that knowledge is validated by the improvement of practice. However, for Fals Borda (1999, 78–80), the focus on praxis, which is the reflection on practice, does not imply that other methods and relevant validity

criteria are not necessary to ensure that the knowledge generated is both rigorous and relevant.

3. *Encounter between subject and subject*: Fals Borda (1999, 80–81) argues that the subject–subject bond as *sentipensantes* (English: feeling-thinking beings), whose diverse knowledge and points of view are considered together, is what makes it possible to define authentic participation as different from manipulative and instrumental forms. This principle of horizontality is reflected in the agreements generated around the questions to be answered and acted upon; in the co-construction of instruments and methods; in the systematic feedback by diverse and accessible means; and in shared action (Fals Borda 1987). In Latin America and Spain, Oscar Jara (2012) and Tomás Rodríguez Villasante (2006) formulated some of the principles most in line with Fals Borda's. Jara put forward a systematization of experiences, an approach for critical analysis and reproduction of experiences from social movements. Rodríguez Villasante and the International Observatory for Citizenship and Sustainable Environment (Observatorio Internacional de Ciudadanía y Medio Ambiente Sostenible, CIMAS) put forward *sociopraxis*, an approach based on participatory methodologies and contributions from network theory, popular environmentalism, and the paradigms of complexity.

Debate and criticism

Nowadays, there are numerous streams that practice and theorize about participatory action research (Bradbury 2015; Dick 2011; Kindon et al. 2007). Some are developed in rural and urban community contexts, close to the roots of participatory action research, while others focus on organizations, educational institutions, and even companies. As it fundamentally conducts research together with people, participatory action research is highly contextual – it is influenced, reconfigured, conditioned, and reinvented within its specific context. Methodology plays a central role, as it determines the forms of participation, the co-construction of knowledge and transformation through praxis. Even though a participatory action research process starts with a methodological proposal and objective, it is redefined and reconfigured during the process, in constant deliberation with the actors and considering the conditions to be transformed.

Current streams are distinguished by a different distribution of weight between the triad research–action–participation. While some focus more on doing participatory research with less practice and shared advocacy (generally driven and sustained by academia), others focus more on reflexive activism from the social base (mostly driven by social movements and civil society organizations).

Another distinguishing element between different streams and practices of participatory action research is participation, due to the central role it plays in the co-construction of knowledge and actions through praxis. In this way, it opposes the dominant forms in which the coloniality of knowledge persists (Lander 2000) and only researchers have the power to determine the problem, agenda, subjects, or objects of research, and the forms of relationship, interpretation, and use of information. In contrast, participation in which all dimensions of a process – from defining the objectives to the implementation of the research process – are decided collectively has the potential to be co-emancipatory, as it changes the power relations between civil society and academia, as well as stimulating transformative learning and knowledge situated in a concrete context, relevant to the people involved. In Freire's words: "The more women and men take an active stance in thematic research, the more they deepen their awareness of reality and make it their own" (1970, 90, own translation).

Participation is a concept that embraces highly diverse practices, which have been categorized by some authors as a ladder (Arnstein 1996) or a wheel (Davidson 1998) of participation. This elasticity in the notion of participation implies that there are participatory action research processes that have lost their emancipatory and transformative nature; some authors even speak of a colonization of participatory action research (Cascante Fernández 2013) as it becomes instrumental within the current system without a critical stance towards the structures and aspirations for its transformation. To distinguish it from approaches with a more instrumental participation, some researchers prefer to change the word participatory for adjectives that highlight the emancipatory aspect of this type of action research. Collaborative Action Research (CARE) seeks collaboration in all aspects of the research process (Casals et al. 2008; García Eiroá and Trigo 2000; Hensler 2023). Collaborative action research brings together diverse people based on the dialogue between different knowledges and practices that generates collective and solidary actions towards a shared goal (Hensler 2023). Critical or transformative action research emphasizes its political character and the need for critical analysis of history and structures in order to transform them (Cascante Fernández 2013).

Participatory action research as a theoretical-practical and epistemological approach has been the target of criticism from more conventional research strands. The following four appear particularly relevant.

1. *Imperative of transformation:* One of the pillars of participatory action research is its focus on transformation, which involves a partnership with marginalized groups and a constant politicization of research. This imperative of transformation has been criticized as saviorism, as well as for the irrelevance of continuing to think of the world in terms of oppressors and oppressed. Today, participatory action research does not focus exclusively on working with mar-

ginalized groups, and it searches to form links with other sectors in a transdisciplinary participatory action research.

2. *Lack of rigor and objectivity*: Because of how engaged researchers are with organizations and processes, there is a highly contextualized and involved vision of the processes that allows them to see certain elements from the inside, but possibly prevents them seeing others that are visible at a distance. Furthermore, the experience-based approach limits the scope to a local or regional level, which cannot be generalized. In addition to this, the methodologies used by participatory action research have been criticized for not meeting scientific, rigorous standards, as they are not experiments that can be replicated but unique experiences and processes under construction, with changing objectives and emergent social learning. Working in a participatory or collaborative manner does not imply that it cannot be orderly and disciplined in keeping records and analyzing. Moreover, tools have been established to balance high subjectivity in the processes, such as methodological, theoretical, and observant triangulation (Arias 2007).
3. *Slowness of participatory action research processes*: Another criticism focuses on the time involved in these processes, as they are medium- to long-term collaborations that, at their foundation, require trust, common languages, and shared goals, among other things. The time required is often in tension with the time marked by conventional academia, which is why participatory action research is criticized as a utopian approach that is unsuitable in current structures.
4. *Forcing participatory methodologies*: From the perspective of decolonial and critical approaches, participatory action research is at risk of imposing methods and forms in certain cultural contexts, insensitive to the different sociocultural realities (Merçon 2018). Cooke and Kothari (2001) titled their book *Participation: The New Tyranny?* as a criticism of the concept and practice that imposes forms of interaction in order to legitimize or fulfill institutional requirements, without recognizing Indigenous and peasant communities' own forms of reflection and action.

Current forms of implementation in higher education

Participatory action research is showing its epistemic and political potential in various higher education experiences. Clear examples of this are three graduate programs at Latin American universities: the Master of Arts in Education for Interculturality and Sustainability at Universidad Veracruzana, Mexico (Maestría en Educación para la Interculturalidad y la Sustentabilidad, MEIS 2019); the Professional Masters in Sustainability together with Peoples and Traditional Lands at the University of Brasilia, Brazil (Mestrado Profissional em Sustentabilidade

Junto a Povos e Terras Tradicionais, MESPT 2019); and the Master of Arts in Interdisciplinary Development Studies at the Universidad del Cauca, Colombia (Maestría en Estudios Interdisciplinarios del Desarrollo; Universidad del Cauca 2022).

The three programs assume the ethical-political and methodological approach of participatory action research. They take the students' knowledge and practices as a starting point and provide tools to deepen critical and self-critical reflection on these practices. In the projects developed by students, research has the function of generating knowledge from and for collective action. The point of departure and arrival is collaboration with communities and collectives mobilized for a fairer, more sustainable world that is open to cultural diversity. Through these links, students learn to dynamize dialogues of knowledge and to establish a constant back-and-forth between transformative action and analytical reflection, in order to enhance the scope of organizations, in terms both of knowledge and of social change and strengthening of collective subjects.

The aim of these graduate courses is not to generate neutral or universal knowledge; students make their positioning explicit and produce situated knowledge with their projects, which acquire meaning for the collective actions that are developed in a given area. Participatory action research thus appears as a strategy for constructing shared meanings in the understanding and practical approach to the processes and problems being experienced, nurturing community and citizen self-management processes, and contributing to the construction of "a world where all worlds fit", to use the expression created by Zapatista communities in southeast Mexico.

The master's program in Sustainability with Traditional Peoples and Territories focuses especially on working with Indigenous, Afro-descendant, and other traditional communities, but it also includes other actors, as do the other two master's degrees (Nogueira and da Silva 2019). The three programs are located in territories characterized by a rich biodiversity and cultural diversity (Mexico, Brazil, and Colombia), and this same richness exists among the students and the processes and people they are linked with – Indigenous, peasants, and Afro-descendants, as well as urban communities and collectives of very different kinds.

These master's degrees share an interest in the flourishing of ontological pluralism, epistemic justice, and transdisciplinarity. They recover ancestral practices and knowledge, and attach great importance to original cosmovision and communalities and to approaches of complexity as indispensable orientations within the challenges of the current civilizational crisis, where the ecosystems that sustain life are deteriorating rapidly and violence and social asymmetries are increasing. The aim is to transform power relations by favoring dialogues and collaboration between actors from different cultural backgrounds, cultivating, as the website of the master's degree in Interdisciplinary Development Studies points out, "new imaginaries that dignify the economy, justice, health and life in general" (Universidad del Cauca 2022, own translation). In the case of the Education for Interculturality and Sustainability program, the objective is to "move towards a perspective of sustainability that implies the conservation and maintenance of the vital cycles of ecosystems, as well as the understanding and appreciation of the close interrelationship that many communities and peoples maintain with nature in all its cultural expressions" (Merçon and Alatorre 2019, 147).

In addition to the decolonial and anti-capitalist approach, there is an anti-patriarchal positioning, which not only implies building gender equity but also strengthening relational ontologies. Diverse types of knowledge come into play, including those that come from sensitivity, sensoriality, emotions, and bodies. The defense of territories begins with caring for bodies; the common goods that we need to protect are both natural goods and the networks of care that we weave on a daily basis.

At the pedagogical and didactic levels, these study programs share many elements, such as how mandatory theoretical subjects, elective courses, and professional internships complement one another. However, each program has specific features. For example, in the Master in Interdisciplinary Development Studies, students and teachers participate together in the so-called "Tramas y Mingas para el Buen Vivir" (Weavings and Mingas for Good Living), where they recover knowledge and practices (traditional or not) that contribute to the cultivation of more harmonious relations among human beings, and between human beings and the rest of nature. The Master in Sustainability together with Peoples and Traditional Lands works with what it calls the Pedagogy of Alternation, "a teaching–learning methodology that combines different formative experiences distributed over different times and places: University Time, which involves carrying out place-based activities in an academic environment, and Community Time, which is carried out (preferably) in a community environment or in the environment where the student works professionally" (MESPT 2019, 16, own translation).

The Education for Interculturality and Sustainability program works with people who are already collaborating in collectives and organizations involved in socio-environmental transformation processes in rural or urban communities (Universidad Veracruzana 2019). The Motor Group, a key element of any participatory action research process, is formed by colleagues from these organizations who are interested in contributing to the reflection–action process. In this way, each student works by linking at least two epistemic communities: the organization with which they are linked and the master's degree.

In general, these graduate programs not only train individual professionals, but also strengthen analytical and political capacities in the collectives and organizations with which they are linked. The impact of student projects is based on the work of the organizations they collaborate with, but is projected as broadly as possible, extending networks through which knowledge and solidarity circulate.

There are also experiences in other parts of the world that have been inspired by participatory action research. For example, in Spain, the International Observatory for Citizenship and Sustainable Environment (Red CIMAS 2022) promotes participatory democracy processes through teaching and the facilitation of participatory processes that integrate different types of knowledge coming from local communities, government, and academia.

In an era marked by overlapping crises and where it can be difficult to visualize solutions, these initiatives cultivate hope, emphasizing the real possibility of changing the world, planting seeds for the future in the daily life of groups, communities, organizations, and institutions. Experiences such as those presented here allow us to see, in a tangible way, the contributions of participatory action research in the 21st century.

References

Arias, Miguel Alvarado. 2007. José Martí y Paulo Freire: aproximaciones para una lectura de la pedagogía crítica. *Revista Electrónica de Investigación Educativa* 9 (1): 1–19.

Arnstein, Sherry. 1969. A ladder of citizen participation. *Journal of the Royal Planning Institute* 35 (4): 216–24.

Bonfil Batalla, Guillermo. 1987. *El México Profundo, una civilización negada.* Mexico: Grijalbo.

Bradbury, Hilary, ed. 2015. *The Sage handbook of action research.* London: Sage.

Casals, Albert, Mercè Vilar, and Jaume Ayats. 2008. La Investigación-Acción Colaborativa: Reflexiones metodológicas a partir de su aplicación en un proyecto de Música y Lengua. *Revista Electrónica Complutense de Investigación en Educación Musical* 5 (4): 1–17.

Cascante Fernández, César. 2013. La investigación-acción crítica y nosotros (que te quisimos tanto). *Revista Interuniversitaria de Formación del Profesorado* 27 (2): 45–63.

Cooke, Bill, and Uma Kothari. 2001. *Participation: The new tyranny?* London: Zed.

Davidson, Scott. 1998. Spinning the wheel of empowerment. *Planning* (3): 14–15.

Dick, Bob. 2011. Action research literature 2008–2010: Themes and trends. *Action Research* 9 (2): 122–43.

Fals Borda, Orlando. 1970. *Ciencia propia y colonialismo intelectual.* Mexico City: Editorial Nuestro Tiempo.

Fals Borda, Orlando. 1987. The application of participatory action-research in Latin America. *International Sociology* 2 (4): 329–47.

Fals Borda, Orlando. 1990. El Tercer Mundo y la reorientación de las ciencias contemporáneas. *Nueva Sociedad* (107): 169–81.

Fals Borda, Orlando. 1999. Orígenes universales y retos actuales de la IAP. *Análisis Político* (38): 73–90.

Freire, Paulo. 1970. *Pedagogía del oprimido.* Montevideo: Tierra Nueva.

García Eiroá, Jesús, and Eugenia Trigo. 2000. Investigación Colaborativa y Formación de Universitarios. *Revista de Educación* (323): 289–318.

González Casanova, Paulo. 2004. *Las nuevas ciencias y las humanidades: de la academia a la política.* Barcelona: Anthropos.

Harper, Douglas, ed. 2023. *Online etymology dictionary.* Available from http://etymonline.com.

Hensler, Loni. 2023. *Territorios en movimiento. Un análisis de procesos participativos para una gestión colaborativa del territorio en Xalapa, México.* Available from https://ru.dgb.unam.mx/handle/DGB_UNAM/TES01000832823.

Jara, Oscar. 2012. *La sistematización de experiencias: práctica y teoría para otros mundos posibles.* San José: Alforja.

Kindon, Sara, Rachel Pain, and Mike Kesby, eds. 2007. *Participatory action research approaches and methods: Connecting people, participation and place.* London: Routledge.

Lander, Edgardo, ed. 2000. *La colonialidad del saber: Eurocentrismo y ciencias sociales. Perspectivas latinoamericanas.* Buenos Aires: Clacso.

Lewin, Kurt. 1946. Action research and minority problems. *Journal of Social Issues* 2 (4): 34–46.

Merçon, Juliana. 2018. Participatory action research and decolonial studies. Critical mirrors. Decolonial education in the Americas: Lessons of resistance, pedagogies of hope. *Latin American Philosophy of Education Journal* (3): 20–29.

Merçon, Juliana, and Gerardo Alatorre. 2019. Educação, Interculturalidade e Sustentabilidade. Uma experiência de pós-graduação no México. *Interethnic@ – Revista de Estudos em Relações Interétnicas* 22 (1): 142–61.

MESPT [Mestrado Profissional em Sustentabilidade junto a Povos e Territórios Tradicionais], ed. 2019. *Projeto Político Pedagógico de Curso.* Available from http://www.mespt.unb.br/images/Documentos/PPPC_final.pdf.

Nogueira, Mônica Celeida Rabelo, and Ana Tereza Reis da Silva. 2019. *Diálogo de Saberes na Pós-Graduação: relato da experiência do Mestrado em Sustentabilidade junto a Povos e Territórios Tradicionais.* Brasilia: Forges, IFB, Universidade de Brasilia

Rahman, M. Anisur. 1984. Asian rural workers' group develop own grassroots methodology. *Convergence* 17 (2): 34.

Rahman, M. Anisur, and Orlando Fals Borda. 1992. La situación actual y las perspectivas de la investigación-acción participativa en el mundo. In *La investigación-acción participativa: inicios y desarrollos*, ed. María Cristina Salazar, 205–30. Madrid: Editorial Popular.

Red CIMAS, ed. 2022. *Observatorio Internacional de Ciudadanía y Medio Ambiente Sostenible.* Available from https://www.redcimas.org.

Swantz, Marja-Liisa. 1975. Research as an educational tool for development. *Convergence* 8 (2): 44.

Universidad del Cauca, ed. 2022. *Master in Interdisciplinary Development Studies.* Available from https://www.unicauca.edu.co/posgrados/programas/maestria-en-estudios-interdisciplinarios-del-desarrollo.

Universidad Veracruzana, ed. 2019. *Plan de Estudios.* Available from https://www.uv.mx/meis/files/2020/06/Plan-de-Estudios-MEIS-2019_web.pdf.

Villasante, Tomas Rodriguez. 2006. *Desbordes creativos: Estilos y estrategias para la transformación social.* Madrid: La Catarata.

Performative Knowledge

Karen van den Berg and Stephan Schmidt-Wulffen

Definition

The term *performative knowledge* implies two different types of discourse; one relates to performativity, while the other concerns epistemology. John Langshaw Austin differentiated the performative function of language from its propositional one: those who say "Yes" during a marriage ceremony *do* marry but they do not *report* on the marriage (Austin 1962, 12). And he remarked that a performative speech act can never be false or true. While Austin used *performative* only for language, by the end of the 1980s, Judith Butler connected it to bodily actions. For Butler, gender is not founded biologically or even ontologically but results from specific social actions (Butler 1988). Performative acts constitute social practice. As Andreas Reckwitz noted, practices are not only bodily behavior but, at the same time, "sets of mental activities" (Reckwitz 2002, 251) – routinized ways of understanding the world. Practices imply "implicit knowledge" (Polanyi 2009). In contrast to propositional knowledge, "performative knowledge" is always embedded in actions. Therefore Donald Schön (1983) also called it "reflection-in-action", highlighting the fact that the seemingly spontaneous acts embody rules but do not refer to a level of reflection anticipating the action. Schön spoke of it also as an "art" (1983, 130). Due to the importance of contingency and the impact of medium and material, performative knowledge is closely related to aesthetic practices.

Background

Alfred Julius Ayer (1952), Ludwig Wittgenstein (1953), and Gilbert Ryle (1959) have contributed significantly to the understanding of knowledge as an activity. Pierre Bourdieu and Anthony Giddens have focused on knowledge inherent in everyday practices (Bourdieu 1977; Giddens 1984). In science, the "pragmatic turn" (Bernstein 2010) redefined "truth". The proposition, when seen as action, was qualified in terms of its potential and possible impact. Knowledge claims were regarded as a product of practice in a community of inquirers, also as fallible, to be improved through continuous testing in action (Hacking 1983; Rheinberger 2010). The laboratory studies of Karin Knorr-Cetina, Bruno Latour, and Hans-Jörg Rheinberger have proven the pragmatic basis of scientific knowledge (Knorr-Cetina 1984; Latour 1988; Rheinberger 2010). As a result, the development from a propositional to a performative concept of knowledge has replaced homogeneity with heterogeneity, absoluteness with contingency, and "academic" with "transdisciplinary" (Gibbons et al. 1994; Schatzky et al. 2001).

The "pragmatic turn" in the sciences also aroused interest in artistic concepts, processes, and forms of production at a time when process-based and ephemeral art forms such as performance and video were becoming paradigmatic within the visual arts. The "dematerialization of the artwork" (Lippard 1973) had already led to a developing rapprochement of aesthetic practice with philosophy and the sciences from the 1960s. Practice-based artistic research – the methodological basis of the new artistic PhDs – soon became part of the repertoire of disciplines such as anthropology or sociology. Scholars confronted the performativity of the humanities and science and encountered the arts as a model. Initially, the "self-commissioning" of modern artists inspired management theory (Boltanski and Chiapello 2007). Peculiarities of aesthetic practice, such as the spontaneous emergence of results and the distinctive nature of materials in the creative process, deepened the understanding of a pragmatic approach to knowledge (Fischer-Lichte 2008).

Performative knowledge presupposes a community of practice (Lave and Wenger 1991), as part of which the learners can observe, compare, act, repeat, and correct themselves. However, a performative understanding of knowledge and knowledge production poses a specific pedagogical problem: while propositional knowledge combines with a teaching concept of explanatory mediation, in which – as Rancière puts it – one intelligence is subordinated to another, performative knowledge does not emerge in this way (Rancière 1991). Here mediation needs demonstration and showing; an already practiced behavior in everyday social life that the "apprentice" observes and that the teacher displays and stages. If performative knowledge is taught, it requires teaching formats that enable active participation in the respective practice. Here, *criticality* might be considered differently: the practice does not imply a reflection in terms of an abstract propositional examination of habitual frames. Transformation does not result from a specific failure that retroacts to a set of mental dispositions. On the contrary, *change* is understood as a central component of practice in general so that learning – without interrupting practice – becomes a continuous element of everyday life.

Debate and criticism

Despite the epistemic change towards a performative understanding of science and its intense theoretical debate, there were few attempts by universities to adopt curricula or to develop a specific pedagogy. The Center for Advanced Visual Studies at the Massachusetts Institute for Technology in Cambridge (US), founded by the artist and designer György Kepes in 1967, invited artists to inspire scientists early on in its existence (Shulman 2017). Faced with a growing demand for creativity and entrepreneurship through the growing importance of immaterial labor, business schools, such as Copenhagen Business School, were expanding their academic curricula in the mid-1990s (Copenhagen Business School 2021). Like many other business schools during this period, Copenhagen Business School imagined the artist as an entrepreneur, a role model for the future businessperson (Guillet de Monthoux 2004). To learn more about artistic practice, artists were invited for dialogue. The encounter with performative knowledge was organized through dialogue and architectural conditioning of experiences (Guillet de Monthoux and Wikberg 2021). Critics argued repeatedly that such an instrumentalization of artistic experience comes at the expense of the very nature of the artistic practice and its autonomy (Holert 2020; Osborne 2014). The interest of Copenhagen Business School in learning from artists, promoted especially by professors like Pierre Guillet de Monthoux, was inspired by Witten/Herdecke University in Germany, founded in 1982 (Guillet de Monthoux 2004, 251). When Witten/Herdecke introduced a *Studium fundamentale*, it also offered practical courses in theater, creative writing, photography, film making, and choral, orchestral, piano, and chamber music, in addition to rhetoric, philosophy, etc. as part of a business studies and a medical studies program to allow students to pursue other interests and place their major discipline in new contexts (van den Berg and Landkammer 2002). Similar to Mezirow's (1978) concept of transformative learning this program promoted active participation of students in a practice instead of "explaining" it – as a more traditional understanding of university training would have it.

These early examples were developed primarily in the context of alternative management training programs, and with no elaborate debate on the didactics of higher education developing from their implementation. It was not until the turn of the millennium, however, that programs merging theory and practice began to professionalize and develop specific methods. In this, two major trends can be observed: On the one hand "design thinking" was inspired by the design practices. It spread rapidly from *d.school* at Stanford University and the management programs developed there (Lawson 2005). On the other hand, there were approaches that emerged from an experimental philosophy, aesthetics, and art practice, such as the *SenseLab* at Concordia University in Canada established in 2004 (Manning 2020; Manning and Massumi 2014).

However, these approaches remain rather marginal for the time being. The promise of a performative concept of knowledge was its innovative dimension (Razzouk and Shute 2012). Although design thinking became a fashion in the fields of management and leadership around the turn of the millennium, its training of performative competencies and creativity techniques was repeatedly criticized as difficult to objectify and offering limited measurability in its effects (Rotherham and Willingham 2009). On the other hand, artists and designers critiqued that the transfer of designerly and artistic practices to further education management runs the risk of its vulgarization or even standardization (van den Berg and Schmidt-Wulffen 2015).

A major problem arising from the transmission of performative knowledge to other practices was in its processes of critical reflection. The kind of meta-reflection called for within the university context poses a specific pedagogical problem: how can one experience the performativity of knowledge without losing its performative character? And without arriving back in the field of propositional knowledge?

Current forms of implementation in higher education

A new dimension of the implementation of performative knowledge can be observed only at the beginning of the second decade of the 21st century. In a wide-ranging reform of studies, Stanford University, for instance, created the *Ways of Thinking/Ways of Doing* program in 2013. It emerged from a document commissioned by the university directorate that recommended a non-disciplinary study model and pursued a more generalist educational concept for undergraduates (Sheehan 2012). "Ways" was developed for undergraduates of all faculties to learn about different disciplines and acquire respective skills. The compulsory bundle of modules "Ways" consists of a total of seven areas of competence: (1) aesthetic and interpretive inquiry; (2) social inquiry; (3) scientific analysis; (4) formal and quantitative reasoning; (5) engaging difference; (6) moral and ethical reasoning; and (7) creative expression. This last competence field comprises performative knowledge (Stanford Undergrad 2023). The program's website states: "Through a combination of instruction and mentoring, Creative Expression (CE) courses offer students significant opportunities to study the creative process and at the same time acquire the requisite skills to 'practice' creative expression themselves" (Stanford Undergrad 2023). The concept allows non-art students to visit the courses of their fellow students studying arts. In this way, law or management students can participate in pottery courses, attend theater classes, or start a film production. However, they will also find courses such as Introduction to Computer Graphics and Imaging, Queered Tech and Speculative Design, Stanford Laptop Orchestra: Composition, Coding, and Performance, or Wild Writing.

The attention to non-propositional epistemologies at Stanford University echoes ideas of the Hasso-Plattner Institute for Design Thinking and also highlights keywords such as "co-creation" and "network thinking". It is remarkable that after the launch of the Creative Expression program in 2015, and with no investment spared, large workshop and studio buildings were built on campus with a stage for theater and performance, sculpture and printmaking workshops, film and media studios, and spaces for music instruction. In terms of didactic development, however, it can be noted that Creative Expression courses primarily opened up existing teaching formats, such as those designed for traditional arts majors, to all undergraduate students.

A different approach in this respect was taken by the Aalto University in Helsinki with its University-Wide Art Studies (Tervo 2020). Aalto University, the result of a 2010 merger between a business school, a design school, and a technical university, initially saw this program as a link between the faculties. Here, too, the idea was to develop a program based neither on classical artistic disciplines nor on a simplified and functionalized concept of creativity but to make available a complex, non-propositional form of knowledge. Within the framework of the university, around 30 courses were offered at the bachelor's and master's levels as of 2014. This course program included more conventional crafts such as painting, pottery, and design techniques, e.g. 3D prototyping, but also equine husbandry, brewing beer, or sausage-making (Tervo 2020). This was not just about classical artistic disciplines, but an "emergent learning" of trying things out and exploring, where the sharing of students' experiences is essential. However, as it was neither an independent administrative unit nor a compulsory part of the teaching program, it fell victim to austerity measures and was discontinued in 2022 (Aalto University 2022).

A program that emerged in exchange with the programs mentioned is the Creative Performance Program of the private Zeppelin University in Germany, developed in 2012, later renamed Creativity and Performance (Schmidt-Wulffen 2022). Teachers and participating artists developed specific didactics based on a post-disciplinary, conceptual notion of art. Artistic practice rather than artworks were at the center, where "practice" was to be experienced as a collection of bodily and mental activities informed by the specific materials and media used. Like Stanford, this program implies the handing down of the experience of artistic production, such as drawing, creative writing, musical improvisation, performance, photography, or film. However, a significant difference between this and the programs discussed earlier is that the university invited internationally known artists, designers, architects, and art-related yoga practitioners to work together in co-teaching formats (Grosser and Kleinmichel 2022; van den Berg and Buck 2017). To strengthen an experience beyond art objects and techniques, students had to combine two disciplines to experience similarity and difference – the characteristics of aesthetic practice and performative knowledge. An accompanying theory course would deliver the conceptual premises of the program and allow for discussion. Alternating annual themes helped to retain the vitality of the program over several years (Zeppelin University 2022).

The development of the course falsified some of the assumptions derived from relevant theories. While they, for example, treat the non-intentionality of aesthetic practice and its foundation in the responsiveness of materials as crucial, the students were concerned with more elementary experiences. It emerged that it was especially important to foster the students' underdeveloped observation skills (van den Berg and Schmidt-Wulffen 2015). They experienced narration in film, photography, or drawing as an essential alternative to their scientific production. The cooperation with fellow students, reinforced even by choreography or other collective practices, became a crucial course experience, hinting at something like swarm intelligence. Improvisation strengthened the feeling of doing it in the right way. Space gained a central role, experienced as a hidden guiding principle connected to the order of an academic institution. Several exercises forced the students to transgress these rules and to invent diverting behavior: participants exercised yoga in a law seminar – with the consent of the lecturer; students had to undergo the painful experience of an outsider in what was later called "affirmative embarrassment" (Schmidt-Wulffen 2022, 207).

Undoubtedly the program hinted at principles of performative knowledge. The economy of time, however, proved to be a problem which also occurred in the aforementioned programs. Embodied knowledge needs exercise and corrections as part of a community of practice. While the program established a community, even addressing its specific collective swarm intelligence, it did not invest enough time for the bodily knowledge to develop (Schmidt-Wulffen 2022). The structuralist method of combining two significant but diverse experiences through participation in two different disciplinary courses tried to "abbreviate" a process that – at an art school, for example – would take years. The limitations of economic and epistemic rules of a university, however, did not allow for a more considerable "investment" to support this process. The evaluation of outcomes in the final presentation at Zeppelin University also demonstrated that students were much better in media familiar from everyday digital communication – like photography or film with mobile phones – than with traditional artistic crafts like drawing, indicating that in a postmedia world, aesthetic practice is not a matter of art alone, but is already generalized into daily practices.

One is more likely today to find performative knowledge in ordinary behavior, which is increasingly aestheticized, than specifically in the arts. The focus on art also raises the question of its role in a globalized culture; but just as there has not yet been a stand-alone debate on integrating performative knowledge into non-arts or sports science courses, it proved impossible to find similar efforts in universities outside Europe and the US. The reason for this might be authors' limited experience. However, it can also be understood as a consequence of the Western tradition, of enlightenment and its division of emotion and intellect, of arts, science, and the everyday. Non-European cultures allow for a much more generalizing approach to embodied knowledge, even in their universities.

It is worth noting that the European programs mentioned above – which include Copenhagen Business School, Witten/Herdecke University, but also Stockholm School of Economics, the University of St. Gallen, and occasionally Leuphana University and Milan's Bocconi University with their art programs – have been in close contact at times to exchange ideas about the integration of artistic practices into university teaching. It became obvious within these meetings, in which the authors of this chapter also participated, that such programs not only pose infrastructural demands and require careful curatorial work, they are also difficult to reconcile with established curricula and spatial conditions. At various network meetings, moreover, it became apparent that it remains controversial whether there is a benefit in offering such programs on a mandatory basis. What speaks in favor of mandatory participation is that these programs should not remain ultimately a destination for those whose sensibilities would always gravitate towards them, but should retain something resistant that is methodically valuable for making systems of thought tangible.

Furthermore, it became evident in the network meetings that aesthetic practice should be organized at universities as neither an alibi for personality development nor as an excursion into an exotic episteme. It makes more sense to recognize the aesthetic practice as an aspect of academic research and to take it into account in the curricula. This presupposes a changed understanding of scientific activity in the sense of a "practice turn" (Schatzky et al. 2001), in which the aesthetic–artistic aspects of the scientific activity itself are recognized. At the center of this changed understanding of scientific productivity is the "unfolding, dispersed, and signifying (meaning-producing) character of epistemic objects" (Knorr-Cetina 1999, 184) that stands in process-based relation to its researching subjects. A new epistemology is emerging here that reaches beyond the sciences into the arts and even everyday practice. Universities should begin to adapt their organization to this epistemology to accommodate a changing society. There is a direct interrelation between pedagogy and the conceptual structure of universities.

References

Aalto University. 2022. *UWAS. University-wide art studies.* Available from https://www.aalto.fi/en/uwas.

Austin, John Langshaw. 1962. *How to do things with words.* Oxford: Oxford University Press.

Ayer, Alfred Julius. 1952. *Language, truth, and logic*. New York: Dover.

Bernstein, Richard. 2010. *The pragmatic turn*. Cambridge: Polity.

Boltanski, Luc, and Eve Chiapello. 2007. *The new spirit of capitalism*. London: Verso.

Bourdieu, Pierre. 1977. *Outline of a theory of practice*. Cambridge: Cambridge University Press.

Butler, Judith. 1988. Performative acts and gender constitution – An essay in phenomenology and feminist theory. *Theatre Journal* 40 (4): 519–31.

Copenhagen Business School, ed. 2021. *Art at CBS*. Available from https://www.cbs.dk/en/about-cbs/art-cbs.

Fischer-Lichte, Erika. 2008. *The transformative power of performance: A new aesthetics*. New York: Routledge.

Gibbons, Michael, Camille Limoges, Helga Nowotny, Simon Schwartmann, Peter Scott, and Martin Trow. 1994. *The new production of knowledge: The dynamics of science and research in contemporary societies*. Los Angeles: Sage.

Giddens, Anthony. 1984. *The constitution of society: Outline of the theory of structuration*. Cambridge: Polity.

Grosser, Benita-Immanuel, and Philipp Kleinmichel. 2022. Rituals, dress code, and floor grid: Yoga practice and artistic knowledge. A conversation between Benita-Immanuel Grosser and Philipp Kleinmichel. In *Art, science & society. The Artsprogram of the Zeppelin University*, eds. Karen van den Berg and Ulrike Shepherd, 211–15. Berlin: Distanz.

Guillet de Monthoux, Pierre. 2004. *The art firm: Aesthetic management and metaphysical marketing*. Stanford, CA: Stanford University Press.

Guillet de Monthoux, Pierre, and Erik Wikberg, eds. 2021. *Economic ekphrasis: Goldin+Senneby and art for business education*. Berlin: Sternberg.

Hacking, Ian. 1983. *Representing and intervening: Introductory topics in the philosophy of natural science*. Cambridge: Cambridge University Press.

Holert, Tom. 2020. *Knowledge beside itself: Contemporary art's epistemic politics*. Berlin: Sternberg.

Knorr-Cetina, Karin. 1984. The fabrication of facts: Toward a microsociology of scientific knowledge. In *The sociology of knowledge*, eds. Nico Stehr and Volker Meja, 223–44. New Brunswick, NJ: Transaction.

Knorr-Cetina, Karin. 1999. *Epistemic cultures: how the sciences make knowledge*. Cambridge, MA: Harvard University Press.

Latour, Bruno. 1988. *The pasteurization of France*. Cambridge, MA: Harvard University Press.

Lave, Jean, and Etienne Wenger. 1991. *Situated learning: Legitimate peripheral participation*. Cambridge: Cambridge University Press.

Lawson, Bryan. 2005. *How designers think: The design process demystified*. London: Routledge.

Lippard, Lucy. 1973. *Six years: The dematerialization of the art object*. New York: Praeger.

Manning, Erin. 2020. *For a pragmatics of the useless.* Durham, NC: Duke University Press.

Manning, Erin, and Brian Massumi. 2014. *Thought in the act: Passages in the ecology of experience.* 4th edition. Minneapolis: University of Minnesota Press.

Mezirow, Jack. 1978. Perspective transformation. *Adult Education Quarterly* 28(2): 100–10.

Osborne, Peter. 2014. The postconceptual condition: Or, the cultural logic of high capitalism today. *Radical Philosophy* 184: 19–27.

Polanyi, Michael. [1966] 2009. *The tacit dimension.* 22nd edition. Chicago: University of Chicago Press.

Rancière, Jacques. 1991. *The ignorant schoolmaster: Five lessons in intellectual emancipation.* Stanford, CA: Stanford University Press.

Razzouk, Rim, and Valerie Shute. 2012. What is design thinking and why is it important? *Review of Educational Research* 82 (3): 330–48.

Reckwitz, Andreas. 2002. Toward a theory of social practices. *European Journal of Social Theory* 5 (2): 243–63.

Rheinberger, Hans-Jörg. 2010. *On historicizing epistemology.* Stanford, CA: Stanford University Press.

Rotherham, Andrew J., and Daniel T. Willingham. 2009. To work, the 21st century skills movement will require keen attention to curriculum, teacher quality, and assessment. *Educational Leadership* 67 (1): 16–21.

Ryle, Gilbert. 1959. *The concept of mind.* New York: Barnes & Noble.

Schatzky, Theodore R., Karin Knorr-Cetina, and Eike von Savigny. 2001. *The practice turn in contemporary theory.* London: Routledge.

Schmidt-Wulffen, Stephan. 2022. Artistic knowledge in university teaching. In *Art, science & society: The Artsprogram of the Zeppelin University,* eds. Karen van den Berg and Ulrike Shepherd. Berlin: Distanz.

Schön, Donald A. 1983. *The reflective practitioner: How professionals think in action.* New York: Basic.

Sheehan, James J. 2012. *The study of undergraduate education at Stanford University.* Available from https://drive.google.com/file/d/1ibg5Dke2rVKQcFzbPY24x4LK8flKCggV/view.

Shulman, Ken. 2017. *Celebrating the Center for Advanced Visual Studies, a pioneer in melding art, science, and tech.* MIT News on Campus and around the World. Available from https://news.mit.edu/2017/celebrating-50-years-since-cavs-center-advanced-visual-studies-founding-0927.

Stanford Undergrad. 2023. *Creative expression (CE).* Available from https://ways.stanford.edu/about/ways-categories/creative-expression-ce.

Tervo, Juuso. 2020. *Learning to unlearn: What could radically creative education be?* Available from http://www.aalto.fi/en/news/learning-to-unlearn-what-could-radically-creative-education-be.

Van den Berg, Karen, and Christina Buck. 2017. On the poetics of measuring space. Appropriating, acting, creating atmospheres. *GAM: Architectural Magazin* 13: 10–23.

Van den Berg, Karen, and Joachim Landkammer. 2002. Das Studium fundamentale an der Privaten Universität Witten/Herdecke. Konzepte, Entwicklungen, Perspektiven. In *Neues Handbuch Hochschullehre*, eds. Brigitte Berendt, Hans-Peter Voss, and Johannes Wildt, 1–25. Stuttgart: Raabe.

Van den Berg, Karen, and Stephan Schmidt-Wulffen. 2015. The politics of artistic knowledge at universities. In *Artistic research in applied arts*, eds. Gabriele Schmid, Harald Gruber, Peter Sinapius, and Rosemarie Tüpker, 159–76. Berlin: HPB University Press.

Wittgenstein, Ludwig. 1953. *Philosophical investigations*. New York: Macmillan.

Zeppelin University, ed. 2022. *Creativity & performance*. Available from https://www.zeppelin-university.com/chairs/kunsttheorie/creative-performance.php.

Personal Sustainability

Oliver Parodi, Christine Wamsler, and Marc Dusseldorp

Definition

During the last decade, the concept of personal sustainability and similar approaches, such as inner transition or inner transformation, have received increasing attention in sustainability science, education, policy, and practice. Personal sustainability is a highly transdisciplinary field and approach that deals with the human being and its relationality in the context of sustainable development. This applies particularly with regard to (1) human beings as bodily, conscious, and rational subjects, (2) their inner worlds, and above all (3) their relationships and interdependencies with the external world (see Parodi and Tamm 2018b).

As a conscious and rational subject (1), the *human being* is seen as a responsible and capable agent in the struggle for sustainable development. The human body is addressed, too, not only in terms of health but also as an essential condition and expression of human life. *Inner worlds* (2) include individual and collective mindsets, values, beliefs, attitudes, worldviews, emotions, and sensations and associated cognitive, emotional, and relational barriers and capacities (Wamsler 2020; Wamsler et al. 2020, 2021; cf. Hunecke 2023). These inner worlds must not only be described intersubjectively or scientifically from the outside, but essentially have to be explored and experienced individually. Those (inner) human dimensions are intrinsically linked to the *"outer" world* (3) in the context of sustainable development in many respects: as sustainable or unsustainable acting and behavior (e.g. consumption, lifestyles); or as affected by outer factors (e.g. climate anxiety); as drivers or barriers for adequate action; or as root causes for sustainability crises and deep leverage points for change – and as such as fundamental to the solutions to the world's greatest challenges (IPCC 2022a, 2022b; Wamsler and Bristow 2022).

Personal sustainability is thus about addressing inner human dimensions to enable a deepening and expansion of human consciousness, awareness and connectedness (to self, others, and nature) and to nourish inner human potential and capacities to care for a better, more sustainable life across individual, collective and system levels (Wamsler et al. 2021, 2022; cf. Hunecke 2023; Parodi 2018).

Personal sustainability thus includes a profound shift in perspectives towards a more relational paradigm, by emphasizing and expanding interdependency and connectedness. It is based on the understanding that strengthening the relation and connectedness to ourselves, and the world we share, is leading to an increasing circle of identity, care, and responsibility, and hence to a more ethical, more prosocial, compassionate life – in alignment with what is needed for an (outer) sustainable development (Wamsler et al. 2021, 2022).

The concept of personal sustainability is immediately related to the concept of sustainable development. At least three relations can be differentiated (Parodi and Tamm 2018a, 2018b). First, personal sustainability is an integrated *part* of sustainable development that complements the current sustainability discourse, which is focused on outer aspects. Second, personal sustainability is a *condition* for sustainable development, as an outer sustainability transformation is not fully feasible without an inner transformation. And third, personal sustainability is an independent *end in itself* of sustainable development, which is important since it would contradict the idea to use personal sustainability as a mere instrument for "outer sustainability" purposes or achieving utilitarian aims.

To gain access to this too often ignored part of the sustainability discourse and efforts, academic concepts and methods, e.g. from psychology, anthropology, philosophy, neuroscience, behavioral economics, education, health sciences, and (micro)phenomenology, but also practical approaches of consciousness and relationship work such as contemplation and meditation techniques, perception exercises, and held conversations are applied. The academic and practical approaches are often interlinked and complementary.

The transdisciplinary character of personal sustainability relates to both the macro- and the micro-level. At the macro-level, relevant knowledge and competencies for personal sustainability are not only coming from scientific fields, but also from implicit and indigenous knowledge or contemplative and wisdom traditions. Acknowledging and integrating them is crucial. On the micro-level, personal sustainability condenses transdisciplinarity in one person: a scientific approach is combined with self-reflection, exploring and personal experience and action. In parallel, approaches with a similar focus, like *inner transformation, inner transition, inner change, personal development,* and *personal spheres of transformation* have been developed.

Background

Sustainable development as global guiding principle has been present in the political acting and scientific discourses for more than 30 years (Dixson-Declève et al. 2022; Schultz et al. 2008; WCED 1987). But, despite wide-ranging actions at transnational, national, and sub-national levels, sustainability problems like poverty, unequal income, climate change, environmental pollution, exploitation of natural resources, the massive loss of biodiversity and fertile grounds are still getting worse and challenges are increasing globally. Policy approaches as well as new technologies have failed so far to generate change at anywhere near the rate, scale, or depth that is needed (IPCC 2022a, 2022b). At the same time, the knowledge required for a sustainable development has increased massively over the past 30 years – in all dimensions: system knowledge, target knowledge, and practical knowledge. From a natural-scientific and technical point of view, we have known what to do for decades. But the gap between knowledge and action is increasing dramatically.

So, what's going wrong? If one looks at (un)sustainable development as a cultural phenomenon, where culture and cultural change is carried out in the interplay between the collective and the individual (Hansen 2011), one can recognize that almost all effort for sustainable development so far has been located at the *collective* side of culture (technology, legislations, rules, economic mechanisms, political strategies, etc.). This focus on collective and outer aspects is part of modern societies' scientific and mechanistic worldview – and, as a result, climate change, loss of biodiversity, and other sustainability problems are generally framed as *outer* – technical or political challenges which are addressed with a "fix-it mindset", and less as a matter of human consciousness, worldviews, associated disconnectedness, and alienation (Leichenko and O'Brien 2019).

While the role of individuals and their inner worlds were initially largely ignored, over the past two decades they have been increasingly considered in the sustainability discourse, but from an external, and instrumental perspective (e.g. nudging). Inner, and relational perspectives, capabilities, and interdependencies remain largely ignored however (Parodi 2011; Wamsler et al. 2022). This, in turn, narrows the possibilities for deeper change that requires tackling the human and inner root causes of global challenges. Put together, personal sustainability involves a change of perspective and as such is not an alternative but a complement and should be an integrative part to the common discourses, theories and practices of sustainable development.

As for etymology, the term *personal sustainability* is translated from the German term *Personale Nachhaltigkeit*. It was invented and introduced in 2008 in the course of the formation of the School of Sustainability at the Karlsruhe Institute of Technology as a search term and working title for new ways of understanding, teaching and practicing sustainable development (Parodi 2011; Parodi and Tamm 2018b). The attribute *personal* seemed to be best suited to describe the intended focus on human, individual, and inner human aspects of sustainable development. Thereby the term *person* describes a human being as a relational individual in its specific character of being conscious, responsible, and able to act reasonably. In this sense, *personal sustainability* is also always to be thought of as *trans*personal sustainability.

Although personal sustainability is a new field of transdisciplinary research, it has antecedents and roots in earlier fields and concepts. At the least, *sustainability science, environmental psychology, behavioral economics, systems theory, human ecology,* and *socio-ecological research* have to be mentioned here. *Deep ecology* (Naess 1972) and *ecopsychology* (Roszak et al. 1995) can be seen as precursors of a kind. Another practice- and change-oriented root lies in the field of *sustainable* or *ethical leadership* and at the interface between arts and sciences in the context of sustainable development. Finally, in the sphere of the ecological movement and community-building there are a lot of efforts that bring together ecology and personality, as well as communal and individual sustainable development (e.g. Joubert and Alfred 2007).

Debate and criticism

Personal sustainability is still an emerging field of transdisciplinary sustainability science and action, and still a search term for a huge field of unexplored phenomena and interdependencies. the publication *Personal Sustainability* (Parodi and Tamm 2018a) was a first important step to grasp and map the research field, and it was accompanied by further advances, reviews, and theoretical developments (Wamsler et al. 2021, 2022). The latter include, for instance, "the inner–outer transformation model" (Wamsler et al. 2022), "the three spheres of transformation model", and conceptual reflections on paradigm shifts in consciousness from an I–I, I–it to an I–We World (Parodi 2018; Siegel 2022). The number of publications on personal sustainability and related discourses is growing rapidly, and researchers' networks like the international *Inner Transition Group* conduct cooperative research and produce collective publications; they also exchange ideas on related teaching tools (Woiwode et al. 2021).

A recent literature review systematizes the current linkages between inner and outer transformation in different research disciplines (Wamsler et al. 2021). It shows that in *psychology,* mental health and related applied sciences (including leadership, personal, and adult development), diverse context-sensitive frameworks have been developed for understanding individuals, their (cognitive) drivers, and the motivations that can influence sustainability. However, they tend to give little consideration to wider societal or systemic issues. Related exceptions come predominantly from the field of environmental psychology.

Contributions in the discipline of *behavioral economics* tend to focus on individuals (or consumers) and the cognitive, motivational, and contextual factors that affect their decisions and choices. Within this context, approaches are limited with respect to: (1) psychological mechanisms; (2) the emphasis on quantitative assessments (mostly via experiments); and (3) their instrumental approach (Wamsler et al. 2021, 4).

Studies from *sustainability science* and *education* tend to emphasize the importance of systems change and the lack of individual agency due to structural constraints. They focus on systems analyses of wider socioeconomic structures, dynamics, and technology, often based on interdisciplinary and mixed-methods approaches. The role of individuals is, in this context, perceived to be of little importance (see *agent–system dichotomy*, Wamsler et al. 2021, 5).

At the same time, there is increasing recognition and associated *systems theory* that inner dimensions are deep leverage points for change (Ives et al. 2020; Wams-ler et al. 2021, 7-8). They are more difficult to influence, but lead to more substantial transformation, as it is from this level that the system's goals, structures, rules, and parameters emerge. Despite the urgent need to better link inner and outer approaches for sustainability and climate action, related knowledge is still scarce and fragmented (Meadows 1999, 7–8).

One central question in the current personal sustainability debate is the relevance of relations, connection, and interdependency. There is mounting evidence that the human story of separation, disconnection, and alienation is the underlying common thread of interlinked social, socioeconomic, and environmental crises and, in general, of today's global unsustainable way of life (Leichenko and O'Brien 2019; Wamsler and Bristow 2022; Wamsler et al. 2021). Separation, dualism, and disconnection form parts of the modern worldview – and are part of the success story of modern civilization: abstract thinking, science, predicting, controlling, and exploiting our environments via technology are achievements that allow and support wealth and security. At the same time, the massive and life-threatening destruction of the human environment and of humans' own basis of life is a direct consequence of this separation. The world seen as a pure object, free for human use and unrestrained access has led to the present excessive overuse and alienation. With the rise of science and technology, humans have become increasingly removed from nature, from each other, and even from themselves (Wamsler and Bristow 2022, 4–11). Climate change and all other sustainability problems can thus also be understood as an unintended – albeit deeply inscribed in culture – consequence, a subconscious manifestation of the globalized disconnected modern way of life, or, pointedly, of human *being* (Wamsler and Bristow 2022, 5).

In consequence, one crucial sustainability challenge is to "know thyself" (Niehaus et al. 2018, 51) to become aware of ourselves, our individual and collective inner worlds, worldviews, emotions, and attitudes. And then, to work on (reestablishing) our relations and connections to the world in and around us. In essence, personal sustainability work essentially means relationship work – regarding how we relate to ourselves, others, and the world at large.

For personal sustainability as a scientific endeavor, an important (transdisciplinary and methodological) question is about how to link and integrate knowledge from the increasing number of studies that look at personal sustainability topics, especially at the linkages between inner and outer (systems) change. However, related approaches are segregated across multiple disciplines that use heterogeneous terminology, with different ontological, epistemological, and ethical underpinnings. In addition, most studies adopt a narrow scope. They look at the link between individual and systems change from a one-directional perspective (Wamsler et al. 2021).

To actualize its transformative potential, personal sustainability has to become more inter- and transdisciplinary, and become common practice (e.g. in the course of relationship work). In addition, it is important to highlight that sustainability is *not only* a scientific endeavor. Engaging with inner human worlds requires introspection, self-perception, and experience that are to a high degree individually and not easily accessible for traditional scientific approaches.

Consequently, personal sustainability is also normative and programmatic, because (1) on the micro-level it is about experiencing and being involved, to perceive, feel, and be as a human being – and not only to think and learn about inner worlds from a scientific mediated third-person perspective; and (2) on a macro-level, following the proverb "You can't solve problems with the same mindset that created them". In fact, current worldviews and scientific approaches are at the root of our unsustainable way of live, and thus we have to challenge them – without ignoring or rejecting them completely.

All in all, despite advances, more interdisciplinary discourses, and theory and method formation, are needed to advance the transdisciplinary field of personal sustainability. This includes exploring aspects, phenomena, and practices of personal sustainability in diverse and new ways. The latter is linked to a call for a "personal sustainability science" that works in a connected way and includes more first-person research and methods such as micro-phenomenology. More investigation into different forms of education and practices is needed (Parodi and Tamm 2018a, Wamsler et al. 2021).

Current forms of implementation in higher education

In the context of transdisciplinary learning, personal sustainability puts emphasis on self-knowledge, relationship work, and people's potential as change agents to support individual, collective, and systems change. Courses in this field differ in their foci regarding: (1) the individual, collective, or system level; (2) their transdisciplinary and transformative substance: more cognition-, experience-, or action-oriented; and (3) their closeness to and occupation with sustainability science and concepts.

A review of evidence-based academic literature by Wamsler et al. (2022) suggests four interrelated categories of practices that can contribute to personal sustainability: (1) contemplative practices and interventions; (2) psychological- and cognitive–behavioral-based interventions; (3) transformative facilitation, communication, and coaching tools; and (4) transformational education and leadership approaches.

Over the past five years, scholars and practitioners have increasingly combined and adapted such practices for personal sustainability to develop transformative education and leadership approaches in higher education. They have combined complexity and systems and design thinking with various practices, and come up with a certain theory and pedagogy for linking inner and outer change (Wamsler et al. 2022). Transformative education is increasingly offered by universities all over the world, but only few explicitly address personal sustainability and inner transformation in a comprehensive way. One example comes from the Waterloo Institute for Social Innovation and Resilience (WISIR, Canada). Their "Decolonial Systems Thinking & Resilience" courses offer a series of seminars and professional development courses that help to foster the capacity for in-depth work and cross-cultural capabilities for broad, systemic change. Through decolonial practices and methodologies, these courses – held by a couple of scientific and Indigenous knowledge holders – support the cultivation of new skills and capacities required for sustainability transformations.

Another example, the "Sustainability and Inner Transformation" course at Lund University Centre for Sustainability Studies (LUCSUS, Sweden), runs every year over three months and includes lectures, seminars, councils, and a practice lab that are designed to explore the role of inner dimensions, to support individual, collective, and systems transformation toward sustainability. Knowledge, tools, and practices from sustainability science, social neuroscience, psychology, behavioral economics, contemplative studies, climate policy integration, and inner–outer transformation theories are systematically integrated. A further recent example from the University of Natural Resources and Life Sciences Vienna (Austria) consists of a course on "The Inner Dimension of Sustainability: The Role of Values, Emotions and Worldviews". Over two weeks, students explore the inner dimension of sustainability on both theoretical and practical levels.

At Karlsruhe Institute of Technology (KIT, Germany) there have been courses and seminars explicitly focused on personal sustainability since 2008. These courses address the individual and systemic level, are self-experience and relation-oriented, and link inner aspects closely to the common theory, concepts, and debates of sustainable development. Further courses at KIT, like transformative project seminars (since 2015), include self-experiments and address inner psychological resources of sustainable development.

In addition, the Principles for Responsible Management Education (PRME), a United Nations-supported initiative, engages worldwide in personal sustainability education and offers related train-the-trainer programs ("The Sustainability Mindset Action Lab").

Aside from higher education, a growing number of private and nonprofit organizations offer or support adult development and leadership courses all over the world that include aspects of personal sustainability. Examples of organizations working in Africa, Asia, Latin America, Europe, the US, or Australia include The Work That Reconnects, Pacific Integral, The Inner Green Deal, The IDG Initiative; RTLWorks, and CChange. And an increasing number of guidelines provide an overview of different practices (Wamsler et al. 2022).

Overall, transdisciplinary settings are particularly fruitful for supporting sustainable development and associated education, as transdisciplinary methods have inherent didactic qualities (Dusseldorp and Beecroft 2012, 11–35). Personal sustainability can and should become an integrated part of related endeavors. Ivanova and Rimanoczy (2022) present examples across five continents and over 150 student voices depicting transformative experiences and shifts in mindsets. Put together, this shows that personal sustainability education is urgently needed and possible.

References

Dixson-Declève, Sandrine, Owen Gaffney, Jayati Ghosh, Jorgen Randers, Johan Rockstrom, and Per Espen Stoknes. 2022. *Earth for all: A survival guide to humanity.* Gabriola Island: New Society.

Dusseldorp, Marc, and Richard Beecroft, eds. 2012. *Technikfolgen abschätzen lehren – Bildungspotenziale transdisziplinärer Methoden.* Wiesbaden: Springer.

Hansen, Klaus P. 2011. *Kultur und Kulturwissenschaft.* Stuttgart: UTB.

Hunecke, Marcel. 2023. *Psychology of sustainability.* Cham: Springer.

IPCC [Intergovernmental Panel on Climate Change]. 2022a. *Climate change 2022: Impacts, adaptation, and vulnerability.* Contribution of Working Group II to the Sixth Assessment Report of the Intergovernmental Panel on Climate Change. Cambridge: Cambridge University Press.

IPCC [Intergovernmental Panel on Climate Change]. 2022b. *Climate Change 2022: Mitigation of climate change.* Contribution of Working Group III to the Sixth Assessment Report of the IPCC. Cambridge: Cambridge University Press.

Ivanova, Ekaterina, and Isabel Rimanoczy, eds. 2022. *Revolutionizing sustainability education.* London: Routledge.

Ives, Christopher D., Rebecca Freeth, and Joern Fischer. 2020. *Inside-out sustainability: The neglect of inner worlds.* Ambio 49 (1): 208–17.

Joubert, Kosha Anja, and Robin Alfred, eds. 2007. *Social key: Beyond you and me. Inspirations and wisdom for building community.* East Meon, Hampshire: Permanent Publications.

Leichenko, Robin, and Karen O'Brien. 2019. *Climate and society: Transforming the future.* Cambridge: Polity.

Meadows, Donella. 1999. *Leverage points: Places to intervene in a system*. Hartland: The Sustainability Institute.

Næss, Arne. 1972. *Shallow and the deep*. Oslo: Inquiry.

Niehaus, Michael, Dirk Schmidt, and Shirli Homburg. 2018. Care of the self and "Bildung" as condition for and result of personal sustainability. In *Personal sustainability: Exploring the far side of sustainable development*, eds. Oliver Parodi and Kaidi Tamm, 51–64. London: Routledge.

Parodi, Oliver. 2011. Personal sustainability – including body and soul. The Karlsruhe School of Sustainability. In *Sustainable development: The cultural perspective*, eds. Gerhard Banse, Gordon L. Nelson, and Oliver Parodi, 223–38. Berlin: Edition Sigma.

Parodi, Oliver. 2018. Sustainable development: A matter of truth and love. In *Personal sustainability: Exploring the far side of sustainable development*, eds. Oliver Parodi and Kaidi Tamm, 65–82. London: Routledge.

Parodi, Oliver, and Kaidi Tamm, eds. 2018a. *Personal sustainability: Exploring the far side of sustainable development*. London: Routledge.

Parodi, Oliver, and Kaidi Tamm. 2018b. Personal sustainability: Exploring a new field of sustainable development. In *Personal sustainability: Exploring the far side of sustainable development, eds.* Oliver Parodi and Kaidi Tamm, 1–17. London: Routledge.

Roszak, Theodore, Mary E. Gomes, and Allen D. Kanner, eds. 1995. *Ecopsychology: Restoring the earth, healing the mind*. San Francisco: Sierra Club Books.

Schultz, Julia, Fridolin Brand, Jürgen Kopfmüller, and Konrad Ott. 2008. Building a 'theory of sustainable development'. *Environment and Sustainable Development* 7(4): 465–82.

Siegel, Daniel J. 2022. *IntraConnected: MWe (Me + We) as the integration of self, identity, and belonging*. New York: Norton.

Wamsler, Christine. 2020. Education for sustainability: Fostering a more conscious society and transformation towards sustainability. *International Journal of Sustainability in Higher Education* 21 (1): 112–30.

Wamsler, Christine, and Jamie Bristow. 2022. *At the intersection of mind and climate change: Integrating inner dimensions of climate change into policymaking and practice.* Available from https://link.springer.com/10.1007/s10584-022-03398-9.

Wamsler, Christine, Niko Schäpke, Carolin Fraude, Dorota Stasiak, Thomas Bruhn, Mark Lawrence, Heike Schroeder, and Luis Mundaca. 2020. Enabling new mindsets and transformative skills for negotiating and activating climate action: Lessons from UNFCCC conferences of the parties. *Environmental Science and Policy* 112: 227–35.

Wamsler, Christine, Gustav Osberg, Walter Osika, Heidi Hendersson, and Luis Mundaca. 2021. Linking internal and external transformation for sustainability and climate action: Towards a new research and policy agenda. *Global Environmental Change* 71: 102373.

Wamsler, Christine, Jamie Bristow, Kira Cooper, Gretchen Steidle, Sara Taggart, Lene Søvold, Jessica Bockler, Tom Oliver, and Thomas Legrand. 2022. *Theoretical foundations report: Research and evidence for the potential of consciousness approaches and practices to unlock sustainability and systems transformation. Report of the UNDP Conscious Food Systems Alliance (CoFSA).* New York: UNDP.

WCED [World Commission on Environment and Development]. 1987. *Our common future*. Oxford: Oxford University Press.

Woiwode, Christoph, Niko Schäpke, Olivia Bina, Stella Veciana, Iris Kunze, Oliver Parodi, Petra Schweizer-Ries, and Christine Wamsler. 2021. Inner transformation to sustainability as a deep leverage point: Fostering new avenues for change through dialogue and reflection. *Sustainability Science* 16 (3): 841–58.

Real-World Lab

Oliver Parodi, Anja Steglich, and Jonas Bylund

Definition

In the pursuit of accelerating and extending sustainability transformations, an experimental turn has occurred in sustainability research since the 2000s. Around the globe, more and more transdisciplinary laboratories and workshops like *living labs, transition labs*, etc. are established in real-world contexts, which foregrounds experimentation for more sustainability (McCrory et al. 2022; Turnheim et al. 2018). However, in Germany, a specific form of transdisciplinary experimental lab was developed during this time: real-world labs (German: *Reallabor*). These models are elaborate and impactful in terms of promoting change in everyday settings, with a focus on transdisciplinary and transformative sustainability research at the intersection of science and society at large. Against the background of increasingly pressing sustainability challenges and a necessary change in science (Schneidewind and Singer-Brodowski 2014; WBGU 2011), the concept of real-world labs found its way into academic research discourse and practice in the early 2010s (Parodi 2011; Schneidewind and Scheck 2013). Real-world labs are places and incubators to develop and research sustainability solutions, or, in a nutshell, to experiment and examine desirable societal futures by scientific means.

Real-world labs can be set to explore a wide range of issues. This can be, for example, regenerative energy supply, socially responsible environmental protection, sustainable consumption, climate protection, or even the sustainable development of a city district. Many examples can be found on the real-world lab network website (2023). The crucial common aspects in how real-world labs are realized is that researchers and problem-owners enter into dialogue, share goals, and collaborate in practice. Researchers and (other) societal actors proceed transdisciplinarily and cooperatively, learn from each other, reflexively minimize risks, and jointly initiate and shape contributions to sustainable development (Bergmann et al. 2021).

In terms of content and concept, the recent term *real-world lab* still remains open to interpretation. Its definition is the subject of scientific and political debate (Schäpke et al. 2018) and does not meet with consensus. However, a widely shared sense of what real-world labs are has emerged in theoretical, conceptual, and practical discourse: a real-world lab describes a transdisciplinary research and development facility and setting that serves to conduct experiments in a spatially delimited societal context, initiate sustainable transformation processes, and support scientific and societal learning processes respectively (Parodi et al. 2017, 80). Although definitions may vary, these aspects are reflected in them (see de Flander et al. 2014). Based on this understanding of real-world labs, a set of nine constitutive core characteristics can be identified (Parodi et al. 2016):

1. *Research orientation*: Real-world labs are research undertakings aimed at generating knowledge, specifically transformative knowledge.
2. *Design and transformativity*: Real-world labs contribute directly to societal change and sustainability transformation. They provide concrete contributions to sustainable development in practice.
3. *Normativity and sustainability*: Real-world labs are normative undertakings; they follow the guiding principle of sustainable development and make their normative starting points transparent.
4. *Transdisciplinarity and participation*: The predominant scientific mode in real-world labs is transdisciplinarity. From a social perspective, participation and co-design are central elements of real-world lab work.
5. *Civil society orientation*: In addition to other non-academic actors (such as local government, companies, schools), civil society actors and inhabitants in particular are also involved.
6. *Model character*: Real-world labs are place-specific and context-bound, but strive for transferability of results and solutions to other contexts, spaces, or scales.
7. *Long-term*: Real-world labs should be set up for as long as possible (ideally a minimum of five years, but potentially several decades), in order to be able to accompany transformation processes both scientifically and in everyday practices and to evaluate them ex post.
8. *Laboratory character and experimental space*: Real-world labs are for experimenting and provide specific (social) spaces, possibilities, and opportunities for joint experimentation.
9. *Education*: Real-world labs are highly condensed reflexive and learning spaces and as such at least implicit educational institutions. If possible, they should integrate educational aspects into their activities.

Ideally, one would only speak of a *real-world lab* when all these characteristics are present. Beyond that, real-world labs are closely related to other kinds of co-creative labs – such as *living labs, transition labs,* or *green urban labs* (Aßmann et al. 2017; JPI Urban Europe 2023). The explanation of the term and the list of constitutive core characteristics can also be used to distinguish real-world labs from related concepts and labs. For instance, living labs are similar to real-world labs regarding characteristics 1, 2, 4, 6 and 8 but not necessarily to the characteristics of 3, 5, 7, and 9.

Etymologically, the term also refers to the decades older term *real-world experiment* and the related notion of "society as laboratory" (German: *Gesellschaft als Labor*) (Krohn and Weyer 1989, own translation). It draws upon the critical discourse on the risks of technically advanced societies and the (inadequately perceived) role of science in the 1980s but turns it into a constructive approach.

Background

The development of real-world labs as hybrid entities at the intersection of research and society ties in with different, partly convergent schools, concepts, and currents in science and society (Parodi et al. 2017). It is the combination of these that bestows on real-world labs their novelty and originality. The central problem background is the increasingly destructive side effects of modern lifestyles and economies, endangering the continued existence of humanity. The concern for a good, humane life for all in the future and the corresponding efforts towards sustainable development (Dixson-Declève et al. 2022; Schultz et al. 2008) is the main motivational and innovation driver for real-world labs. Historically, the idea and first implementations of real-world labs originate from transformative sustainability research (Schneidewind and Singer-Brodowski 2014; Wiek and Lang 2016).

The urgency of societal crises, such as climate change and its effects on ecosystems, habitats, and societies, also makes clear the need for action in sustainability research and calls on the research community to move from knowledge to action. Thus, real-world labs can be seen as a contemporary practical and applied form of sustainability research. With the transformative approach of real-world labs, transdisciplinary research has been expanded to the effect that the goal is no longer only to gain knowledge – the production of *knowledge* for sustainable development – but also to develop practical impulses: contributions to sustainable development in the course of the research. These take place in the form of (transdisciplinary) experiments. Real-world labs are both a specific case of applying transdisciplinary research *and* its further development. With their direct design mission, real-world labs leave the sphere of conventional academic research and become a force to change and reshape societies. They are at the same time knowledge producers as well as practical actors outside the academic context and in this respect *trans-scientific*. This ambivalence does not imply that they are non-scientific. However, they do *not only* proceed scientifically – which in turn gives rise to specific potentials, but also to challenges.

Thus, real-world labs operate also in the tradition of those forces of inter- and transdisciplinarity that try to broaden, renew, and reform research and science

(Bergmann et al. 2021). In general, their aim is to bring the cognitive processes of research closer to the issue at hand in order to be able to grasp and describe them more accurately. In the case of real-world labs, the issue is the transformation of an unsustainable society into a sustainable way of life and economy. Real-world lab research – and also teaching – approaches these transformation processes from an inside perspective. Real-world lab researchers are part of the change and gain their knowledge from an active and participating perspective. Real-world labs and the idea of the "Great Transformation" (Polanyi 2001) introduced into scientific and political debates, develop in parallel and relate to each other (WBGU 2011, 2016). As "institutions of change" (Parodi 2019, 8, own translation), real-world labs are intended to support, research, and accelerate the transformation of settlements in particular (WBGU 2016).

Alongside sustainable development, transdisciplinarity, and transformative research, the democratization of science forms another ideal root of the real-world labs. Against the background of a democratic society, in real-world labs the knowledge-producing process of research is more closely linked to the legitimized subject of knowledge: the population, the citizens, and the diversity of social actors. Thus, participation, the involvement and co-determination of many actors – if possible from the beginning – is an essential part of real-world lab work (Parodi et al. 2018). As many levels of participation as possible, from information to consultation, cooperation, and empowerment are to be realized (Meyer-Soylu et al. 2016; Parodi et al. 2018). Science communication and bidirectional knowledge transfer play a central role in the work of the real-world labs: not only research should be communicated to different actors in popularized or non-technical jargon. Similarly, impulses from non-academic actors should be equally incorporated into the real-world lab activities in order to generate scientific and societal resonance and effectiveness (Steglich et al. 2020). Furthermore, a democratization of research takes place directly through an (equal) participation of non-academic actors in the entire transdisciplinary process: from agenda setting to co-design and co-production to the utilization of the results.

The orientation on civil society also expresses the democratization of science. Extensive and far-reaching cooperation between science and private sector actors, especially in technology and product development, is widespread. The inclusion of citizens, (local) government, and civil society groups (including nongovernmental organizations) in real-world lab work broadens the social base of people participating in science. In this process, civil society grows into a new role that is crucial for transformation: through the direct participation in knowledge production made possible by real-world labs, civil society is recognized as a full partner in research.

Historically, both the term and the first real-world labs launched originated in transformative sustainability research: Schneidewind and Scheck (2013) introduce the term in the context of a transformative science, which not only conducts research on transformation processes but also actively supports them. Real-world labs have developed into a research format in their own right that methodically condenses and practically concretizes transformative sustainability research. From the very beginning, real world labs – directly linked to urban development, urban research, and urban transformation – were conceived as a framework for societal research, transformation, and learning processes (de Flander et al. 2014). Already at the beginning of the real-world lab discourse, the above characteristics 1, 2, 3, 4, 8 are programmatically linked to the real-world lab concept. Almost at the same time, first proto-real-world labs emerged apart from the conceptual debate, such as the *District Future – Urban Lab* (2023; and see Parodi 2011), whose objective is the sustainable transformation of an existing urban district in Karlsruhe by means of a long-running transdisciplinary and participatory process (Parodi et al. 2018). To date, the establishment and spread of the term and concept have been accompanied by a semantic diversification and partial reinterpretation of the term.

Debate and criticism

Are real-world labs really new? The concept, discourse, and practice of real-world labs are undoubtedly new, especially in the context of academic research. However, real-world labs build upon many strands of discourse and practice, some of them decades old, combine them, and develop them further (JPI Urban Europe 2023). Since real-world labs are essentially about the concretization, operationalization, and ideally institutionalization of transdisciplinary sustainability research, the debates about real-world labs focus predominantly on aspects of transdisciplinarity and transformative research. Some of the real-world lab characteristics are controversial, such as the question of whether and how real-world labs are to be aligned with the guiding principle of sustainable development (for a critical appraisal, see Defila and Di Giulio 2018).

The term combines and merges *reality* and *laboratory* and thus points to a immanent epistemic tension between, on the one hand, the highly controlled environment of a laboratory, with which the attempt is made to create a stringent framework for the production of knowledge out of delimited experiments, and, on the other hand, the non-academic everyday practices full of complexity and contingency. In order to achieve impact, real-world labs must be based in everyday life settings and are therefore context-bound. This in turn makes it difficult for them to generate transferable knowledge. In this respect, real-world labs often run the risk of merely producing case studies while comparative analysis between them still needs to be done, bar a few rare occasions.

The role of the participating researchers in real-world labs is sharpened by the dual objective and strong proximity of research and design. Thus, in addition to being the "honest broker" and "issue advocate" (Grunwald 2019, 170–76), the academic researcher is potentially also present in the role of designer, mediator, and process organizer. Hence, in addition to (individual) conflicts between roles and interests, the researcher involved may be overtaxed. However, some of these role conflicts are due to the hitherto unbalanced sponsorship of real-world labs.

Ideally, the flow of funds and sponsorship corresponds to the types of real-world lab activities. Hence, real-world labs should be supported and financed by research funding *and* other areas of society, but in reality, they have so far mainly been financed by the research funding system (in contrast to other labs internationally financed from a diverse range of funders; see Bylund et al. 2022). This distortion of funding and sponsorship hinders the development of transformative potential and discredits real-world labs as research ventures. Thus, it is not the task of academic research to shape society directly or to pursue societal transformation. A real-world lab financed exclusively by research funding misuses research resources for design purposes. On an individual level, the scientific sponsorship of real-world labs leads to conflicts and double burdens, as researchers have to manage research as well as design (processes). The necessary, often costly non-scientific activities such as the conception, initiation, and support of transformation processes, event organization, communication, etc. are rarely rewarded in the academic research system.

A major potential of real-world labs is to institutionalize them as actors of change (Karvonen 2018; WBGU 2016). Real-world labs, which – similar to engineering or natural science laboratories – would be established for 30, 50, or 100 years, would be a true innovation in the science system and would entail new framework conditions and unprecedented possibilities for transdisciplinary and transformative research. Transformation processes could be stimulated, accompanied, and researched over the long term, and evaluated ex post. So far, with few exceptions, real-world labs have been designed as research projects with durations and funding periods of about three years. However, far-reaching social and cultural transformation processes do not take place within a few years, but take decades. In addition, setting up a real-world lab can take a great deal of time and money: exploring the social context, identifying relevant actors, building trust and networks, acquiring real-world lab skills and premises, synchronizing research, practice, and teaching. The multifaceted and intense set-up work in a project lasting only a few years becomes disproportionate to the actual experimental work and its evaluation.

Current research questions for the further development of real-world labs are "What impact do real-world labs really have?", "How can we scale them up?", and "How can real-world labs contribute to a to sustainable technology transformation in a reflexive and responsible manner?" (Parodi et al. 2022).

Current forms of implementation in higher education

Real-world labs aim to implement and perpetuate scientific and social learning (Singer-Brodowski et al. 2018), and are thus didactic undertakings per se. Through the coupling of research, teaching, and practice, as well as the involvement of many often very different kinds of actors (Parodi et al. 2018; Steglich et al. 2020), real-world labs form rich learning environments and are, at least implicitly, educational institutions (Beecroft 2018). Education can occur at all scales: from individuals to groups, organizations, and society at large. Dialogue, resonance, and reflexivity are central aspects of learning. As a framework for transdisciplinary and transformative research, they enable precisely those didactic aspects of transdisciplinary and transformative research.

The forms of didactic implementation in the real-world lab are as diverse as the constellations of topics and actors within the lab or its social and spatial contexts. They range from self- and group experiments, conventional or transformative seminars incorporating student projects, service learning, lecture series, training courses, practical or scenario workshops, to self-experience of personal sustainability (Parodi and Tamm 2018) or serious gaming forms such as planning or learning games in which the actors involved swap roles (Beecroft 2018). In addition to the obvious and often established connection with university teaching (see Beecroft 2018; Steglich et al. 2020), real-world lab educational activities are also finding their way into primary, secondary, and vocational schools. An established link between real-world lab research and university teaching can be found for example at Leuphana University Lüneburg, ETH Zurich, TU Berlin, and the Karlsruhe Institute of Technology, where lab activities are integrated in the accompanying studies in sustainable development as transdisciplinary student projects. Related labs are also being directly integrated into higher education outside Europe, such as Cité-ID in Montreal, Canada. As real-world labs are showing signs of transformative potential and impact, other universities are currently expanding their offerings related to this approach.

Overall, real-world labs, in their pursuit of sustainability and with their core characteristics, offer a universal and flexible framework that can be applied around the world. However, they only work if they are adapted to and integrated in the local social and cultural context.

References

Aßmann, Katja, Markus Bader, Fiona Shipwright, and Rosario Talevi, eds. 2017. *Explorations in urban practice: Urban School Ruhr Series.* Barcelona: Dpr-Barcelona.

Beecroft, Richard. 2018. Embedding higher education into a real-world lab: A process-oriented analysis of six transdisciplinary project courses. *Sustainability* 10 (10): 3798.

Bergmann, Matthias, Niko Schäpke, Oskar Marg, Franziska Stelzer, Daniel J. Lang, Michael Bossert, Marius Gantert, Elke Häußler, Editha Marquardt, Felix M. Piontek, Thomas Potthast, Regina Rhodius, Matthias Rudolph, Matthias Ruddat, Andreas Seebacher, and Nico Sußmann. 2021. Transdisciplinary sustainability research in real-world labs: Success factors and methods for change. *Sustainability Science* (16): 541–64.

Bylund, Jonas, Johannes Riegler, and Caroline Wrangsten. 2022. Anticipating experimentation as the "the new normal" through urban living labs 2.0: Lessons learnt by JPI Urban Europe. *Urban Transform* 4 (8).

De Flander, Kathleen, Ulf Hahne, Harald Kegler, Daniel Lang, Rainer Lucas, Uwe Schneidewind, Karl-Heinz Simon, Mandy Singer-Brodowski, Matthias Wanner, and Arnim Wiek. 2014. Resilience and real-life laboratories as key concepts for urban transition research. 12 theses. *GAIA* 23 (3): 284–86.

Defila, Rico, and Antonietta Di Giulio. 2018. Reallabore als Quelle für die Methodik transdisziplinären und transformativen Forschens – eine Einführung. In *Transdisziplinär und transformativ forschen. Eine Methodensammlung Band 1*, eds. Rico Defila and Antonietta Di Giulio, 9–35. Wiesbaden: Springer VS.

District Future. 2023. *District future – Urban lab.* Available from https://www.quartierzukunft.de/en.

Dixson-Declève, Sandrine, Owen Gaffney, Jayati Gosh, Jorgen Randers, Johann Rockström, and Per Espen Stoknes. 2022. *Earth for all – A survival guide to humanity.* Gabriola: New Society.

Grunwald, Arnim. 2019. *Technology assessment in practice and theory.* London: Routledge.

JPI Urban Europe. 2023. *Guidelines for urban labs.* Available from https://jpi-urbaneurope.eu/wp-content/uploads/2016/09/URB@EXP-Guidelines.pdf.

Karvonen, Andrew. 2018. The city of permanent experiments? In *Innovating climate governance: Moving beyond experiments*, eds. Bruno Turnheim, Paula Kivimaa, and Frans Berkhout, 201–15. Cambridge: Cambridge University Press.

Krohn, Wolfgang, and Johannes Weyer. 1989. Die Gesellschaft als Labor: Die Erzeugung sozialer Risiken durch experimentelle Forschung. *Soziale Welt* 40 (3): 349–73.

McCrory, Gavin, Johan Holmén, Niko Schäpke, and John Holmberg. 2022. Sustainability-oriented labs in transitions: An empirically grounded typology. *Environmental Innovation and Societal Transitions* (43): 99–117.

Meyer-Soylu, Sarah, Oliver Parodi, Helena Trenks, and Andreas Seebacher. 2016. Das Reallabor als Partizipationskontinuum – Erfahrungen aus dem Quartier Zukunft und Reallabor 131 in Karlsruhe. *TATuP* 25 (3): 31–40.

Parodi, Oliver. 2011. *Quartier Zukunft – Labor Stadt*. Available from https://www.itas.kit.edu/pub/v/2011/paro11a.pdf.

Parodi, Oliver. 2019. Wider eine Engführung des Reallabor-Konzepts. Ökologisches *Wirtschaften* (34): 8–9.

Parodi, Oliver, Marius Albiez, Richard Beecroft, Sarah Meyer-Soylu, Alexandra Quint, Andreas Seebacher, Helena Trenks, and Colette Waitz. 2016. Das Konzept "Reallabor" schärfen. *GAIA* 25 (4): 284–85.

Parodi, Oliver, Richard Beecroft, Marius Albiez, Alexandra Quint, Andreas Seebacher, Kaidi Tamm, and Colette Waitz. 2017. The ABC of real-world lab methodology – From "action research" to "participation" and beyond. *Trialog* 126/127 (3-4): 74–82.

Parodi, Oliver, and Kaidi Tamm, eds. 2018. *Personal sustainability: Exploring the far side of sustainable development.* London: Routledge.

Parodi, Oliver, Colette Waitz, Monika Bachinger, Rainer Kuhn, Sarah Meyer-Soylu, Sophia Alcántara, and Regina Rhodius. 2018. Insights into and recommendations from three real-world laboratories: An experience-based comparison. *GAIA* 27 (1): 52–59.

Parodi, Oliver, and Anja Steglich. 2021. Reallabor. In *Handbuch Transdisziplinäre Didaktik,* eds. Tobias Schmohl and Thorsten Philipp, 255–65. Bielefeld: transcript.

Parodi, Oliver, Paula Bögel, Richard Beecroft, Andreas Seebacher, Felix Wagner, and Julia Hahn. 2022. Reflexive sustainable technology labs: Combining real-world labs, technology assessment and responsible research & innovation. *Sustainability* 14 (22): 15094.

Polanyi, Karl. 2001. *The great transformation: The political and economic origins of our time.* Boston, MA: Beacon.

Real-world lab network. 2023. Available from https://www.reallabor-netzwerk.de.

Schäpke, Niko, Franziska Stelzer, Guido Caniglia, Matthias Bergmann, Matthias Wanner, Mandy Singer-Brodowski, Derk Loorbach, Per Olsson, Carolin Baedeker, and Daniel J. Lang. 2018. Jointly experimenting for transformation? Shaping real-world laboratories by comparing them. *GAIA* 27 (1): 85–96.

Schneidewind, Uwe, and Hanna Scheck. 2013. Die Stadt als "Reallabor" für Systeminnovationen. In *Soziale Innovation und Nachhaltigkeit. Innovation und Gesellschaft,* ed. Jana Rückert-John, 229–48. Wiesbaden: Springer.

Schneidewind, Uwe, and Mandy Singer-Brodowski. 2014. *Transformative Wissenschaft: Klimawandel im deutschen Wissenschafts- und Hochschulsystem.* Marburg: Metropolis.

Schultz, Julia, Fridolin Brand, Jürgen Kopfmüller, and Konrad Ott. 2008. Building a "theory of sustainable development": Two salient conceptions within the German discourse. *IJESD* 7 (4): 465–82.

Singer-Brodowski, Mandy, Richard Beecroft, and Oliver Parodi. 2018. Learning in real-world laboratories: A systematic impulse for discussion. *GAIA* 27 (1): 23–27.

Steglich, Anja, Grit Bürgow, and Angela Million. 2020. Optimising aquaculture in urban agriculture. In *Achieving sustainable urban agriculture*, ed. Han Wiskerke, 303–30. Cambridge: Burleigh Dodds Science.

Turnheim, Bruno, Paula Kivimaa, and Frans Berkhout. 2018. *Innovating climate governance: Moving beyond experiments*. Cambridge: Cambridge University Press.

WBGU [Wissenschaftlicher Beirat Globale Umweltveränderungen]. 2011. *World in transition – A social contract for sustainability.* Flagship Report. Available from https://www.wbgu.de/fileadmin/user_upload/wbgu/publikationen/hauptgutachten/hg2011/pdf/wbgu_jg2011_en.pdf.

WBGU. 2016. *Humanity on the move – Unlocking the transformative power of cities. Flagship Report.* Berlin: WBGU.

Wiek, Arnim, and Daniel J. Lang. 2016. Transformational sustainability research methodology. In *Sustainability science*, eds. Harald Heinrichs, Pim Martens, Gerd Michelsen, and Arnim Wiek, 31–41. Berlin: Springer.

Research Integrity

Marie Alavi and Tobias Schmohl

Definition

Research integrity refers to the conduct of research in accordance with accepted research norms and practices as well as ethical values. It is helpful to examine the term's constituent parts in order to define this concept more precisely.

In general, *research* refers to a methodologically sound and critically minded scientific investigation: It is a process of exploring the new or unknown according to methodological standards usually derived from domain-specific contexts and under the premises of objectivity, reproducibility, and reliability. Students and research professionals must adhere to their discipline's standards, codes, and guidelines as well as those of their institutions (universities, businesses, or research organizations) and their respective governments' laws, rules, and regulations. The importance of disseminating findings to other researchers (*science to science*), practitioners who apply their findings in practice (*science to business*), or to the general public (*science to society*) is also growing. Given the societal, ecological, political, and economic significance of research findings, one of the most important requirements to all research actors, fields, and disciplines is "to ensure the highest levels of integrity in research" (OECD 2007, 1), to maintain society's trust in research, to protect the reputation of researchers and institutions, to ensure the reproducibility of research results, and to prevent fraud and misconduct (Science Europe 2017, 2-5).

The word *integrity* is composed by the Latin prefix *in-* (not) and the verb *tangere* (to touch). Its adjectival form, "integer", refers to the state of an entity that is "in no way touched, affected, altered [or] corrupted" (ten Have et al. 2021, 641). In this sense, research integrity consequently carries moral implications: There is widespread agreement that research should be conducted ethically and in conformity with the established norms in order to generate honest and reliable research results. In this context, the term "integrity" refers to the disposition of researchers to conduct research in accordance with appropriate ethical, legal, and professional frameworks, obligations, and standards. This relates to several levels of accountability, including those of individual researchers, their institutions, (potential) clients, the government, and the general public.

Between individuals, structures, and systems, research is conducted on a multilateral scale. Therefore, all parties involved in or influencing the research process are expected to generate scientific knowledge in an honest, valid, and trustworthy manner, ideally independent of ideological, economic, or political motives. Individual researchers, collaborative teams (institutional, inter-institutional, international, corporate), research performing organizations (such as universities and non-university research institutions), research funding organizations, scientific journals and publishers are parties involved in research. Moreover, public authorities, "university administrations, ethical review institutions, [and] legislation" (Helgesson and Bülow 2023, 118) are expected to establish the conditions necessary for the achievement of research integrity in its entirety.

Research integrity encompasses both external and internal research norms: external norms in the form of laws or regulations, guidelines, codes, or rules that guide the conduct of researchers, and internal norms in terms of internalized standards or desired behaviors. The research integrity framework addresses research *actors'* behavior and responsibility (a, c), as well as the conduct and impact of the *research itself* (b, c).

(a) Research integrity is the *attitude* of research actors to conduct research in an accountable, equitable, and reliable manner within the context of generally accepted scientific common sense and a scientific ethos (in the sense of a research habitus; Steneck 2006, 55). Different research integrity codes of conduct provide a normative framework for individual researchers, research collectives, institutions, and higher education. General codes cover content which is relevant across disciplines, such as the research environment, training, supervision and mentoring, research procedures and ethics structures, data practices and management, research collaboration, dissemination, publication, reviewing, and integrity breaches, and share (to varying degrees) fundamental principles to be applied in all research contexts, such as honesty, accountability, reliability, respect and others (WCRI 2010; WCRI 2013; ALLEA 2017), while discipline-specific, institution-specific, or region-specific codes focus on specialized aspects and needs of institutions.
(b) Research integrity also refers to both the research process (*conduct*) and the reporting or publishing (*dissemination*) of scientific information. As a result, it includes systematic procedural guidelines for responsible conduct of research to ensure the quality of the methodological process (Wilder et al. 2022, 206).
(c) Finally, research integrity is closely related to *research ethics,* which refers to the ethical responsibility of research and research actors (Steneck 2006, 56). Research ethics emphasizes behavior, attitude, values, and virtues. Therefore,

research integrity and research ethics should be understood holistically as complementary perspectives that relate to a situation, problem, or strategy to varying degrees.

Research integrity transcends disciplinary, thematic, cultural, and national boundaries by incorporating all of these factors and multiple viewpoints. Therefore, a transdisciplinary approach to research integrity education offers a promising means of effectively communicating pertinent aspects of research integrity across disciplines.

Background

Research integrity is not an enshrined construct, but has been a topic of ongoing activities of the research community for two decades. It spans national, institutional, and disciplinary boundaries in a "global effort to foster integrity" (Steneck et al. 2017, 3). Numerous guiding documents, conferences, studies, or educational resources have emerged in this context. Fifty-two nations are currently represented at the World Conferences on Research Integrity, held in Lisbon (2007), Singapore (2010), Montreal (2013), Rio de Janeiro (2015), Amsterdam (2017), Hong Kong (2019), and Cape Town (2022), as a result of initial work by the United States Office of Research Integrity and two members of the Organisation for Economic Co-operation and Development, Canada and Japan. The European Code of Conduct for Research Integrity (ESF and ALLEA 2011), which was intended to serve as an umbrella standard for Europe (revised and published in its final version in 2017), or the Montreal Statement on Research Integrity (3rd World Conference on Research Integrity) are some of the concrete reference documents for research standards that have emerged primarily in the last decade. The conclusions of the Council on research integrity (Council of the European Union 2015) was another significant document because it placed an emphasis on ethical principles in addition to integrity. This was followed by the Hong Kong principles (Moher et al. 2019), which emphasize researchers' behavior. On the global landscape, there are several other guidelines. To implement them on the institutional level, they are either imposed on institutions by government mandates or by making their establishment a requirement to receive funding. However, most higher education institutions implement the standards on the basis of their commitment (Steneck 2006, 67), with the demand for easy access and dissemination among their researchers, employees, and students emphasized.

These efforts paved the way for further initiatives promoting and safeguarding research integrity and ethics and creating an interplay of various types of expertise and methodologies. Among such initiatives are science-led infrastructures like the ombuds system as well as reproducibility networks, research integrity offices and research ethics committees, which serve as platforms for the dialogue between organizations and professionals and provide advisory services for the investigation of misconduct or ethical aspects. In addition, several guiding, mentoring, and training initiatives as well as studies thereof have emerged. The different instructional, methodological, and content-related programs for professionals, educators, students at different levels of qualification, institutions, policymakers, and ideally industry stakeholders show the different angles from which we can approach a comprehensive culture of research integrity and meet the current needs of specific communities.

The reason for many initiatives regarding research integrity are breaches in research. The research community places particular emphasis on serious misconduct like *fabrication* (creating data and reporting them as if they were real), *falsification* (the manipulation, modification, withholding, or elimination of data), and *plagiarism* (taking others' statements, data, ideas with inadequate or no citation of the source; Bouter 2020, 2364). Moreover, so-called *questionable research practices* exist in the gray area between scientifically desirable practices and those to be rejected (Fanelli 2009, 1). Among many others, such practices include *cooking* (giving ordinary observations extraordinary character), *mining* (highlighting a discovered statistically significant relationship as the true intention of the analysis), selective reporting or citing (only if it meets one's own expectations), etc. (Bouter 2020, 2364). Predatory and hijacked journals (Abalkina 2022) and AI-based paraphrasing tools or text generators pose new challenges to research integrity.

Debate and criticism

The need for trustworthy and high-quality research is the main driver in the establishment of research integrity, while persistence of misconduct and questionable research practices demonstrates that research integrity is not firmly established in the field. Especially the latter are considered to be remarkably detrimental due to their greater prevalence (Bouter et al. 2016, 2363). John et al. even assume "that some questionable practices may constitute the prevailing research norm" (2012, 524). Therefore, there is a great need for a multi-perspective approach of enabling, empowering, mentoring, and training across disciplines, actors, and institutional structures. These include the implementation of codes of conduct as guiding documents for institutions and the monitoring of their observance, the establishment of peer review systems, the introduction of ethics committees, the conduct of misconduct investigations, and strengthening the position of whistleblowers (e.g. ORI 1995). In addition, the concept of *open science* represents an opportunity to make research findings "more traceable and verifiable" (Haven et al. 2022, 2), also for the benefit of citizens as recipients of research results (Priess-Buchheit et al. 2020, 30).

Kalichman points out that research integrity involves "socialization, incentives, and culture" (2016, 785). All actors are therefore called upon to work together in order to develop a structural, institutional, financial, and political environment that is conducive and stimulating for a responsible conduct of research (Sørensen et al. 2021, 2). The most problematic criticism, however, is that the research system creates incentives that work against research integrity such as funding and publication pressures (Bouter 2020, 2364) and that there are situations when the opportunities and obligations of each research actor are not entirely transparent or well-defined (Horbach and Halffmann 2017, 1464).

Consistent efforts to intensify training are therefore required from the educational sector. However, the effectiveness of interventions is the subject of ongoing research. The various requirements of research integrity in single disciplines (for example, data management or ethics codes in medical sciences versus humanities) raise the question whether a generic approach of the training or a more specialized, discipline-specific one should be chosen (Sørensen et al. 2021, 2), which didactic method is the most productive, and which content should be taught. Another aspect is the challenge to transmit or evoke values and an inner orientation in the sense of a scientific habitus, in addition to teaching concrete knowledge (such as rules or methods). This challenge can be addressed through a transdisciplinary approach, where learners can gain experience in other contexts, be confronted with insights from other disciplines or new theories, and thus enrich their inner attitude towards their research activity.

Since research results stemming from the private sector also affect society, it is necessary to investigate the extent to which research integrity is taught and practiced in the economic realm. Given that academic research integrity faces the behavioral, institutional, and infrastructural challenges outlined above, and that research, its funding, and its dissemination involve numerous, potentially very diverse interests, the assessment of responsible conduct of research in industry and the private sector is a crucial issue. This leads to the conclusion that research integrity education should be accessible not only to students, but also to researchers and educators at all career stages. A certain level of training should be available as well to non-researchers (such as funders, reviewers, journals, policy makers; Fanelli 2019, 5, 11) who are involved in the research process in different contexts. Considering that society's ability to understand research outcomes is a prerequisite for the trust in science, citizens should also be educated in research integrity (Priess-Buchheit et al. 2020, 30).

Current forms of implementation in higher education

The claim that "integrity in research should be developed in the context of an overall research education program" (Institute of Medicine and National Research Council 2002, 84) is highly pertinent and an ongoing effort of educational institutions. Key findings from the training program Path2Integrity (2019–2022) show that students at lower qualification levels are less motivated to engage in research integrity training lacking relevance for them (Valeva et al. 2022, 530). Whereas it is precisely during their university studies that learners are introduced to research activities and required to apply scientific methods such as literature review, responsible elaboration, and report of data, but also to be accountable and open for critical reflection (Steneck 2006, 56).

While universities offer various courses on individual research integrity topics (such as scientific writing, research methods, etc.), there are several educational programs, mostly developed in the academic context, which are applicable at an international level and which enrich the educational landscape both thematically and methodologically. These include, for example, the toolbox Standard Operating Procedures for Research Integrity, which assists institutions in developing a Research Integrity Promotion Plan (SOPs4RI) or the wiki-platform The Embassy of Good Science, which serves as a repository for comprehensive information on educational resources worldwide. By 2020, Pizzolato et al. have collected 237 mostly online and freely accessible domain and non-domain-specific educational resources consisting of videos, online (self-processing or collaborative) courses, textbooks or case study collections covering primarily misconduct-related content, followed by publication ethics, data management, and others such as research procedures or collaborative working. Examples of these are VIRT2UE, INTEGRITY, and Path2Integrity. While all of them make their training available worldwide, VIRT2UE addresses mainly educators from all disciplines and aims to strengthen the learners' attitudes towards research integrity. INTEGRITY is a modular training mainly for high school students, stimulating their critical awareness by mostly socially relevant (real or fictive) cases that are considered to be interesting for this target group (from activism to usage of mobile devices). Path2Integrity offers various modular and dialogue-based learning materials for citizens, undergraduates, graduates, and early career researchers that address the content fields of the European Code of Conduct for Research Integrity and are thus applicable across disciplines. Formal and informal learning pathways (learning materials, campaign, and role models) are used to address learner awareness. Additionally, it offers several evaluation instruments like a feedback sheet or a pre- and post-test survey in order to assess possible learning gains (Zollitsch et al. 2021).

Several recommendations and findings from training conducted can be gleaned from the available literature. Nonetheless, they do not constitute universal guidance and must be evaluated in light of the particular context and learning objective. Katsarov et al. (2022, 951) outline that voluntary courses have a more positive impact on learning outcomes. The much-needed learner attitude for research integrity, however, is partially hampered by learners with lower qualification levels who question the relevance of research integrity for them (Valeva et al. 2022, 530). The design and effectiveness of interventions, in addition to the question of an (effective) instructional approach, depend on the different statues of learners, such as prior knowledge, research experience, skills, and qualifications.

It is noteworthy that most of the trainings (especially the latter two) tend towards a learner-centered dialogical approach, allowing learners to experience research integrity through role-play, discussion, storytelling, or the presentation of concrete research-relevant scenarios (Hermeking and Priess-Buchheit 2022, 112). This, together with the cross-disciplinary topics and the aspect of formal and informal learning (Path2Integrity), is a useful example of how education in research integrity through a transdisciplinary teaching–learning setting is an effective way to teach and learn many aspects of research integrity.

References

Abalkina, Anna. 2022. Publication and collaboration anomalies in academic papers originating from a paper mill: Evidence from a Russia-based paper mill. *arXiv* 2112.13322: 1–30.

ALLEA [All European Academies]. 2017. *The European code of conduct for research integrity*. Revised edition. Berlin: ALLEA. Available from https://allea.org/code-of-conduct/.

Bouter, Lex M. 2020. What research institutions can do to foster research integrity. *Science and Engineering Ethics* 26: 2363–69.

Bouter, Lex M., Joeri Tijdink, Nils Axelsen, Brian C. Martinson, and Gerben ter Riet. 2016. Ranking major and minor research misbehaviors: Results from a survey among participants of four World Conferences on Research Integrity. *Research Integrity and Peer Review* 1 (17): 2–8.

Council of the European Union. 2015. *Council conclusions on research integrity*. Available from https://data.consilium.europa.eu/doc/document/ST-14853-2015-INIT/en/pdf.

ESF and ALLEA [European Science Foundation and All European Academies]. 2011. *European Code of Conduct for Research Integrity*. Available from https://www.allea.org/wp-content/uploads/2015/07/Code_Conduct_ResearchIntegrity.pdf.

Fanelli, Daniele. 2009. How many scientists fabricate and falsify research? A systematic review and meta-analysis of survey data. *PLoS ONE* 4 (5): 1–11.

Fanelli, Daniele. 2019. *MLE on research integrity – training and education.* Available from https://ec.europa.eu/research-and-innovation/en/statistics/policy-support-facility/mle-research-integrity.

Haven, Tamarinde, Gowri Gopalakrishna, Joeri Tijdink, Dorien van der Schot, and Lex M. Bouter. 2022. Promoting trust in research and researchers: How open science and research integrity are intertwined. *BMC Research Notes* 15 (302): 2–5.

Helgesson, Gert, and Wiliam Bülow. 2023. Research integrity and hidden value conflicts. *Journal of Academic Ethics* 21: 113–23.

Hermeking, Noémie, and Julia Priess-Buchheit. 2022. What's integrity got to do with it? Second-year experiences of the Path2Integrity e-learning programme. *FACETS* 7: 110–25.

Horbach, Serge P. J. M., and Willem Halffman. 2017. Promoting virtue or punishing fraud: Mapping contrasts in the language of "Scientific Integrity". *Science and Engineering Ethics* 23: 1461–85.

Institute of Medicine and National Research Council. 2002. Promoting integrity in research through education. In *Integrity in scientific research: Creating an environment that promotes responsible conduct*, ed. Institute of Medicine and National Research Council, 84–111. Washington, DC: The National Academies Press.

John, Leslie K., George Loewenstein, and Drazen Prelec. 2012. Measuring the prevalence of questionable research practices with incentives for truth telling. *Psychological Science* 23 (5): 524–32.

Kalichman, Michael. 2016. Research integrity: Introduction. In *Handbook of academic integrity*, ed. Tracey Bretag, 785–87. Singapore: Springer.

Katsarov, Johannes, Roberto Andorno, André Krom, and Mariëtte van der Hoven. 2022. Effective strategies for research integrity training – A meta-analysis. *Educational Psychology Review* 34: 935–55.

Moher, David, Lex Bouter, Sabine Kleinert, Paul Glasziou, Mai Har Sham, Virginia Barbour, Anne-Marie Coriat, Nicole Foeger, and Ulrich Dirnagl. 2020. The Hong Kong Principles for assessing researchers: Fostering research integrity. *PLOS Biology* 18(7): 1-14.

OECD [Organisation for Economic Co-operation and Development]. 2007. *Best practices for ensuring scientific integrity and preventing misconduct.* Paris: OECD. Available from https://www.oecd.org/sti/inno/40188303.pdf.

ORI [Office of Research Integrity]. 1995. *Guidelines for institutions and whistleblowers: Responding to possible retaliation against whistleblowers in extramural research.* Available from https://ori.hhs.gov/ori-guidelines-institutions-and-whistleblowers-responding-possible-retaliation-against.

Pizzolato, Daniel, Shila Abdi, and Kris Dierickx. 2020. Collecting and characterizing existing and freely accessible research integrity educational resources. *Accountability in Research* 27(4): 195–211.

Priess-Buchheit, Julia, Arja R. Aro, Iliyana Demirova, Dirk Lanzerath, Pavel Stoev, and Nicolaus Wilder. 2020. Rotatory role-playing and role-models to enhance the research integrity culture. *Research Ideas and Outcomes* 6: e53921: 2–49.

Science Europe. 2017. *Seven reasons to care about integrity in research*. Available from https://www.scienceeurope.org/media/42sphgqt/20150617_seven-reasons_web2_final.pdf.

Sørensen, Mads P., Tine Ravn, Ana Marušić, Andrea Reyes Elizondo, Panagiotis Kavouras, Joeri K. Tijdink, and Anna-Kathrine Bendtsen. 2021. Strengthening research integrity: Which topic areas should organisations focus on? *Humanities and Social Sciences Communications* 8(198): 1–15.

Steneck, Nicholas H. 2006. Fostering integrity in research: Definitions, current knowledge, and future directions. *Science and Engineering Ethics* 12 (1): 53–74.

Steneck, Nicholas H., Tony Mayer, Melissa S. Anderson, and Sabine Kleinert. 2017. The origin, objectives, and evolution of the world conferences on research integrity. In *Scientific Integrity and Ethics in the Geosciences*, ed. Linda C. Gundersen, 3–14. Hoboken: Wiley.

Ten Have, Henk, and Maria do Céu Patrão Neves. 2021. Integrity concept. In *Dictionary of global bioethics*, eds. Henk ten Have and Maria do Céu Patrão Neves, 641. Cham: Springer.

Valeva, Milena, Petya Dankova, and Julia Priess-Buchheit. 2022. Students' mindsets on research integrity – A cross-cultural comparison. *FACETS* 7: 528–42.

WCRI [World Conference on Research Integrity]. 2010. *Singapore Statement on research integrity*. Available from https://www.singaporestatement.org/downloads/main-website/singapore-statements/223-singpore-statement-a4size/file.

WCRI [World Conference on Research Integrity]. 2013. *Montreal Statement on Research Integrity in Cross-Boundary Research Collaborations*. Available from https://www.wcrif.org/downloads/main-website/montreal-statement/123-montreal-statement-english/file.

WCRI [World Conference on Research Integrity]. 2019. *Montreal Statement on Research Integrity in Cross-Boundary Research Collaborations*. Available from https://www.wcrif.org/downloads/main-website/montreal-statement/123-montreal-statement-english/file.

Wilder, Nicolaus, Doris Wessels, Johanna Gröpler, Andrea Klein, and Margret Mundorf. 2022. Forschungsintegrität und Künstliche Intelligenz mit Fokus auf den wissenschaftlichen Schreibprozess. In *Verlässliche Wissenschaft. Bedingungen, Analysen, Reflexionen*, eds. Katharina Miller, Milena Valeva, and Julia Priess-Buchheit, 203–23. Darmstadt: wbg.

Zollitsch, Linda, Nicolaus Wilder, and Julia Priess-Buchheit. 2021. The development of a four-tier test to evaluate research integrity training. *Open Research Europe* 1(90): 1–32.

Research-Based Education

Tibor Koltay and László Z. Karvalics

Definition

Research-based education, leaning on Elliott Seif's (2021) outstanding summary, is "a specific approach to classroom teaching that places less emphasis on teacher-centered learning of content and facts and greater emphasis on students as active researchers", preparing them to be lifelong inquirers and learners. Research-based education orients students' performance through five stages, beginning with active search for and then use of multiple resources in order to explore important, relevant and interesting questions and challenges that lead to clarification and identification of a personal research topic. Gradually, they "find, process, organize and evaluate information and ideas ... learn how to read for understanding, form interpretations, develop and evaluate hypotheses, and think critically and creatively". During this research process they can also understand how to generate contexts, setting a hierarchy of research questions. Through sharing the results, students are developing "communication skills through writing and discussion" (Seif 2021).

From a student's perspective it is common to talk about research-based education (also as inquiry-based learning, research-intensive learning), which "should help motivate them to become experts in their self-described field. And the more often a student gets a taste of what it feels like to be an expert, in however small a concept, the more they will want that feeling later on in life" (Wolpert-Gawron 2016). Again from a teacher's perspective, it is necessary to call the "other side" research or inquiry-based instruction or research-engaged teaching. From a complex educational and organizational culture perspective the overall framework is regularly designated as research-based environment.

Research-based education gradually consolidated as a well-founded methodological-didactical direction, developing special "teaching the teachers"-type courses, and came closer to citizen science initiatives, channeling open scientific crowdsource projects into higher education. This development could progressively improve lifelong learning, transforming its certain terrains to lifelong research (Z. Karvalics 2013). Lifelong research is an extension of research-based education, transplanting "research skills and academic productivity in a feasible and sustainable approach" to the post-university life of the former students (Himelhoch et al. 2015, 445). However, research-based education is not about supporting the reproduction of scientific reinforcements, nurturing new scholars – it is about producing scientifically literate generations from higher education.

Background

It is a common supposition that children are natural-born scientists, since the process by which children turn experience into knowledge is identical to the process that we call scientific knowledge, produced by scientists (Holt 1989 1989). An example of this thinking is described by Alison Gopnik (2012), who found that "young children, in their play and interactions with their surroundings, learn from statistics, experiments and from the actions of others in much the same way that scientists do" (Yaffe 2022, 10). Francis (2012) evaluates this popular approach as a mixture of truth and falsehood, as researchers' thinking has its sources not only in their natural curiosity and mental plasticity, but it is a learned skill. However, as Shanahan (2011) underlines it, science is not just a grown-up version of a child's curiosity.

> While children have the fertile beginnings, becoming a scientist requires that they learn and skillfully practice many abstract skills that are far from intuitive. When students struggle with scientific thinking later in life it isn't because they have unlearned or lost the ability, it's because they (for any number of reasons) did not get to take the next steps to developing those skills and understandings (Shanahan 2011).

This mission has never been completed in elementary and secondary schools. Basic disciplinary science education is a fundamental feature, sometimes with its advanced discovery-based forms which challenged and changed the traditional instruction-based pedagogical culture (Mirzoyan 2021). However, the research environment is simply simulated or emulated into talent management solutions and forms. Now, it is not the child prodigies or wunderkinds, but "child scientists" (McCartney 2011) are the best proofs that there would be several reasons to teach scientific literacy and research skills already to age groups of 12–18 and 8–12. When wanting to complete this unfulfilled mission, it is the role of higher education to provide and practice elementary research skills for all – from the birth of modern universities in the late 19th century.

In medieval universities science making was on the periphery. After the Humboldtian turn, "research (as a process of searching for truth) became a system-forming element of university education, since students, interacting with teachers, acquired not only formal knowledge, but also certain value imperatives that formed their professional vocation and personal position" (Islomovich 2021, 75). To be able to compose a dissertation became a ritual precondition of graduation, as a metaphor of an intellectual initiation process. Science today is apparently not a privilege of the small number of elected, outstanding people. Moreover, there are enormous differences between three different levels of research-based education practice.

The main goal of the introductory (typically undergraduate) level is to foster student awareness and motivation, making them familiar with scientific thinking, mediating research-related forms of literacy and skills. The outcome is some learnt elements of the scientific method, picking up as many abilities from the research literacy complex as students can. On the medium level, the challenge is to be able to use these skills, capacities, and sensitivities for an inquiry-based practice, solving a research problem while acquiring new disciplinary knowledge in a given field. It is nothing else than the rediscovery of existing scientific knowledge, while students are completing micro-research cycles and learning intensively. Finally, at a high level, the students can become producers of new scientific knowledge. Unfortunately, today this training philosophy is not a strict requirement for research-based education practitioners.

Healey and Jenkins (2018, 54) draw a distinction between four overlapping ways in which students may engage with research. The first one is the *research-led level*, where students learn about current disciplinary research. This is followed by (2) the *research-oriented level*, where they develop research skills; (3) the *research-based level*, where the focus is on undertaking research; and (4) the *research-tutored level*, where they engage in discussions on current research. These "stratifications" come into view in the same way in the most popular typology of citizen science, composed by Haklay (2012), reflecting the level of scientific profundity of personal involvement in research processes from the simple *crowdsource* logic, followed by *distributed intelligence* practices, reaching the *participatory science* stage, and finally the *extreme citizen science* projects, where full-value contribution is expected and required from the group members.

In order to reinvent the research-based education environment, it is necessary to accept that education and research are equally important and bridge divides between research and teaching staff. These staff members should excel in both research and teaching. Positive attitudes towards research by students should be strengthened among staff and students. Resources to do research must be available for students, among others by involving libraries in teaching information literacy to students. There should be opportunities and incentives for teachers for further development of their research-based teaching competence and excellence, including the creation of opportunities for dissemination of successful practices. This cannot be done without recognizing teaching excellence and monitoring the growth of research-based teaching. Inter alia, introducing an undergraduate student research award may help in achieving goals that can be solidified if there is more research initiated and financed on the nexus between teaching and research, as well as of research-based teaching and learning in particular (Dekker and Wolf 2016, 10–11).

Debate and criticism

The analysis of research-based education and learning became a popular scholarly field after the millennium. "More than half of the studies were published since 2010, [which] suggests an increasing interest in disseminating the outcomes of incorporating RBL [research-based learning] practices in Higher Education courses" (Camacho et al. 2017, 4192). However, in the forge of the discourse today is living dialogue, debate, and shared experience of the practitioners through textbooks, special reports, methodology exchange forms, blogposts, and comments. This semi-formal ecosystem of ideas gives account of the advantages and disadvantages of research-based methods.

Many teachers mention the greater interest of students during the whole learning and activity cycle, generating more attention, emphasis, engagement, and ambition, discussing the key topical issues in an open way. It is often highlighted that intensive problem-solving focus and the acquired teamwork routine prepares for real-world situations like few others. Practitioners recurrently testify that the retrieval, recall, and reuse of information in the afterlife is strongly supported by research-based forms of education, enhancing long-term knowledge retention (Lindsey et al. 2014).

The list of disadvantages begins with the ambiguous and shaky feedback. The lack of proper assessment creates confusion and anxiety, and the standardized testing performance is missing or poor quality. The risk of students' embarrassment and reluctance is high. Slow thinkers, introverted students, and the ones with learning disabilities are not prepared for the flexibility and freedom assigned to this kind of activity. A majority faces difficulties in collaboration, teamwork, and a culture of sharing results. Their overall readiness and responsibility are characteristically low. The lack of reinforcement flow is also a typical problem. The first steps can be attractive and alluring, but later, unravelling the higher skills of participants' than what is needed for, even an eager student quickly become bored. The lack of such skills can be undermining, and can easily frustrate the students. Unprepared teachers produce disorganized teaching, with vague requirements, and a lack of guidance and task personalization. The result is a sloppy classroom.

The methodology of research-based education was developed in multifarious ways by researchers and practitioners. The long-standing, literature-hunting desk research was stepwise enhanced by varied quantitative and interpretive disciplinary methods (Slater et al. 2015), and action research in education (Efron and Ravid 2017). One of the most intensive and engaging forms of action research is community-based research, which is "collaborative and change-oriented and finds its research questions in the needs of communities", combining "classroom learning with social action in ways that can ultimately empower community groups to address their own agendas and shape their own futures. At the same time, it emphasizes the development of knowledge and skills that truly prepare students for active civic engagement" (Strand et al. 2003, 1).

Methodologies were also enriched by disciplinary endeavors. As a part of the development of general art-based research methodologies (Leavy 2020), universities were early adapters, building the experiences fruitfully into their curricula, creating an independent field, art-based research in education (Cahnman-Taylor and Siegesmund 2017). Alongside the design thinking paradigm, its approaches and considerations were transferred easily to research-based education praxis, as design-based research in education (Philippakos et al. 2021).

The latest frontier is challenge-based research and learning. Challenges are for competing student groups to solve problems. Today it seems to be one of the most efficient and motivating frameworks for learning while solving real-world problems through research. This method develops "student transversal competencies, knowledge of sociotechnical problems, and collaboration with industry and community actors" in a versatile way (Gallagher and Savage 2020). This approach is obviously popular in *research-intensive universities*, where the increase of knowledge flow and the production of new knowledge across diverse disciplines are deeply embedded in the education practice (Njuguna 2015), thus directly creating value for society. There are mingled versions of these methods, too. It is easy to mix art-based platforms with other approaches, and design thinking regularly meets with challenge-based projects (Charosky et al. 2018).

Current forms of implementation in higher education

Healey and Jenkins (2018, 67) collected comparative examples of implemented policies and cases in different higher education institutions all over the world. The updated and improved list below also represents further possibilities of investigation to discover and apply new and new best practices, although there are not many venues where the organized exchange of experiences takes place.

Table 1. International RBE practices

Higher Education education provider	Institutional approach
University of Adelaide, Australia	Small group discovery experience
Dublin City University, Ireland	Challenge-based learning
Humboldt University, Berlin, Germany	Research-based education
Kingston University, London, UK	Promoting and reinforcing a research-based education environment to STEM undergraduate students
Leiden University, Netherlands	Fostering students' awareness of research
University of Lincoln, UK	Student as producer
Maastricht University Netherlands	Extending problem-based learning to research-based learning
McMaster University, Canada	Problem-based and inquiry-based learning
Miami University, US	Student as scholar
Massachusetts Institute of Technology, US	Undergraduate research opportunity program
Olin College of Engineering, US	Group project-based entrepreneurial engineering design projects
Quest University, Canada	Research-based education
Roskilde, Denmark	Problem-oriented project-based learning
University College, London, UK	Research-based education and the connected curriculum
University of Delaware, US	Providing a discovery-oriented environment
Carl von Ossietzky University, Oldenburg, Germany	Research-based teaching and learning as a guideline for developing various degree programs, modules, and individual courses.

The involvement and curriculum-based development of varied literacies also fruitfully supports the goals of research-based education. From these kinds of literacies, *information literacy* appeared early and became a fundamental one that is not restricted to textual information, but relates to digital content, data, and images; thus it is not a stand-alone concept, but overlaps with other literacies (CILIP 2018, 3).

Media literacy, especially in its critical from is similarly fundamental, as it focuses on trustworthy media content, and considers how messages are constructed (Funk et al. 2016).

The newest entry in this group of literacies is *data literacy*. According to one of its definitions, it aims at enabling individuals to access, assess, manage, handle, and use data (Calzada Prado and Marzal 2013, 126). Citizens' critical and active agency is paramount when society's datafication and decision-making, driven by algorithms, has become normalized. One of the enabling factors of data literacy is data citizenship that underlines critical and active agency that takes account of society's datafication and decision-making, driven by algorithms. It is divided into three components: (1) *Data thinking*, i.e. citizens' critical understanding of data collection and data economy; (2) *Data doing*, e.g. everyday engagements with data, including using and deleting it in an ethical way; (3) *Understanding the digital economy*, i.e. how algorithms work and who is funding social media platforms (Carmi et al. 2020, 10). Nevertheless, *visual literacy* (*visuacy*) can greatly help in providing confident and attractive representation of the used data.

Illustrating these new triggers of research-based education, the following table provides an overview of emblematic courses from leading universities. In the future, the number and plurality of these kinds of literacy-focused and data culture-related improvements will expectedly spread in higher education.

Table 2. International literacy practices

Higher education provider	Institutional approach
Rutgers University, US	Producing media literacy-based interventions for active involvement in creating secondary school substance abuse prevention messages.
Stanford University, US	Two curricula (Beyond the Bubble; Reading Like a Historian), based on media literacy and information literacy, and directed to contextualize and corroborate historical texts and stimuli.
University of Nebraska-Lincoln (UNL), Columbia University/ NASA, US	Supporting secondary school students to use authentic climate models and understand epistemic dimensions of climate science, relying on data literacy.

By relying on inquiry-based and research-intensive learning, supported by varied literacies, research-based education is meant to provide engaging research opportunities that are well incorporated in learning activities and well supervised by teaching staff (Van der Rijst 2017). This quality research-based education promises varied types of transformative learning experience for a wide range of students. It will play a growing role in education and, simultaneously, in the production of new scientific knowledge, while building more future-proof universities. Research-based education is not just a way to refresh education practice with stronger student motivation, but also outlines a new, community-driven culture of doing science.

References

Carmi, Elinor, Simeon J. Yates, Simeon J. Eleanor Lockley, and Alicja, Pawluczuk. 2020. Data citizenship: Rethinking data literacy in the age of disinformation, misinformation, and malinformation *Internet Policy Review* 9 (2): 1–22.

Cahnmann-Taylor, Melissa, and Richard Siegesmund. 2017. *Arts-based research in education: Foundations for practice.* New York: Routledge

Calzada Prado, Javier, and Miguel Ángel Marzal. 2013. Incorporating data literacy into information literacy programs: Core competencies and contents. *Libri*, 63 (2): 123–34.

Camacho, Maria H., Martin Valcke, and Katherine Chiluiza. 2017. Research based learning in higher education: A review of literature. INTED 2017 Proceedings 11th International Technology, Education and Development Conference, Valencia, Spain, 6-8 March, 2017. 4188–97.

Charosky, Guido, Luciana Leveratto, Lotta Hassi, Luciana Leveratto, Lotta Hassi, Kyriaki Papageorgiou, Ramon Llull, Juan Ramos-Castro, and Ramon Bragós. 2018. Challenge based education: An approach to innovation through multidisciplinary teams of students using design thinking. *XIII Technologies Applied to Electronics Teaching Conference (TAEE)*, La Laguna, Spain, 2018, 446-53.

CILIP. 2018. *CILIP definition of information literacy.* Available from https://infolit.org.uk/ILdefinitionCILIP2018.pdf.

Dekker, Henk, and Sylvia Walsarie Wolff. 2016. *Re-inventing research-based teaching and learning.* Available from https://bl.curriculumdesignhe.eu/wp-content/uploads/2015/08/Dekker-H.-Walsarie-Wolff-S.-2016-Re-inventing-Research-Based-Teaching-and-Learning.pdf.

Efron, Sara Efrat, and Ruth Ravid 2017. *Action research in education: A practical guide.* New York: Guilford.

Francis, Matthew. 2012. *Children are not "natural" scientists.* Available from https://galileospendulum.org/2012/11/15/children-are-not-natural-scientists.

Funk, Steven S., Douglas Kellner, and Jeff Share. 2016. Critical media literacy as transformative pedagogy. In *Handbook of research on media literacy in the digital age*, eds. Melda Yildiz and Jared Keengwe, 318–48. Hershey, PA: IGI Global.

Gallagher, Silvia Elena, and Timothy Savage. 2020. *Challenge-based learning in higher education: An exploratory literature review.* Available from https:/www.tandfonline.com/doi/abs/10.1080/13562517.2020.1863354Gopnik, Alison. 2012. Scientific thinking in young children: Theoretical advances, empirical research, and policy implications. *Science* 337 (6102): 1623–27.

Haklay, Muki. 2012. Citizen Science and volunteered geographic information: Overview and typology of participation. In *Crowdsourcing geographic knowledge*, eds. Daniel Sui and Michael Goodchild, 105–22. Dordrecht: Springer.

Healey, Mick, and Jenkins, Alan. 2018. The role of academic developers in embedding high-impact undergraduate research and inquiry in mainstream higher education: Twenty years' reflection. *International Journal for Academic Development* 23 (1): 52–64.

Himelhoch, Seth, Sarah Edwards, Mark Ehrenreich, and Philip Luber. 2015. Teaching lifelong research skills in residency: Implementation and outcome of a systematic review and meta-analysis course. *Journal of Graduate Medical Education* 7 (3): 445–50.

Holt, John. 1989. *learning all the time.* Reading, MA: Addison-Wesley.

Islomovich, Ismoilov Temurbek. 2021. Academic profession and university in the context of the historical role of higher education. *International Journal of Economics, Finance and Sustainable Development* 3 (3): 73–77.

Leavy, Patricia. 2020. *Method meets art: Arts-based research practice.* 3rd edition. New York: Guilford.

Lindsey, Robert V., Jeffery D. Shroyer, and Michael C. Mozer. 2014. Improving students' long-term knowledge retention through personalized review. *Psychological Science* (25) 3: 639–47.

McCartney, Melissa. 2011. Child scientists. *Science* 331 (6016): 379.

Mirzoyan, Vera. 2021. Discovery-based learning: Definition, principles, and techniques. Available from https://uteach.io/articles/discovery-based-learning-definition-principles-and-techniques.

Njuguna, James. 2015. *What is the purpose of research-intensive universities?* Cranfield: Cranfield University.

Philippakos, Zoi A., Emily Howell, and Anthony Pellegrino, eds. 2021. *Design-based research in education: Theory and applications.* New York: Guilford.

Seif, Elliott. 2021. *Research based learning: A lifelong learning necessity.* Available from https://www.solutiontree.com/blog/research-based-learning-a-lifelong-learning-necessity/

Shanahan, Marie-Claire. 2011. *Students don't lose their ability to think scientifically.* Available from https://mcshanahan.wordpress.com/2011/09/22/students-dont-lose-their-ability-to-think-scientifically

Slater, Stephanie J., Timothy F. Slater, Inge Heyer, and Janelle M. Bailey. 2015. *Discipline-based education research: A guide for scientists.* Scotts Valley, CA: CreateSpace.

Strand, Kerry J., Sam Marullo, Nick Cutforth, Randy Stoecker, and Patrick Donohue. 2003. *Community-based research and higher education.* New York: Jossey-Bass.

Van der Rijst, Roeland. 2017. The transformative nature of research-based education: A thematic overview of the literature. In *Research-based learning: Case studies from Maastricht University. Professional learning and development in schools and higher education*, eds. Ellen E. Bastiaens and Jeroen van Merriënboer, 2–23.

Cham: Springer. Wolpert-Gawron, Heather. 2016. *What the heck is inquiry-based learning?* Available from https://www.edutopia.org/blog/what-heck-inquiry-based-learning-heather-wolpert-gawron

Yaffe, Philip. 2022. Workings of science: Is science limited to the sciences? *Ubiquity* (22) 3: 1–11.

Z. Karvalics, László. 2013. From scientific literacy to lifelong research: A social innovation approach. In *Worldwide commonalities and challenges in information literacy research and practice,* eds. Kurbanoglu, Serap, Esther Grassian, Diane Mizrachi, Ralph Catts, and Sonja Špiranec, 126–33. Cham: Springer.

Science Communication

Konstantin S. Kiprijanov and Marina Joubert

Definition

Science communication is a dynamic field of practice and research that deals with the communication of scientific knowledge to audiences that are typically outside academic institutions. Science communication is widely recognized as a broad term that encompasses a wide variety of actors and formats, and includes a spectrum of activities that range from informal to strategic in nature (Bucchi and Trench 2021). As such, science communication does not constitute a clearly defined discipline, but can be conceptualized – in the words of Bucchi and Trench – as "an inherently, even joyously, interdisciplinary field" (2021, 2).

Historically speaking, *science* is derived from the Latin word *scientia* for knowledge, understanding, and learning (Onions 1966, 797). Today, the term refers to a practice of systematic production and organization of specialized knowledge by means of specific methods and strict quality standards; it is simultaneously a system for stabilizing said knowledge (Bauchspies et al. 2006, 5–6; Mittelstraß 2010). This chapter employs the term science in a broad sense that is not limited to the natural sciences such as physics or geology, but "has a much broader meaning and includes all the academic specialties, including the humanities" (Hansson 2021). Communication refers to the "practice of producing and negotiating meanings" (Schirato and Yell 1997, x).

In order to do justice to the dynamics and diversity of this multi-faceted field of practice and research, science communication is understood as a broad concept that includes "all forms of communication focused on scientific knowledge or scientific work, both within and outside institutionalized science, including its production, content, use, and effects" (Schäfer et al. 2015, our translation). In this way, knowledge is not simply transferred from one person or community to the other, but it is rather negotiated, mediated, and transformed in a mutual exchange of ideas, opinions, and values. In this chapter, a distinction is made between science communication *teaching* and science communication *training*. For this purpose, *science communication training* is defined as practical courses that are typically offered over one or more days to researchers who are interested to advance their science communication skills. In contrast, *science communication teaching* is defined as academic courses offered by universities and other higher education institutions. Many of these courses are presented as dedicated postgraduate diplomas, Master's courses or PhD programs, but some universities offer modules in science communication as part of degree courses at undergraduate and postgraduate levels.

Although science communication constitutes an essential part of the processes of knowledge production (Horst et al. 2017; Secord 2004) and exchange (Jensen and Gerber 2020), and plays an essential role in transdisciplinary processes, the complex relationship between transdisciplinarity and science communication is yet to be studied from a comprehensive and systematic perspective (Wang 2019). Individual studies, however, offer some promising insights and lay the groundwork for further inquiry. For instance, Mercer-Mapstone and Kuchel (2017) conceptualize the entirety of science communication as a transdisciplinary field due to the inclusion of different actors and the traversing of disciplinary and professional boundaries. Others emphasize that science communication provides the communication skills that are essential for shaping transdisciplinary research processes (Kalmár and Stenfert 2020; Misra and Lotrecchiano 2018; Wang et al. 2019). In addition, Burns et al. (2003, 193) suggest that the act of communication itself has a transdisciplinary dimension, as "the need to explain complex issues in lay terms can lead to new perspectives on a topic and a deeper understanding of the field by the professional".

Background

The multifaceted nature of science communication stems from the complex socioeconomic background that has determined the evolution of the field within distinct national and cultural settings. Scholars of different branches of knowledge have been sharing their knowledge in different contexts and languages for millennia (Gordin 2015, 35–40; Secord 2004). Many traditional approaches to science communication included elements of entertainment – such as music and art – as well as listening and dialogue. The diverse histories, cultural roots, and trajectories of science communication are reflected in Gascoigne et al. (2020). Despite the existing national idiosyncrasies, it is possible to trace a common historical trajectory (Gascoigne et al. 2020) along a number of major historical landmarks, such as the growth of the print market; the professionalization, specialization, and institutionalization of academic research; the emergence and proliferation of mass media and, more recently, social media (Bucchi 2008; Dawson and Topham 2020; Kiprijanov 2021). For a comprehensive overview of the history of science communication from the 18th to the 20th century, see Knight (2006), as well as Gascoigne et al. (2020) and the contributions in volume 16, number 3 (2017) of the *Journal of Science Communication*.

In the wake of the institutionalization of academic research, professionalization, and internal and external differentiation of academic disciplines at the beginning of the twentieth century (Mittelstraß 2010), researchers came to believe that science was too complicated for the general public to understand. This idea corresponds to the *deficit model* (also called *deficit-diffusion model*) of science communication, which was the predominant model until the 2000s. This model describes science communication as a linear transmission of specialized knowledge from a small number of experts to an allegedly ignorant mass audience. The transmission occurs from areas of high concentration among experts to areas of low concentration among audiences. Controversies and misunderstandings are attributed to a lack of scientific literacy among the public. Comparable to Shannon and Weaver's (1949) mathematical model, the deficit model distinguishes between active communicators (senders) and passive recipients who lack any agency.

The deficit model served as the ideological foundation and justification of the *Public Understanding of Science* movement that gained momentum in the United Kingdom during the 1980s and 1990s with the aim of promoting public interest in, and awareness of, the natural sciences (Bucchi 2008). While the dissemination of information about new advances in science is a legitimate and useful activity, the critical flaw of the deficit model is the logical fallacy and incorrect assumption that providing more information and better explanations will lead to more public support for and increased public trust in science.

Reflections on the motivations behind efforts to improve the so-called public understanding of science, as well as concerns about the efficacy of these campaigns, prompted the beginning of systematic science communication research. Criticism from social scientists led to revision of the assumptions and goals of the deficit model. Today, a variety of competing or complementary science communication models exist, as described by Trench (2008), and Schmid-Petri and Bürger (2020), and it is clear that these approaches (or models) are interdependent and often overlap in science communication practice (Brossard and Lewenstein 2010).

According to Horst et al. (2017, 883), the existing models can be placed on a continuum that ranges from *deficit* (one-way, elitist, and fact-oriented) to *dialogue* (two-way or interactive, participatory, and reflective). Dialogue and participation models acknowledge the communication needs and preferences of specific audiences and prioritize meaningful and "mutually supportive relationships between research and society ... through high levels and varied forms of interaction between the two" (Burchell et al. 2017, 200). These forms are also used in transdisciplinary processes, such as participatory research and citizen science. These models reflect current societal demands for more transparency (Weingart et al.

2021) and a "general participatory/collaborative opening of the science system" (Schrögel and Humm 2020, 488).

As science communication became more professionalized and institutionalized, the demand for professionals in the field began to grow, and science communication became a flourishing industry (Davies and Horst 2016). Universities responded by offering a growing number and range of degree programs, as has been documented by a number of scholars (e.g. Massarani et al. 2016; Trench 2012; Turney 1994). For example, Schiele and Gascoigne (2020) document how university-based courses in science communication started to emerge in the 1980s (with one precursor program in the United States in 1960), spreading to countries around the world since then.

Debate and criticism

Proponents of science communication teaching programs use a number of motivations to justify the need for this type of offering in higher education. Motivations include professional capacity building, development of evidence-based policy around public participation in science, provision of authentic educational experiences involving academics and communication professionals, as well as the contributions of this type of program to the employability of students (Longnecker 2014; Longnecker and Gondwe 2014; McKinnon and Bryant 2017; Ramani and Pitrelli 2007).

Already in 1994, Turney pointed out that science communication courses on offer in the United Kingdom in the early 1990s varied from those focusing purely on skills to those with a more theoretical (or big-picture) approach. He argued for the inclusion of theory into science communication education in order to deliver more effective communicators who benefit from lasting intellectual resources, as well as to guard against courses that merely teach students to promote science. Since then, scholars and educators have debated the recurring question about to what extent students need to know the theoretical underpinnings of the science communication skills they learn and how much practical experience should be included in science communication teaching (e.g. Baram-Tsabari and Lewenstein 2017; Mercer-Mapstone and Kuchel 2017).

Overall, there is agreement that the content of science communication programs needs to address both theory and practice, striving for a balance and a productive interface between these components (Baram-Tsabari and Lewenstein 2017; Longnecker 2014; Mercer-Mapstone and Kuchel 2017; Mellor 2013). In general, these authors argue that a theoretical foundation is essential for students to be able to understand and apply the knowledge base, and to identify relevant evidence that can inform and enhance their practice. Practical work, on the other hand, is essential to consolidate what they have learnt and to prepare them for the world of work. Practical expertise can be brought into these courses via guest lectures by industry experts, as well as by work placements and internships. For instance, Bray et al. (2012) emphasize the need for students to develop self-awareness around their own scientific values and science communication objectives, as well as the contexts in which they operate, necessitating a broad understanding of the societal implications of science, rather than a focus on technical media skills.

Academic programs in science communication cannot afford to be static. In order to prepare students to cope with the increasing complexities of science communication and fast-changing media ecosystems, courses must be flexible and respond to changing circumstances and the evolving needs and expectations of students and future employers (Fähnrich 2020, 3; Ramani 2009, 2). Science communication students must be prepared for the complexities of communication around increasingly contested topics that are rooted in science but have social, moral, or ethical dimensions and are often heavily politicized, such as climate change, biotechnology, stem cell research, and artificial intelligence. In addition, societal challenges such as science skepticism and dwindling trust in democratic institutions call for a critical reflection on the relationship between science and society, as well as a new and more inclusive approach to knowledge production (Schrögel and Humm 2020). Students must therefore be equipped with competencies that will be required to navigate the controversies, uncertainties, and polarized debates around science and its applications in society.

Teaching in this field must also keep up with fast-evolving science communication ecosystems and landscapes, characterized by a general move away from mainstream media where journalists were the traditional gatekeepers, towards online and social media where everyone can communicate and comment on science (Fähnrich 2020, 1). Digital science communication channels are characterized by fragmented audiences, online hostility and concerns around mis- and disinformation, and social media may disrupt scientific standards and challenge the authority of science. Jointly, these trends demand a continuous reassessment of the theoretical and practical content taught within academic science communication courses.

In the 1990s, higher education and research institutions around the globe embarked on a journey of transformation toward fostering positive societal impact beyond economic goals. This ongoing transformation, reflected by concepts such as *Third Mission* (Trencher et al. 2014) and *Quadruple Helix* (Carayannis and Campbell 2009), is accompanied by a shift toward a mode of knowledge production that surpasses the boundaries of established academic disciplines (Scholz 2020). Science communication is understood to be an essential and integral part of this process, as dialogue-focused and participation-oriented activities play a key role in knowledge and research exchange through engaging a wide range of stakeholders from outside the academic domain (Jensen and Gerber 2020; Leshner and Scheufele 2017). Scholars also point out that, following the idea of Responsible

Research and Innovation, participatory science communication – understood as joint knowledge production – can make a significant contribution to strengthening innovation processes, and to developing solutions to challenges that affect society as a whole (Loroño-Leturiondo and Davies 2018).

Current forms of implementation in higher education

Academic programs in science communication around the world share some common approaches and characteristics, but also significant variations in content, goals, learning outcomes, and delivery (Trench 2012; Trench and Bucchi 2021). This plurality originates from different national and institutional contexts, and the unique views of the individuals who champion these programs. However, despite the interdisciplinary nature and diversity of the field, it is argued that there are some core topics that should be covered in a science communication degree program (Longnecker and Gondwe 2014).

A number of handbooks related to science communication constitute evidence that the field is maturing and encompasses a range of core topics that should be considered by teachers in the field. These books provide a valuable starting point for curriculum developers and are useful guides for educators and students (Bucchi and Trench 2021; Jamieson et al. 2017; Leßmöllmann et al. 2020; Van Dam et al. 2020). In addition to the field delineation and guidance provided by these handbooks, several science communication scholars have generated ideas and topics for a core science communication curriculum (e.g. Baram-Tsabari and Lewenstein 2017; Bray et al. 2012; Mercer-Mapstone and Kuchel 2017). Further reports by scholars such as Costa et al. (2019), Fähnrich (2020), Gascoigne et al. (2020), Hong and Wehrmann (2010), Longnecker (2016), Longnecker and Gondwe (2014), and Massarani et al. (2016) have suggested several core topics to be considered when new science communication programs are designed. These include: (1) Social studies of science, including the history, sociology, and philosophy of science and science communication; (2) Media studies and communication science, theory, and strategies; (3) Behavioral studies, including persuasive communication; (4) Education studies and learning theories associated with informal learning; (5) Public participation in science and the co-creation of scientific knowledge, including citizen science; (6) Evaluation of science communication materials and projects; (7) Research methodologies relevant to science communication scholarship; and (8) Real-world experiences for students via industry placements or institutional internships.

Despite consensus on the value of agreeing on core concepts, experienced science communication teachers recognize that most science communication programs will not be able to include all these topics. They therefore advise curriculum designers of new programs to consider local needs and priorities carefully (Longnecker and Gondwe 2014). New degree programs in the field should be clear about their learning outcomes and strengths, and should use their uniqueness as a way to attract relevant students (Hong and Wehrman 2010).

Although science communication plays an essential role in transdisciplinary processes, there is little systematic research into the specific relationship between transdisciplinarity and science communication (Du Plessis 2012; Wang 2019), and only a few studies explore this question from the perspective of science education (Arber 1999; Mercer-Mapstone and Kuchel 2017). There is, however, a number of practical examples that illustrate the effective integration of transdisciplinary approaches into science communication courses. These include programs at the Karlsruhe Institute of Technology (KIT), the Universidad Nacional Autónoma de México (UNAM), and the University of Leeds. The latter offers joint honors degrees consisting of a STEM subject (e.g. biology or physics), instruction in history or philosophy of science, and an integrated research project in science communication (University of Leeds 2021).

Science communication has been an integral part of modern science for over 200 years and contributes significantly to the circulation and targeted exchange of knowledge between academia, politics, society, and business. Today, the presence of academic education in science communication is recognized as a key indicator of the maturing of the field and its associated infrastructure in different national and regional contexts. However, empirical research into the relationship between transdisciplinarity and science communication in the context of higher education is still lacking. Here, prospective studies might benefit from existing scholarship in neighboring areas, most notably in sustainability studies, and Research and Innovation.

References

Arber, Werner. 1999. Transdisciplinarity in science education and in science communication. In *Thirteenth labor*, eds. Eric Chaisson and Tae-Chang Kim, 15–22. Amsterdam: Gordon and Breach.

Baram-Tsabari, Ayelet, and Bruce V. Lewenstein. 2017. Science communication training: What are we trying to teach? *International Journal of Science Education Part B* 7 (3): 285–300.

Bauchspies, Wenda, Jennifer Croissant, and Sal Restivo. 2006. *Science, technology, and society*. Malden, MA: Blackwell.

Bray, Belinda, Bev France, and John K. Gilbert. 2012. Identifying the essential elements of effective science communication. *International Journal of Science Education Part B* 2 (1): 23–41.

Brossard, Dominique, and Bruce Lewenstein. 2010. A critical appraisal of models of public understanding of science. In *Communicating science: New agendas in communication*, eds. LeeAnn Kahlor and Patricia Stout, 11–39. New York, NY: Routledge.

Bucchi, Massimiano. 2008. Of deficits, deviations and dialogues. In *Handbook of public communication of science and technology*, eds. Massimiano Bucchi and Brian Trench, 57–76. Abingdon: Routledge.

Bucchi, Massimiano, and Brian Trench. 2021. Rethinking science communication as the social conversation around science. *Journal of Science Communication (JCOM)* 20 (3): Y01.

Burchell, Kevin, Chloe Sheppard, and Jenni Chambers. 2017. A "work in progress"? *Research for All* 1 (1): 198–224.

Burns, Terry, John O'Connor, and Susan Stocklmayer. 2003. Science communication: A contemporary definition. *Public Understanding of Science* 12: 183–202.

Carayannis, Elias, and David Campbell. 2009. "Mode 3" and "quadruple helix". *International Journal of Technology Management* 46 (3/4): 201.

Costa, Enrico, Sarah Davies, Suzanne Franks, Aaron Jense, Roberta Villa, Rebecca Wells, and Ruth Woods. 2019. *Deliverable 4.1. Science communication education and training across Europe.* Available from https://ec.europa.eu/research/participants/documents/downloadPublic?documentIds=080166e5cc700100&appId=PPGMS.

Davies, Sarah, and Maja Horst. 2016. *Science communication: Culture, identity and citizenship.* London: Palgrave Macmillan.

Dawson, Gowan, and Jonathan Topham. 2020. Introduction: Constructing scientific communities. In *Science periodicals in nineteenth-century Britain: Constructing scientific communities*, eds. Gowan Dawson, Bernard Lightman, Sally Shuttleworth, and Jonathan Topham, 1–32. Chicago: University of Chicago Press.

Du Plessis, Hester. 2012. *Science communication and transdisciplinarity within an African context.* Paper presented at the Law – Criminal Justice and M-I-T Disciplinary Research Conference, 13–15 August 2012. Magaliesberg, ZAF. Available from https://repository.hsrc.ac.za/handle/20.500.11910/2998.

Fähnrich, Birte. 2020. Deliverable 3.1. *Analysis of the status quo and demands for science communication training.* Available from http://www.rethinkscicomm.eu/wp-content/uploads/2020/06/D3.1-Report-on-analysis-of-status-quo-and-requirements-in-focus-countries.pdf.

Gascoigne, Toss, Bernard Schiele, Joan Leach, and Michelle Riedlinger, eds. 2020. *Communicating science: A global perspective.* Acton: Australian National University Press.

Gordin, Michael. 2015. *Scientific babel.* Chicago: University of Chicago Press.

Hansson, Sven Ove. 2021. Science and pseudo-science. In *The Stanford encyclopedia of philosophy (Fall 2021 edition)*, ed. Edward N. Zalta. Available from https://plato.stanford.edu/archives/fall2021/entries/pseudo-science.

Hong, Chao-Ping, and Caroline Wehrmann. 2010. *Do science communication university programs equip students to become professionals?* Paper presented at the 11th

International Conference on Public Communication of Science and Technology (PCST), 6–9 December 2010. New Delhi: MK Patairiya.

Horst, Maja et al. 2017. Reframing science communication. In *Handbook of science and technology studies*, eds. Ulrike Felt, Rayvon Fouché, Clark Miller, and Laurel Smith-Doerr, 881–907. 4th edition. Cambridge, MA: MIT Press.

Jamieson, Kathleen Hall, Dan Kahan, and Dietram Scheufele, eds. 2017. *Oxford handbook on the science of science communication*. Oxford: Oxford University Press.

Jensen, Eric A., and Alexander Gerber. 2020. Evidence-based science communication. *Frontiers in Communication* 4: 201.

Kalmár, Éva, and Hanneke Stenfert. 2020. Science communication as a design challenge in transdisciplinary collaborations. *JCOM* 19 (4): C01.

Kiprijanov, Konstantin S. 2021. Wissenschaftskommunikation. In *Handbuch Transdisziplinäre Didaktik*, eds. Tobias Schmohl and Thorsten Philipp, 383–93. Bielefeld: transcript.

Knight, David. 2006. *Public understanding of science: A history of communicating scientific ideas*. London: Routledge.

Leshner, Alan, and Dietram Scheufele. 2017. *Communicating science effectively: A research agenda*. Washington, DC: National Academies Press.

Leßmöllmann, Annette, Marcelo Dascal, and Thomas Gloning, eds. 2020. *Science communication*. Berlin: De Gruyter.

Longnecker, Nancy. 2014. *Science communication at UWA*. Perth: University of Western Australia.

Longnecker, Nancy. 2016. An integrated model of science communication – more than providing evidence. *JCOM* 15 (5): Y01.

Longnecker, Nancy, and Musa Gondwe. 2014. Graduate degree programmes in science communication. In *Communicating science to the public: Opportunities and challenges for the Asia-Pacific region*, eds. Leo Tan Wee Hin and R. Subramaniam, 141–60. Dordrecht: Springer.

Loroño-Leturiondo, Maria, and Sarah Davies. 2018. Responsibility and science communication. *Journal of Responsible Innovation* 5: 170–85.

Massarani, Luisa, Elaine Reynoso, Sandra Murrielo, and Ayelen Castillo. 2016. Science communication postgraduate studies in Latin America. *JCOM* 15 (5): A03.

McKinnon, Merryn, and Chris Bryant. 2017. Thirty years of a science communication course in Australia. *Science Communication* 39 (2): 169–94.

Mellor, Felicity. 2013. Twenty years of teaching science communication. *Public Understanding of Science* 22 (8): 916–26.

Mercer-Mapstone, Lucy, and Louise Kuchel. 2017. Core skills for effective science communication. *Journal of Science Education Part B* 7 (2): 181–201.

Misra, Shalini, and Gaetano Lotrecchiano. 2018. Transdisciplinary communication. *Informing Science* 21: 41–50.

Mittelstraß, Jürgen. 2010. Wissenschaftstheorie und Wissenschaftsgeschichte. *N.T.M.* 18: 431–36.

Onions, Charles Talbut. 1966. *Oxford dictionary of English etymology*. Oxford: Oxford University Press.

Ramani, Donato. 2009. Master in science communication: An overview. *JCOM* 8 (1): C01.

Ramani, Donato, and Nico Pitrelli. 2007. The output for the master's degree in science communication at SISSA of Trieste. *JCOM* 6 (01): L01.

Schäfer, Mike, Silje Kristiansen, and Heinz Bonfadelli. 2015. Wissenschaftskommunikation im Wandel. *Wissenschaftskommunikation im Wandel*, eds. Heinz Bonfadelli, Mike Schäfer, and Silje Kristiansen, 10–42. Köln: Herbert von Halem.

Schiele, Bernard, and Toss Gascoigne. 2020. The timelines: A broad-brush analysis. In *Communicating science: A global perspective*, eds. Toss Gascoigne, Bernard Schiele, Joan Leach, and Michelle Riedlinger, 15–50. Canberra: Australian National University Press.

Schmid-Petri, Hannah, and Moritz Bürger. 2020. Modeling science communication. In *Science communication*, eds. Annette Leßmöllmann, Marcelo Dascal, and Thomas Gloning, 105–22. Berlin: De Gruyter.

Scholz, Roland W. 2020. Transdisciplinarity. *Sustainability Science* 15 (4): 1033–49.

Schrögel, Philipp, and Christian Humm. 2020. Science communication, advising, and advocacy in public debates. In *Science communication*, eds. Annette Leßmöllmann, Marcelo Dascal, and Thomas Gloning, 485–513. Berlin: De Gruyter.

Shannon, Claude, and Warren Weaver. 1949. *The mathematical theory of communication*. Urbana, IL: University of Illinois Press.

Secord, James. 2004. Knowledge in transit. *Isis* 95: 654–72.

Schirato, Tony, and Susan Yell. 1997. *Communication and cultural literacy*. Sydney: Allen & Unwin.

Trench, Brian. 2008. Towards an analytical framework of science communication models. In *Communicating science in social contexts: New models, new practices*, eds. Donghong Cheng, Michel Claessens, Toss Gascoigne, Jenni Metcalfe, Bernard; Schiele, and Shunke Shi, 119–35. Dordrecht: Springer.

Trench, Brian. 2012. Vital and vulnerable: Science communication as a university subject. In *Science communication in the world: Practices, theories and trends*, eds. Bernard Schiele, Michel Claessens, and Shunke Shi, 241–57. Dordrecht: Springer.

Trench, Brian, and Massimiano Bucchi. 2021. Global spread of science communication institutions and practices across continents. In *Handbook of public communication of science and technology*, eds. Massimiano Bucchi and Brian Trench, 97–113. 3rd edition. London: Routledge.

Trencher, Gregory, et al. 2014. Beyond the third mission. *Science and Public Policy* 41 (2): 151–79.

Turney, Jon. 1994. Teaching science communication: Courses, curricula, theory and practice. *Public Understanding of Science* 3 (4): 435–43.

University of Leeds, ed. 2021. *Module and programme catalogue: 2020/21 selected undergraduate programme index*. Available from http://webprod3.leeds.ac.uk/catalogue/programmesearch.asp?L=UG&Y=202021&F=P&D=all&S=history+science&A=all.

Van Dam, Frans, Liesbeth de Bakker, Anne Dijkstra, and Eric Jensen, eds. 2020. *Science communication: An introduction*. Singapore: Hackensack.

Wang, Jue. 2019. *Science-practice interaction in transdisciplinary research*. Weikersheim: Margraf.

Wang, Jue, Thomas Aenis, and Tuck Fatt Siew. 2019. Communication processes in intercultural transdisciplinary research. *Sustainability Science* 14: 1673–84.

Weingart, Peter, Marina Joubert, and Karien Connoway. 2021. Public engagement with science. *PloS One* 16 (7): e0254201.

Science Shop

Martine Legris and Frank Becker

Definition

A science shop is defined as a collaborative space where communities, nonprofit organizations, researchers, and students work together to address socially relevant issues and problems (Frickel and Moore 2006). It emerged in this form in Europe in the 1970s, offering independent and participatory research support to civil society organizations (CSOs) willing to develop a research project in response to a particular concern.

A science shop is not a commercial shop, but an entry point into the university for anyone outside academia who is looking for answers based on a scientific approach to a problem. However, a single, uniform format that could easily be transferred does not exist. Some science shops operate as an integral part of a university, while others run as a social enterprise, an association, or a civics initiative. The members of a science shop – usually researchers, students, or academic staff and representatives of nongovernmental organizations (NGOs) – mediate the research process so that civil society, academics, and students become co-researchers or co-contributors. Two conditions quoted in the literature as criteria for science shop projects are: (1) The civil society organization does not bear the costs of the research; (2) neither the science shop nor the cooperating institution pursue commercial interests (Stewart 1988).

Research participation requires a collaborative approach among hybrid groups consisting of research professionals and civil society actors, such as NGOs, grassroots organizations, and residents. Members of a science shop act as mediators in the research process, enabling civil society, academics, and students to become co-researchers or co-contributors. Various stages of the research process require a form of cultural translation. Science shops support to translate civil society's concerns into research questions and provide participatory engineering. This mediation goes beyond dialogue or facilitation. The members of a science shop bridge gaps and create meaning where differences in vocabulary, experience, or knowledge make collaboration difficult.

Science shops depend on their context of emergence and are diverse in terms of the themes they address, the type of support they provide, their institutional position, and their governance. Nonetheless they share common characteristics. Based on a true co-production of knowledge between researchers and organized civil society actors, science shops produce the knowledge that democracies need to address today's social, health, and environmental challenges.

Civil society organizations reach out to science shops with an issue related to their activities, and the science shop, in return, provides support to partners throughout the research process, which typically occurs in several phases (Blangy et al. 2018). All phases involve both civil society organizations and academics, including students. The organizational process may differ from one science shop to another – depending on its institutional design, as science shops may be independent units or university-based – but it follows a common path: (1) gathering the request, followed by (2) translating it into a scientific issue, which is a critical point and the main added value of the support, (3) identifying an academic team to address the subject, (4) conducting the research work based on the co-production of knowledge, (5) returning the results to the actors and disseminating them widely, and (6) concluding with a reflexive phase of research evaluation that enables the system's improvement. Today, more than 50 science shops and comparable intermediaries exist in Europe, Tunisia, and Canada (Living Knowledge 2023).

Background

Following a long tradition of public engagement in research, science shops are part of a movement to redress a divide that emerged at the beginning of the 19th century. This global divide separated the scientific and academic spheres from all other kinds of knowledge, whether experiential, know-how, or user-based knowledge, among others. The construction of a disciplinary structure of science and the belief in technical progress (Habermas 1970) led to a high specialization of knowledge, excluding lay people from the margins of scientific endeavor. However, a countermovement emerged to adapt scientific research to the needs of society rather than the other way around. Several crises and epidemics highlighted the need for a more inclusive and systemic approach to knowledge production (as argued by Morin 1992).

Although a historical overview of science shops does not exist, scholars (Fischer et al. 2004; Millot 2019) agree on decisive turning points. The first wave of science shops started in the Netherlands in the 1970s. The phenomenon then spread throughout Europe due to a positive political and institutional climate. In contrast, the 1990s saw the closure of many European science shops, and in the 2000s even the Netherlands witnessed the demise of several historic science shops.

Only in 2010s were several new science shops established. In 2001, the European Union launched its Science and Society Program (European Commission 2001) to strengthen the societal impact of research. This program funded large research projects, which in turn triggered the rise of new science shops across Europe. Initially, a European network was established, today known as the *Living Knowledge Network*, which efficiently supports science shops in fostering public engagement and participation. Subsequently, many changes in the European governance of research have occurred. Some of them have favored the spread of science shops in European countries, such as the promotion of responsible research and innovation, participatory research, and open science (Rodriguez et al. 2013).

At the same time, a vision of technical democracy, or the empowerment of the public to participate democratically in scientific decision-making, emerged among researchers from different disciplines. This was partly due to the dramatic consequences of past scientific decisions, such as nuclear technology, genetically modified organisms, changes in occupations and jobs, and pollution (Beck 1992; Feenberg 1999). More recently, various trends and theories (Voorberg et al. 2015) have emerged (co-production, co-creation, etc.) to link different practices and methodologies, resulting in a semantic blurring. One remaining question concerns the new challenges science shops are facing. Rather than adapting to a new environment, and possibly reinventing themselves, science shops may initiate the cultural translation and the interdisciplinary process, while providing collaboration frames during the research projects. They also provide meta-analysis of the participatory dimension of the research and improve reflexivity.

Debate and criticism

When addressing societal challenges through a scientific approach, the crucial question is whether the knowledge generated is relevant to the lifeworld solution of the problem at hand. This is facilitated by a non-reductionist approach to the problem and the contextualization of research. Science shops contribute to this development through their constitutive bottom-up approach. In order to remain relevant in research and society, it is also crucial to reflect on their internal quality of intermediation and cultural translation capacity. If science shops are understood as a relevant link between civil society and science, the question of funding their cultural translation services between the logics of the science enterprise and civil society requires attentiveness. Interdependencies may arise from the forms of funding that affect both the interaction between society and science and the contribution of science shops to academic education.

Science shops find themselves in a difficult position between the need to find new social decision-making processes and the pressure of social closure movements (Koppetsch 2019, 34). The tried and tested social negotiation processes of the past no longer seem sufficient to deal with the increased complexity of contemporary society (Latour 2018, 106). To what extent can or should the scientific community be experimentally involved in recent developments such as citizens' assemblies (e.g. Ireland) and major debates (e.g. France)? Phenomena such as populism (Inglehart and Norris 2016) can arise from the uncertainties resulting from the apparent mismatch between societal complexity and existing processing capacities, social closure movements, and tendencies towards greater hierarchization of societies. A reinterpretation and instrumentalization of the dialogical working principle of science shops can be the result. Science shops have to consider this essential aspect of the social ecosystem in which they operate. Such a reflection is equally important for established (as in Europe and North America) and emerging (e.g. in Africa) science shops, as well as for students using science shops as an academic resource: What kind of funding supports a science shop? What decision-making processes govern the interaction between civil society and science? How are civil society representatives involved in the evaluation of outcomes?

In the environment of a university-based science shop, the claim of a university's Third Mission have gained in importance. Transfer, transdisciplinarity, and participation provide a limited ecosystem for academic activities of young academics seeking tenure. At the same time, the social closure movements mentioned above are leading to a change in the capabilities, function, and reputation of science shops. As intermediators, science shops are seen as boundary crossers and disruptive forces at the same time. Neoliberal trends such as "the top-down implementation of competition and market principles under the aegis of New Public Management (NPM) in higher education and science has led not to more but to less professional freedom for those concerned" (Koppetsch 2020, 18, own translation).

The contemporary benefits of science shops are linked to their capacity to deepen and enlarge their main mission. Climate change and major social events as Covid 19 (Latour 2018) request a systemic review of science shops' modus operandi. Six aspects can contribute to enhance collaboration on eye-level: (1) The research question or underlying problem may affect *individuals*. It requires major attention to clarify who are the people that are actively involved in the project. (2) Science shops involve *students* and universities need to develop specific curricula and procedures for them. (3) The *leadership* for the research project can be taken by anyone. To widen the circle of participants or co-researchers, outreach efforts can be made to individuals and groups who may not have been involved initially. (4) An intended *co-research approach* involves collaboration between different actors. This approach aims to facilitate the equal participation of all stakeholders and to ensure that the research is guided by their collective knowledge and expertise. (5) Research projects may involve both *research and implementation*. The balance between these two aspects will depend on the specific aims and objectives of the project. Action and research can be thought of in the same systemic perspective to avoid getting caught up in narrow perspectives that lead to poor experiments. (6) Science shops approach researchers and students to offer them fair cooperation with *nongovernmental organizations*. It is possible to use a transdisciplinary team even if students are supposed to belong to one discipline.

Although solutions to some basic challenges have been identified, embedding them to academic routines and organizational procedures has proved difficult (Schlierf and Mayer 2013). Two key challenges remain: First, research in general and participatory research specifically is often constrained by a lack of time, as co-producing knowledge requires significant time investment. Second, a shortage of human resources, partly due to a lack of institutional recognition and dedicated funding, hinders successful cooperation (Bammer et al. 1992; Legris 2012).

Current forms of implementation in higher education

For budding academics, science shops provide a hands-on learning environment. In universities, the ability to recruit and train academics and students in participatory research and knowledge co-production is crucial for the implementation of science shops in higher education. To achieve this goal, specific training programs are to be designed and integrated into the curriculum, with community-based learning or service learning being the most common form (Ferrari and Chapman 1999; Hyde and Meyer 2004). In independent science shops, links with higher education institutions are established through personal networks, European projects, hiring of undergraduate assistants, and other means.

Students involved gain new knowledge and skills from the process of conducting research on real-life problems. They become familiar with the concept and practice of social responsibility, enhance their professional standing, for example by publishing the research and participating in conferences. In addition, their dissertations serve a purpose. The results of their research are often used to bring about meaningful change.

Science shops provide students with a support system that they can turn to for help with any problems or questions that arise during their research. Nevertheless, students may find it difficult to meet academic standards. New communication and pedagogical skills are needed to deal with multiple stakeholders. In terms of students' experiences, Rao et al.'s (2004) analysis of students' reports suggest that not all of them benefited or were equally satisfied with the experience. Some said they had problems interacting with the community in the field; others did well in the field but had problems with the research itself.

The link between science shops and transdisciplinarity is two-fold: First, science shops engage in cultural translation by promoting reflexivity in the collaboration between civil society and the scientific community. Second, they provide students with an opportunity to engage in negotiation processes on various levels, which they may not otherwise experience in their studies.

Science Shops appear as a mostly Western product of the dialog between science and society. Despite the lack of exhaustive survey, most of them are located today in Europe and Canada.

Figure 1. Geographical distribution of science shops in 2023. Source: own illustration based on data of Living Knowledge (2023), SYNO (2023), and own research.

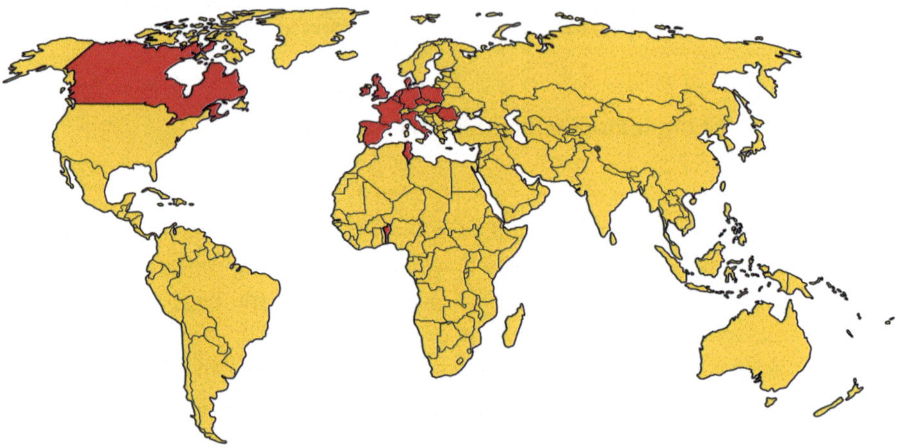

A typology of science shops emerges from their organizational and institutional structure. (1) *University-based* science shops are the original form. In the 1970s, for instance, almost all Dutch universities maintained science shops that were fully integrated into the university structure (Dixon 1988). This integration provided them with stability. They were funded by the universities and other higher education actors. Today, this model is followed by the science shops of the University of Lille and TU Berlin, for instance. (2) A second group of science shops are *nongovernmental associations*, e.g. the Bay Zoltan science shop in Hungary. These institutions often charge for their services or apply for funding. (3) A third group of science shops belongs to the *private sector*. In France, for instance, science shops tended to be independent of universities (Dixon 1970), but they still relied heavily on public funding. Some organizations now accept requests from larger CSOs and commercial companies.

Two cases from Europe and one from Canada may illustrate the specific functionality of science shops.

1. The *FloFauMe* project aims to promote intergenerational measures for preserving urban biodiversity through citizen science cooperation between the Berlin district of Lichtenberg, the environmental NGO NABU, and the science shop of TU Berlin. It provides additional competences and supports cultural translation. Citizens are testing their hypothesis that planting large trees reduces heat in the city, and contribute to establishing a measurement network. FloFauMe involves other stakeholder groups and citizens in DIY workshops to build measuring devices.
2. The science shop at *Lille University* operates on a research ethic that promotes dialogue between different types of knowledge and partners in a relationship of parity and mutual recognition of knowledge. Its focus on participatory research, where the co-production of knowledge between researchers and actors is central, is a distinctive feature. The process begins with the demand from civil society organizations, which is translated into a research problem in collaboration with researchers from relevant disciplines and social actors.
3. The Research Shop at the University of Guelph, Canada, fosters collaborative and mutually beneficial community–university partnerships. Several programs are running, including a community-engaged teaching and learning program (supporting the design of university courses), a program on knowledge mobilization (to support campus-identified dissemination needs), and, more recently, the Guelph Lab. Projects in the Research Shop are undertaken by a small team of student research assistants, supervised and mentored by project managers, all under the supervision of the Research Shop Coordinator. For each project, a work plan is developed involving the community partner, students, and Research Shop manager, to agree the timeline, deliverables, and responsibilities.

In conclusion, the main outputs of science shops include the democratization of expertise, the co-creation of knowledge between researchers and communities, and the promotion of social justice and sustainability (Frickel and Moore 2006; Wibeck et al. 2022). Through science shops, community members, students, and researchers work together to develop research projects aimed at creating social change and promoting democratic, equal, and participatory practices Overall, science shops need to face many challenges while maintaining their inner characteristics favoring the common public good and peer-to-peer relationships.

References

Bammer, Gabriele, Merrelyn Emery, Linda Gowing, and Jennifer Rainforth. 1992. Right idea, wrong time: The Wisenet science shop 1988–1990. *Prometheus* 10 (2): 300–10.

Beck, Ulrich. 1992. *Risk society: Towards a new modernity*. Thousand Oaks, CA: Sage.

Blangy, Sylvie, Bertrand Bocquet , Fiorini Cyril, Fontan Jean-Marc, Martine Legris, and Christian Reynaud. 2018. *Recherche et innovation citoyenne par la Recherche Action Participative*. Available from https://www.openscience.fr/Recherche-et-innovation-citoyenne-par-la-Recherche-Action-Participative.

Chevalier, Jacques M., and Daniel J. Buckles. 2013. *Participatory action research: Theory and methods for engaged inquiry*. New York: Routledge.

Dixon, Bernard. 1970. *What is science for?* New York: Scientific Book Club.

Dixon, Bernard. 1988. Selling research, and it pays. *British Medical Journal* 297 (6660): 1416.

European Commission. 2001. *Science and society action plan*. Available from https://eur-lex.europa.eu/legal-content/EN/TXT/PDF/?uri=CELEX:52001DC0714.

Feenberg, Andrew. 1999. *Questioning technology*. London: Routledge.

Ferrari, Joseph R., and Judith Chapman. 1999. *Educating students to make a difference: Community-based service learning*. Philadelphia: Haworth.

Fischer, Corinna, Loet Leydesdorff, and Malte Schophaus. 2004. Science shops in Europe, the public as stakeholder. *Science and Public Policy* 31 (3):199–211.

Frickel, Scott, and Kelly Moore. 2006. *The new political sociology of science: Institutions, network and power*. Madison: University of Wisconsin Press.

Habermas, Jürgen. 1970. Technology and science as "ideology". In *Towards a rational society: Student protest, science and society*, ed. Jürgen Habermas, 81–121. Boston: Beacon Press.

Hyde, Cheryl, and Megan Meyer. 2004. A collaborative approach to service, learning, and scholarship. *Journal of Community Practice* 12: 71–88.

Inglehart, Ronald F., and Pippa Norris. 2016. *Trump, Brexit, and the rise of populism: Economic have-nots and cultural backlash*. Available from https://ssrn.com/abstract=2818659.

InSpires project open platform. 2023. *ISGlobal, Private Foundation*. Available from https://app.inspiresproject.com.

Koppetsch, Cornelia. 2019. *Gesellschaft des Zorns – Rechtspopulismus im globalen Zeitalter*. Bielefeld: transcript.

Koppetsch, Cornelia, 2020. *Rechtspopulismus als Protest. Die gefährdete Mitte in der globalen Moderne*. Hamburg: VSA.

Latour, Bruno, 2018. *Down to earth: Politics in the new climatic regime*. Available from http://www.bruno-latour.fr/node/754.html.

Legris, Revel Martine. 2012. *The practice of CSO participation: Deliverables of the Consider Project*. Available from https://www.consider-project.eu/activities/wp2-the-practice-of-cso-participation.

Leydesdorff, Loet, and Janet Ward. 2005. Science shops: A kaleidoscope of science–society collaborations in Europe. *Public Understanding of Science* 14 (4): 353–72.

Living Knowledge [The International science shop Network]. 2023. *History of science shops*. Available from https://livingknowledge.org/science-shops/elementor-1260.

Millot, Glen. 2019. *Boutiques des Sciences – La recherche à la rencontre de la demande sociale*. Available from https://www.eclm.fr/livre/boutiques-des-sciences.

Morin, Edgar. 1992. *Method: Towards a study of humankind. Volume 1. The nature of nature*. New York: Peter Lang.

Mulder, Henk, and Gerard Straver. 2015. *Strengthening community–university research partnerships: Science shops in the Netherlands*. Available from https://research.rug.nl/en/publications/strengthening-community-university-research-partnerships-science-.

Rao, Pamela, Thomas Arcury, and Sara A. Quandt. 2004. Student participation in community-based participatory research to improve migrant and seasonal farmworker environmental health: Issues for success. *Journal of Environment Education* 35 (2): 3–15

Rodríguez, Hannot, Erik Fisher, and Daan Schuurbiers. 2013. Integrating science and society in European Framework Programmes: Trends in project-level solicitations. *Research Policy* 42 (5): 1126–37.

Schlierf, Katharina, and Morgan Meyer. 2013. Situating knowledge intermediation: Insights from science shops and knowledge brokers. *Science and Public Policy* 40 (4): 430–41.

Stewart, John. 1988. Science shop in France: A personal view. *Science as Culture* 1 (2): 52–74.

SYNYO. 2023. *SciShops vision*. Available from https://project.scishops.eu.

Voorberg, William H., Victor J. J. M. Bekkers, and Lars G. Tummers. 2015. A systematic review of co-creation and co-production: Embarking on the social innovation journey. *Public Management Review* 17 (9), 1333–57.

Wibeck, Victoria, Karin Eliasson, and Tina-Simone Neset. 2022. Co-creation research for transformative times: Facilitating foresight capacity in view of global sustainability challenges. *Environmental Science & Policy* 128: 290–98.

Scientific Knowledge

Hildrun Walter and Kerstin Kremer

Definition

Scientific knowledge has gained renewed attention in the so-called knowledge-based society today. Challenges and crises highlight the question of what scientific knowledge is and what science can achieve. Especially in inter- and transdisciplinary contexts, one should be aware of the potential of scientific knowledge, but also of the reductionism inherent in a scientific approach.

Defining what scientific knowledge is, what it means when someone claims to know something based on evidence and what this knowledge implies, is part of scientific and philosophical reflections. These questions that originated in ancient Greece were discussed by numerous philosophers, such as Aristotle, Bacon, or Popper, thereby developing the rules of what is today accepted and applied as a scientific approach. Therefore, knowledge acquired via adhering to these rules must be accepted as scientific knowledge (Kuhn and Vessuri 2016, 11).

The term *scientific knowledge* has, regarding its etymology, a double name. *Knowledge* is an Old English word describing the fact of being acquainted with a thing or a familiarity gained by experience (Oxford University Press 2022a). The attribution *scientific* relates to the Latin *scient-, sciēns,* present participle of *scīre,* which means "to know", so knowledge as opposed to belief (Oxford University Press 2022b).

Scientific knowledge is gained via a specific process which adheres to the conditions of science, here used in the embracing notion including non-empirical sciences such as mathematics, law, philosophy, linguistics, and history. The UNESCO (2018, 5) defines science as the

> enterprise whereby humankind, acting individually or in small or large groups, makes an organized attempt, by means of the objective study of observed phenomena and its validation through sharing of findings and data and through peer review, to discover and master the chain of causalities, relations or interaction; brings together in a coordinated form subsystems of knowledge by means of sys-

> tematic reflection and conceptualization; and thereby furnishes itself with the opportunity of using, to its own advantage, understanding of the processes and phenomena occurring in nature and society.

There is no final definition of scientific knowledge, the term is still frequently discussed within and between different scientific disciplines. It encompasses empirical, theoretical, and hermeneutical approaches towards a systematic knowledge acquisition. However, researchers and students should be aware of the various conceptions of knowledge implied in different scholarly cultures. Within transdisciplinary learning contexts, the different cultures of knowledge provide the potential for a more holistic reflection, and therefore more comprehensive understanding. Persons who intend to collaborate in a transdisciplinary research project or educational task should reflect on and communicate their sources of knowledge, its reliability, and limitations. Conscious discussion may enhance mutual understanding for different approaches, documentation, and methodologies in scientific knowledge acquisition and therefore may prevent possible misconceptions (Pohl et al. 2021, 18–19). In addition, this reflection may help to identify common points of contact and complementary additions to the various disciplinary levels of knowledge, thereby enhancing mutual learning.

Background

The conception of scientific knowledge and how scientists gain knowledge is part of epistemology. Epistemology deals with the origin, nature, and limits of human knowledge (Stroll and Martinich 2022). Philosophers, historians, and science sociologists contribute to the subject. The modern scientific system can be divided into three areas.

First, science as a *system of knowledge*. Scientists use terms to specify the origin, conditions, and reliability of statements. They differ between hypotheses – tentative explanations that need to be tested by further investigation; scientific laws – statements that describe the relationship between certain variables under given conditions; and theories – well-substantiated, overarching explanations of natural phenomena. Further, scientific knowledge is classified using disciplinary typologies, grouping items or concepts based on commonalities they share, like the taxonomy in biological classification. These typologies are developed by scholars forming a thought collective (Fleck 1979, 99), which means that they share a framework of ideas, cultural customs, and experiential knowledge.

Second, science as an *organized process*. In their research, scholars apply different methodological approaches including empiricism, analytical methods, and hermeneutics. Empiricism involves careful observation, applying rigorous skepticism about interpretation of observations. It involves formulating hypotheses via induction, experimental testing of deductions and refinement (or elimination) of the hypotheses based on the experimental findings. Not all steps take place in every procedure to the same degree and in the same order. Researchers use analytical methods to reveal type, structure, and function of an object by breaking it down into its components and describing their relationships on a theoretical basis. Hermeneutics describe the theory and practice of interpretation. It reflects the nature, scope, and validity of statements, for example, inherent in texts or observations. Thereby, it can help an understanding of how problems are defined and situate them in a societal and historical context.

Third, science as *cultural achievement*. Scientific knowledge is organized in institutions with specific rules and values, and with a societal role and responsibility. For Europe, this culture has its origins in ancient Greece and, through knowledge exchanges, such as e.g. with the Arab world, has developed over the centuries into modern science as we experience it today. Science in the scholarly sense described above is today applied universally.

The history of scientific knowledge is complex, and a multitude of perspectives and notions exist. However, it is possible to highlight key developments that have led to the emergence of the modern science system. The Greco-Roman ancient world represents a distinct cultural area that produced significant scientific advances and is today regarded as the cultural origin of European science traditions. The first period of scientific history in ancient Greece was characterized by developments in research methods, the formation of rules and systematization, the observation of the course of diseases or the study of order in nature (Merlin 2014, 16–21). Natural philosophers from this period, like Socrates, Plato, or Aristotle, engaged in the earliest known forms of what is today recognized as rational scientific knowledge acquisition. Aristotle's inductive-deductive method used a cyclic process of inductions from observations to infer general principles and deductions from those principles to check against further observations to continue the advance of knowledge systems. Based on Aristotle's work, in the Middle Ages the scientific systematics of scholasticism were developed including, for example, the work of Thomas Aquinas and William of Ockham, elaborating the proof of evidence via disputation (Marrone 2006, 32–37). Assertions were developed based on assumptions, which were then tested for arguments for and against this assertion with the help of logical considerations.

However, during the 16th century English humanists began to value practical knowledge more than solely theoretical consideration, thereby rejecting scholastic disputation (Gaukroger 2001, 6–15). The pursuit of practical knowledge was one major aim Bacon followed (Gaukroger 2001, 9–10, 101). With the help of his commitment to observation and experiment, empiricism became a central part in the reform of natural philosophy towards modern science. This era of scientific revolution achieved "facts, principles, laws, hypotheses, and theories [being] subject to objective judgment in the light of empirical evidence" (Wenning 2009, 12). Furthermore, humanists, such as Nicholas of Cusa and Bacon, recognized the unique role of the researcher, which was a remarkable step in raising awareness of the cultural dimension within science and scientific knowledge. Bacon proposed rules of conduct for researchers, claiming the requisite of good sense and behavior in observation and experiment (Bacon 1620; Gaukroger 2001, 12). With his publication *Discourse on the Method,* Descartes is also recognized as pioneer of the development of modern natural science, especially for emphasizing the significance of doubt or skepticism as an essential attitude for scientific reasoning (Descartes 1637, part two).

In the 17th and 18th centuries, the development of more formalized processes of knowledge creation through empiricism or mathematical reconstruction led to an enormous increase in scientific data that had to be collected and ordered. The amount of available information increased the pressure to treat data selectively, depending on scientific criteria. Furthermore, experimental settings allowed scholars to construct their research around specific subjects and phenomena. Scholars communicated concepts and methods that were more specific to these subjects, finally leading to specialized journals and communities. Through this specialization, the disciplines in the modern sense emerged around 1800. They became institutionalized in scholarly associations and universities. This further structured knowledge formation and conception, its distribution in research and teaching, and its application. The enormous growth of science stimulated by the disciplinary development forced the system to further structuring and internal differentiation, and, therefore, to a multiplication of disciplines. Scholars had to focus their attention on a specific field, thereby leading to an increase in specialization. Although the loss of scientific unity was perceived by scholars themselves, it was not until around 1970 that public attention to environmental protection and technological developments fueled the debate on inter- and transdisciplinarity (Weingart 2010). The previous accumulation of knowledge and techniques within delimited disciplines now allows complex problems to be viewed from different perspectives.

Debate and criticism

The major challenges facing society today are characterized by a multi-layered nature and complex underlying causal chains. Their complexity does not allow for solutions developed within one discipline (Mittelstraß 1987, 154–55), despite the profound stock of disciplinary knowledge available. Researchers need to unify different knowledge perspectives in order to address these challenges in a transformative manner. While the natural sciences can provide insights into laws and relationships, the humanities can offer reflective perspectives and elucidate the cultural embeddedness of observations. Transdisciplinary research thereby represents a complement to disciplinary research, not a replacement. It builds on multidisciplinary research, so addressing the same problem within different disciplines, and by final sharing of results, it combines findings within a common context. It also builds on interdisciplinary research, which means a close interaction between different disciplines in terms of transferring methods and knowledge at an early stage, as well as close cooperation throughout the research process. Furthermore, transdisciplinary approaches also often involve societal actors to integrate their knowledge and perspective (Lawrence et al. 2022, 44–48). A successful integration of these different types of knowledge and practices in such collaborative processes can lead to mutual learning, a more holistic perception of issues, and synergistic, innovative approaches in searching for solutions to societal problems. This integration can only be achieved when the scientists themselves become aware of the properties of disciplinary knowledge, the processes of its development, and its boundaries. Higher education should therefore help to reflect the relevance of mono-, multi-, inter-, and transdisciplinarity to complex problems. Teachers and students should become aware that many disciplinary perspectives exist and that they are not static, but are evolving based on progress in their own as well as in other disciplines (Vereijken et al. 2022, 6). The comprehension of basic scientific concepts and a solid understanding of the epistemological characteristics that are part of disciplinary knowledge are essential baselines for transdisciplinary problem-solving. Here, the concept of Nature of Science could serve as an example for educational implementation.

Nature of Science encompasses an understanding of the epistemic, historical, social, and cultural reach of scientific knowledge as well as an understanding of scientific reasoning and methods. It further reflects the values and norms to justify scientific claims (Heering and Kremer 2018). Since the 1960s, Nature of Science was increasingly taken up by science educators and was then central in the debate from the 1990s (Heering and Kremer 2018, 105; Lederman et al. 2002, 498). Communicating the overarching ideas in science that hold true in several disciplines are today seen as superior outcomes for science education (Lederman et al. 2013, 138–39). One educational aim within Nature of Science is to convey that "scientific knowledge is tentative, empirical, theory-laden, partly the product of human inference, imagination and creativity, and socially and culturally embedded" (Lederman et al. 2002, 499). The authors furthermore underline the importance of teaching the distinction between observation and inference, the lack of a universal method within science, and the functions and relationships between theories and laws in science. Scholars from different disciplines, however, still controversially discuss conceptions of Nature of Science.

Today, educators all over the world accept the comprehensive understanding of Nature of Science as a goal to be achieved in the science classroom and informal educational settings (Allchin 2011, 519; Lederman et al. 2013, 138). Several studies investigated how the explicit reflection of Nature of Science during education supported its understanding, whereby longitudinal studies showed only short-term gains (summarized in Cullinane and Erduran 2022, 2). Alternatives to the rather general "consensus view" of Nature of Science (Lederman et al. 2013, 138) were discussed. Allchin's (2011) "Whole Science" framework highlights dimensions that shape the reliability of science, so that students are empowered for personal and public decision-making. Erduran and Dragher (2014) presented their "Family Resemblance Approach" that provides perspectives on similarities and unique differences of the discipline-specific Nature of Science, such as for chemistry, physics, and biology.

Considering not only the potential but also the limits of scientific knowledge, helps us realize the significance of the plurality of knowledge sources and to recognize other types of knowledge, like indigenous or practitioners' knowledge (Tengö et al. 2014, 579). Indigenous knowledge refers to knowledge with different forms of legitimation and tradition, e.g. through generations of naturalistic observation, place- and community-based insight. From the 1990s onwards, members of the research community have called for the recognition of other cultures of knowledge besides the "Standard Account". This considers knowledge as scientific in the sense of Western culture based on ancient Greek and European heritage (Cobern and Loving 2001, 52–56). As summarized by Cobern and Loving (2001, 54), movements such as multiculturalism (Stanley and Brickhouse 1994), post-colonialism (Rigney 2001), and post-modernism (Lyotard et al. 1995) enabled new epistemological perspectives on the relationship between science, culture, and the 'Standard Account' itself. The conception of scientific knowledge in the standard account helps us to perceive its inherent nature, since it is clearly defined within its disciplinary boundaries and is integrated into educational systems all over the globe. This helps us to place and relate scientific knowledge worldwide in a similar manner, with all the constraints mentioned above. The perception of other knowledge forms, e.g. indigenous knowledge, may be limited due to different cultural backgrounds (Sidik 2022). It is essential to recognize the value and the potential inherent in the diversity of different knowledge forms for more holistic approaches, better ways of social inclusion and their huge potential to support societal transformation. The integration of diverse knowledge systems in transdisciplinary education and research can be challenging (Tengö et al. 2014, 581–82). Therefore, Nature of Science could function as an informative guide for the integration of other knowledge forms. The communication and discussion of their nature in transdisciplinary projects may help to integrate them in a respectful, valid, and synergistic manner.

Current forms of implementation in higher education

In modern science education, the understanding of Nature of Science is a critical component (Khishfe 2022). Therefore, three basic approaches to Nature of Science contextualization are proposed and can be adopted for higher education. Understanding the Nature of Science can be promoted (1) through the integration of case studies from the history and philosophy of science, (2) through the consideration of the mutual influence of science, technology, and society using contemporary cases, as well as (3) through the reflection of individual experimental-research activities using inquiry-based cases (Allchin et al. 2014; Kremer 2008).

The history of science can provide effective Nature of Science contexts. It allows the scientific process and the tentativeness of scientific knowledge to be addressed. It provides insight into the subjective and cultural dimension of science. Another advantage is that the case is already completed in time and thus can help in understanding the evolution of scientific knowledge and the interplay with society. For example, Paraskevopoulou and Koliopoulos (2011, 943) developed a teaching intervention about the dispute between Millikan and Ehrenhaft about the existence of the elementary electrical charge. Douglas Allchin (2012) provides a collection of historical cases for the reflection of Nature of Science and the interplay between science and social and political circumstances.

Using contemporary cases that show the relationship between science and society offer insights into open and controversial debates. Science education researchers contextualized Nature of Science using current socio-scientific issues (Khishfe 2022). For example, press articles or interviews with scientists on the Covid-19 pandemic can serve as material for discussion. In order to be able to bridge the holistic, people-oriented, contextual, social, and personal life-world image and the analytical and objective scientific image that are both part of socio-scientific issues, Zeyer (2022, 5–6) proposes a "Two-Eyed Seeing" method for science teaching. By switching between the two images which can stand side by side, this stereoscopic view provides a more encompassing picture of the world.

The contextualization using inquiry-based cases builds on the conception that Nature of Science can be better understood by actively constructing such knowledge. When the learners engage in a scientific inquiry and reflect on this process, they may gain insight in the nature of the scientific process. Science communication research shows that citizen science has the potential to improve Nature of Science knowledge and attitudes as well as inquiry skills among participants (Peter et al. 2021). Thus, the design and participation in citizen science settings is another promising scenario for transdisciplinarity in higher education.

Teaching scientific knowledge in transdisciplinary contexts in higher education needs both – teaching discipline-specific concepts and knowledge, and teaching ways to transcend disciplinary boundaries to connect complementary fields. In this sense, Baumber et al. (2020, 396) provide a case study focusing on the development and implementation of a four-year curriculum for the Bachelor of Creative Intelligence and Innovation at the University of Technology Sidney. Students first follow three years of disciplinary education and then accomplish a joint fourth year. The curriculum employs a transdisciplinary learning approach based on addressing complex real-world challenges through collaboration and mutual learning across disciplines and with a variety of industry, government, and community partners.

The contextualized reflection of the role of scientific knowledge in historical and contemporary research cases as well as during personal inquiry experiences provides an inevitable foundation for transdisciplinary knowledge formation processes in higher education settings, as it can clarify the contributions, the limitations, as well as the social and political embeddedness of different disciplines.

References

Allchin, Douglas. 2011. Evaluating knowledge of the nature of (whole) science. *Science Education* 95 (3): 518–42.

Allchin, Douglas. 2012. Resource center for science teachers using sociology, history and philosophy of science. Available from http://shipseducation.net.

Allchin, Douglas, Hanne Møller Andersen, and Keld Nielsen. 2014. Complementary approaches to teaching nature of science: Integrating student inquiry, historical cases, and contemporary cases in classroom practice. *Science Education* 98 (3): 461–86.

Bacon, Sir Francis. 1620. *Novum organum. Aphorisms – Book I: True suggestions for the interpretation of nature* (ed. Joseph Devey, 1902). New York: P. F. Collier & Son.

Baumber, Alex, Giedre Kligyte, Mieke van der Bijl-Brouwer, and Susanne Pratt. 2020. Learning together: A transdisciplinary approach to student–staff partnerships in higher education. *Higher Education Research & Development* 39 (3): 395–410.

Cobern, William W., and Cathleen Loving. 2001. Defining "science" in a multicultural world: Implications for science education. *Science Education* 85 (1): 50–67.

Cullinane, Alison, and Sibel Erduran. 2022. Nature of Science in preservice science teacher education: Case studies of Irish pre-service science teachers. *Journal of Science Teacher Education*. Available from https://www.tandfonline.com/10.1080/1046560X.2022.2042978.

Descartes, René. 1637. *A discourse on the method of correctly conducting one's reason and seeking truth in the sciences*. Leiden: Ian Maire.

Erduran, Sibel, and Zoubeida R. Dragher, 2014. *Reconceptualizing the Nature of Science for science education: Scientific knowledge, practices and other family categories*. Dordrecht: Springer.

Fleck, Ludwik. 1979. *Genesis and development of a scientific fact.* Chicago: University of Chicago Press.

Gaukroger, Stephen. 2001. *Francis Bacon and the transformation of early-modern philosophy.* Cambridge: Cambridge University Press.

Heering, Peter, and Kerstin Kremer. 2018. Nature of Science. In *Theorien in der naturwissenschaftsdidaktischen Forschung,* eds. Dirk Krüger, Ilka Parchmann, and Horst Schecker, 105–19. Berlin: Springer.

Khishfe, Rola. 2022. Improving students' conceptions of nature of science: A review of the literature. Available from https://link.springer.com/10.1007/s11191-022-00390-8.

Kremer, Kerstin. 2008. Zufällig nobelpreiswürdig! Unterrichtsbeispiel: Magengeschwür als Infektionskrankheit. *Praxis der Naturwissenschaften – Biologie in der Schule* 57 (2): 16–19.

Kuhn, Michael, and Hebe Vessuri. 2016. The misery of defining what scientific knowledge is – and what not. In *Contributions to alternative concepts of knowledge,* eds. Michael Kuhn and Hebe Vessuri, 9-20. Stuttgart: Ibidem press.

Lawrence, Mark G., Stephen Williams, Patrizia Nanz, and Ortwin Renn. 2022. Characteristics, potentials, and challenges of transdisciplinary research. *One Earth* 5 (1): 44–61.

Lederman, Norman G., Fouad Abd-El-Khalick, Randy L. Bell, and Renée Schwartz. 2002. Views of nature of science questionnaire: Toward valid and meaningful assessment of learners' conceptions of nature of science. *Journal of Research in Science Teaching* 39 (6): 497–521.

Lederman, Norman G., Judith S. Lederman, and Allison Antink. 2013. Nature of science and scientific inquiry as contexts for the learning of science and achievement of scientific literacy. *International Journal of Education in Mathematics, Science and Technology* 1 (3): 138–47.

Lyotard, Jean F., Robert Harvey, and Mark S. Roberts. 1995. *Toward the postmodern.* Atlantic Highlands, NJ: Humanities.

Marrone, Steven P. 2003. Medieval philosophy in context. In *The Cambridge companion to medieval philosophy,* ed. Arthur S. McGrade. Cambridge: Cambridge University Press.

Merlin, Hope. 2014. *The history of science,* ed. Hope Merlin and Nelson Sá. New York: Britannica Educational Publishing.

Mittelstraß, Jürgen. 1987. Die Stunde der Interdisziplinarität? In *Interdisziplinarität. Praxis – Herausforderung – Ideologie,* ed. Jürgen Kocka, 152–58. Frankfurt am Main: Suhrkamp.

Oxford University Press, ed. 2022a. *Oxford English dictionary. Knowledge,* N. https://www.oed.com/view/Entry/104170?rskey=lW5fkv&result=1.

Oxford University Press, ed. 2022b. *Oxford English dictionary. Science,* N. https://www.oed.com/view/Entry/172672?redirectedFrom=science.

Paraskevopoulou, Eleni, and Dimitirs Koliopoulos. 2011. Teaching the Nature of Science through the Millikan–Ehrenhaft dispute. *Science & Education* 20(10): 943–60.

Peter, Maria, Tim Diekötter, Kerstin Kremer, and Tim Höffler. 2021. Citizen science project characteristics: Connection to participants' gains in knowledge and skills. *PLoS ONE* 16 (7): e0253692.

Pohl, Christian, Julie Thompson, Sabine Hoffmann, Cynthia Mitchell, and Dena Fam. 2021. Conceptualising transdisciplinary integration as a multidimensional interactive process. *Environmental Science and Policy* 118(3): 18–26.

Rigney, Lester. 2001. A first perspective of Indigenous Australian participation in science: Framing Indigenous research towards Indigenous Australian intellectual sovereignty. Kaurna *Higher Education Journal* (7): 1-13.

Sidik, Saima M. 2022. Weaving Indigenous knowledge into the scientific method. *Nature* 601 (7892): 285–87.

Stanley, William B., and Nancy W. Brickhouse. 1994. Multiculturalism, universalism, and science education. *Science Education* (78): 387–98.

Stroll, Avrum, and Aloysius P. Martinich, 2022. *Epistemology*. In Britannica [database online]. Available from https://www.britannica.com/topic/epistemology.

Tengö, Maria, Eduardo S. Brondizio, Thomas Elmqvist, Pernilla Malmer, and Maria Spierenburg. 2014. Connecting diverse knowledge systems for enhanced ecosystem governance: The multiple evidence base approach. *AMBIO* (43): 579–91.

UNESCO. 2018. *Recommendation on science and scientific researchers*. Available from https://unesdoc.unesco.org/ark:/48223/pf0000263618.locale=en.

Vereijken, Mayke W. C., Sanne F. Akkerman, Susan F. te Pas, Iris van der Tuin, and Manon Kluijtmans. 2022. *"Undisciplining" higher education without losing disciplines: Furthering transformative potential for students*. Available from https://www.tandfonline.com/doi/full/10.1080/07294360.2022.2156482.

Weingart, Peter. 2010. A short history of knowledge formations. In *The Oxford handbook of interdisciplinarity*, eds. Robert Frodeman, Julie Thompson Klein, Carl Mitcham, and J. Britt Holbrook, 3–14. New York: Oxford University Press.

Wenning, Carl J. 2009. Scientific epistemology: How scientists know what they know. *Journal of Physics Teacher Education Online* 5 (2): 3–15.

Zeyer, Albert. 2022. Teaching two-eyed seeing in education for sustainable development: Inspirations from the Science|Environment|Health pedagogy in pandemic times. *Sustainability* 14 (6343): 1–12.

Scrum

Maren Heibges, Katharina Jungnickel, and Markus A. Feufel

Definition

Scrum is a project management method. Transdisciplinary group learning and collaborative research, especially communication and coordination, can be fostered by adapting the method. The term Scrum originally derives from the rugby ball game, where it refers to a tight huddle when the game restarts (Cervone 2011, 19), and has entered into the (project-)management literature through an article by economists Takeuchi and Nonaka (1986), where they describe new, more flexible and incremental approaches to product development they observed in Japan and the USA in the mid-1980s.

In project management, Scrum's main features are formalized forms of communication and coordination, which require relatively little documentation, a strict meeting rhythm, and particular management roles and artifacts (e.g. a progress board) to manage work tasks (Schwaber and Sutherland 2020). Within this rather rigid framework, Scrum promises flexible, non-hierarchical, and self-organizing work processes thanks to built-in, self-correcting feedback loops for small teams of five to ten people (e.g. Pope-Ruark 2012).

Specifically, project management with Scrum relies on three defining features (Preußig 2015; Schwaber and Sutherland 2020; Shalloway et al. 2010). The first defining feature relates to cyclical workflows, that is, meetings that are repeated after fixed time intervals, which are called *sprints* in Scrum terminology. In contrast to classical project management, Scrum specifies most planning details "on the way". Rather than planning years ahead, which is time-consuming and makes it hard to adapt to new circumstances (Pope-Ruark 2017, 10–13), Scrum needs just a broadly defined general goal at the outset, which is spelled out in detail and revised if needed as the project evolves. This self-corrective function is at the root of the cyclical workflow described above, ultimately enabling adaptive work processes, often referred to as *agile* project management (e.g. Wintersteiger 2015, 20).

Depending on the timeframe of a specific project, work cycles or *Sprints* typically last between one to six weeks and consist of the following meeting formats, which, depending on work cycle length and project size, can each be adapted in length:

- *Planning meeting or Sprint planning:* In this meeting, the project team specifies the tasks necessary for completing the work package that has been prioritized for the upcoming work cycle.
- *Review meeting or Sprint review:* During this format the outcomes of the previous work cycle are reviewed by the team and relevant stakeholders. This meeting aims to evaluate project progress.
- *Introspection meeting or Sprint retrospective:* This is an introspective format, where the team evaluates the past work cycle. Main outcomes of this meeting are decisions about what should be changed in the following work cycle.
- *Update meeting or daily Scrum:* Unlike the other meetings, which are scheduled once every work cycle, this meeting takes place every day. It is only about 15 minutes long and conducted as a stand-up meeting. Each team member identifies what has been completed since the last update, what task they will take on next, and whether there are any obstacles to completion. It serves to synchronize the team members and identify any obstacles.

Once meeting timetables have been specified by the team, they cannot be changed for the ongoing work cycle. This practice points to one essential aspect of the Scrum framework: extremely strict time-keeping referred to as "time-boxing" (e.g. Fowler 2019, 75–76) applied to all Scrum meetings. For instance, if a team runs into problems midway into a work cycle, then the planned timeframe of the work cycle cannot be extended. Instead, the following work cycle will be planned with the past delays in mind. In other words, milestones are more flexible in Scrum and timeframes are more rigid than in classical project management.

The second key component of Scrum is the use of artifacts, such as an (analog or digital) whiteboard to track project progress. This whiteboard contains all tasks identified for a work cycle during the planning meeting (e.g. on sticky notes) and visualizes task progress by the position of these notes in one of three predefined columns, labeled "to do", "in progress", and "done".

The only more conventional documentation format used in Scrum is a project logbook, called *Product Backlog,* which is a list of all work packages to be completed to achieve the overall project goal. Importantly, these work packages are not yet planned out in detail but only roughly defined and prioritized. Over time, as the project progresses, the work packages are broken down into individual tasks during the planning meetings. In the logbook, the goals of individual work packages are usually described by one or two simple sentences, called a *User Story,* to facilitate a shared understanding between team members. A similar communication technique is used when team members define and agree on a short and easy-to-understand *Definition of Done* for each individual task that has been identified as part of a work package.

The third defining feature of Scrum is, in contrast to classical approaches, that it designates two types of project manager roles (Schwaber and Sutherland 2020, 5–7). There is a project facilitator, referred to as the *Scrum Master,* and a project planner, who is called the *Product Owner* in Scrum terminology. The project facilitator runs all meetings, is responsible for solving problems that hinder the workflow, and makes sure that the team complies with the Scrum formats. This role necessitates dominant facilitation techniques to ensure that the Scrum framework stays intact, and it entails significant amounts of troubleshooting to ensure the work processes run smoothly. The project planner, in comparison, focuses on maintaining and refining the project goals and priorities in the logbook by planning and revising necessary next steps *beyond* the current work cycle. The person assuming this role is also in charge of communicating about the project with external stakeholders, such as academic partners or clients.

Background

By the mid-1990s, Schwaber and Sutherland introduced Scrum as a project management framework for software development, which has been updated several times since (Schwaber and Sutherland 2020). From the outset, the design of Scrum has been driven by the desire to operationalize principles of agile management, such as flexibility, transparency, and incremental improvement (Beck et al. 2001). Scrum's ability to foster self-correcting workflows (in other words, its potential to foster agility) is one of its most acclaimed strengths (Wintersteiger 2015). This flexibility is contrasted with more classical approaches to project management, most notably the *Waterfall Model* (e.g. Thesing et al. 2021). Waterfall planning leads to a detailed and fixed project schedule, for instance through highly specified project milestones, which are identified in the planning phase even before the start of the project.

Today, most software is developed within an agile framework (e.g. Krawczyk-Bryłka and Krawczyk 2019), which is based on different forms of implementation, such as Kanban, Extreme Project Management, and Scrum as the most popular method (e.g. Cervone 2011, 19). Given the success of Scrum in the domain of software and product development, it is now being promoted as a general method for project management in many areas of project work, also for higher education and research (Pope-Ruark 2017). However, the literature suggests that Scrum is thus far rarely used in academia (e.g. Fernandes et al. 2021, 4). Most papers discussing applications of Scrum in higher education are low-profile proceedings from international computer science or engineering conferences (e.g. Hicks and Foster 2010; Linos et al. 2020; Ochoa et al. 2021; Persson et al. 2011), with only a few exceptions stemming from educational science and discussing academic applications of Scrum in a non-technical academic setting (e.g. Fernandes et al. 2021; Pope-Ruark 2012, 2017).

Despite the dearth of research, and although Scrum was originally developed to coordinate project management for monodisciplinary teams of software developers (e.g. Schwaber and Sutherland 2020, 5), the literature and the authors' facilitation experience with Scrum suggests that the method can be adapted to create and maintain collaborative accountability (i.e. a shared understanding of goals, processes, and data) in transdisciplinary learning and research contexts. In particular, Scrum can synchronize academic collaboration – acutely so for transdisciplinary teams that are marked by diverse bodies of knowledge and skill sets – through its formalized communication techniques and a focus on creating and maintaining shared (visual) representations of collaborative workflows on progress boards. In other words, Scrum can guide the communication and organization necessitated by transdisciplinary group learning and research.

Debate and criticism

Scrum is mostly criticized for potential shortcomings in oversight, because no detailed milestone roadmap or waterfall project plan is followed (e.g. Cervone 2011, 22; for a general debate see Serrador and Pinto 2015). This is particularly problematic for project management in academia as most funding agencies require standard waterfall plans and reporting. Some practitioners of Scrum also criticize the framework for introducing a certain "breathlessness" into project work, referring to the repetitive work cycles and the metaphorical and literal emphasis on sprinting through fixed time intervals. For academic projects, this may not provide enough time to think and focus on in-depth analyses. Furthermore, Scrum language is rather inaccessible to those unfamiliar with the framework, hampering the creation of a common language and common ground, which already is a challenge for inter- and transdisciplinary team work in academia. Finally, the Scrum philosophy or mindset – which is emphasized as one of its most central parts (e.g. Sloan 2015) – might be perceived as too dogmatic by academic teams and may lead to resistance rather than effective work processes. There are also voices, however, which underline the (politically) progressive potential of the Scrum mindset to foster non-hierarchical work environments and a communicative work and feedback culture (e.g. Pope-Ruark 2017, 15–22).

Project facilitation via Scrum is increasingly in demand for its potential to coordinate distributed teams (e.g. Fowler 2019; Krawczyk-Bryłka 2017) – increasingly so in the wake of the worldwide trend towards remote work during and after the Covid-19 pandemic (Henke et al. 2022). Generally, Scrum seems to function just as well, or even more effectively, for distributed and hybrid teams (Sutherland et al. 2007). However, more technology-prone or introvert team members may fare better in a digital (Scrum) environment than technology-averse or extrovert team members, who may prefer analog interactions (e.g. Grelle and Popp 2021). Similarly, team members who interact in highly diverging intensities with digital progress boards require increased coordinative efforts (Hidalgo 2019). Finally, digital facilitation, for Scrum or other formats, requires more planning and active chairing than face-to-face facilitation to avoid what has become known as "Zoom fatigue" (e.g. Nesher Shoshan and Wehrt 2022).

On the positive side, Scrum has the potential to foster team work and systemize the management of group learning, research, and administration (e.g. Hidalgo 2019; Pope-Ruark 2017). For the university classroom, there is evidence that the transparent and ritualized communication and planning and the clear role allocation in Scrum may foster a collaborative learning environment and reduce student anxiety related to group dynamics and the rejection of group projects (Allan 2016; Fernandes et al. 2021; Pope-Ruark et al. 2011).

Especially in transdisciplinary academic contexts, where students, external stakeholders, and researchers do not share the same knowledge and background, formalized communication techniques may help to keep misunderstandings at bay and to synchronize goals and achievement strategies. For instance, by collectively negotiating a *Definition of Done* in a transdisciplinary group, possibly diverging expectations about the goals ("what do we need to do"), the methodology ("how do we get there?"), and the medium of the final product ("how do we (re-) present our results") can be identified – and clarified if needed – as a basis for effective team work.

Finally, the lack of agility in higher education tends to be a general obstacle to introducing Scrum to academia: Typically, higher education is not particularly flexible when it comes to changing research goals or curricula to meet new requirements. For example, in funding applications, project goals usually need to be mapped out using Gantt charts and underpinned with concrete project milestones, even if the project is to be managed with Scrum. And both the university classroom and the academic research group are typically marked by steep hierarchies in contrast to the flat hierarchies envisioned for effective cooperation and coordination in Scrum teams (Schwaber and Sutherland 2020).

Current forms of implementation in higher education

There are currently two forms of implementation of Scrum in higher education: (1) to help manage transdisciplinary research (*Scrience*) and (2) to facilitate collaborative and transdisciplinary learning (*EduScrum*).

To manage transdisciplinary research groups, Scrum provides a systematic approach to collaboration and communication, which is used to some extent in IT or engineering departments and rarely in domains outside of this context (but see Hidalgo's 2019 case study on using Scrum in a UK policy research center). To provide insights into the more general forms of implementing Scrum to manage research, the authors of this chapter therefore share their own practical experience with devising, introducing, and maintaining Scrum to manage transdisciplinary research projects (see Speiser et al. 2023). Given that the developers of Scrum, Schwaber and Sutherland (2020, 13), are adamant that their framework should *not* be called Scrum if it was changed in any way, the authors refer to their implementation as *Scrum for Science* or *Scrience*. However, the Scrum framework can and should be maintained when implementing *Scrience*.

To account for the fact that *Scrience* is not only about practice and products (as the original Scrum) but also about knowledge, education, and discourse, *Scrience* requires an additional "science meeting" to formalize discussions of scientific theories, findings (e.g. newly collected data), and thesis work. The science meeting should be scheduled in the middle of every work cycle. To account for the busy work schedules of academics, which often involve more than one project at a time, work cycles should be extended (e.g. to about three to four weeks) to create the right meeting density that fits into typical faculty or student schedules (see also Hidalgo 2019, 17). For similar reasons, the update meetings should be scheduled weekly rather than daily (see also Baham 2019, 142; Ochoa et al. 2021, 4).

For *Scrience* to succeed, a second critical feature is that members of academic teams possess, on the one hand, unique and diverse expertise instead of similar skills as originally envisaged in the Scrum framework (Hidalgo 2019, 18–19) and may assume multiple roles on the other. To account for the uneven distribution of expertise, it is recommended to treat each task *as if* it was for a generalist team to ensure transparency, team synchronization, and consensus, even if certain tasks realistically can only be completed by an individual team member. To avoid role confusion, the role of project facilitator (Scrum Master) should be assigned to a senior team member with authority in the team rather than to a student assistant, and with sufficient resources to regularly chair the meetings (see also Baham 2019, 150). For similar reasons, the Principial Investigator (that is, the person in charge of the research project) should take on the role of project planner (Project Owner), who coordinates and prioritizes work packages with the overall project goal in mind. If the Principal Investigator does not have the time to be actively involved in project planning, this role can also be performed by a team member capable of the necessary scientific oversight. However, the Principal Investigator must then accept that this practice reduces his or her role to a project stakeholder, who is only occasionally consulted for input by the chosen project planner (see also Fowler 2019, 74).

Finally, given that *Scrience* is a new form of project management that most team members will be unfamiliar with, every project needs an extended introductory phase during which all team members are familiarized with the main concepts and are given the opportunity to practice the steps involved and make mistakes (also see Pope-Ruark 2012, 165–67). To facilitate transdisciplinary communication, it is recommended to use accessible language instead of potentially inaccessible Scrum terminology for the different components of *Scrience*, such as speaking of 'work cycles' instead of 'Sprints' – as already demonstrated throughout this chapter. During the introductory phase, the team members should also decide on a meeting rhythm and work cycle length and commit to keeping this schedule for at least three to six months before revising it if necessary. As for the original Scrum, Scrience works best for groups of four to ten members. Larger teams may be divided into separate smaller groups with different thematic focus if needed (e.g. on technology development versus study design; also see Schwaber and Sutherland 2020, 5).

To manage collaborative and transdisciplinary learning, Scrum's most common form of implementation is in the context of applied engineering and computer science courses around the world, where students work on development projects and learn Scrum as a project management tool on the side (Baham 2019). This practice is sometimes referred to as *EduScrum* (e.g. Fernandes et al. 2021; Neumann and Baumann 2021). EduScrum has also been used successfully outside of engineering curricula and with the broader goal of enabling project-based, collaborative, and transdisciplinary student learning in the academic classroom. Pope-Ruark (2012, 2017) has shown, for instance, that Scrum can be used to manage student projects in the humanities just as much as in engineering or programming courses.

The focus of EduScrum mostly lies on equipping students with Scrum as a skill, which they will need later in industry and to manage practical course work effectively (e.g. Kudikyala and Dulhare 2020). In this context, the lecturer often assumes the role of the project facilitator (e.g. Persson et al. 2011, 63; Pope-Ruark 2017, 171–74), that is the role of the Scrum Master, whereas project planning is done by a student team member or by an external stakeholder from industry. EduScrum may thus be an effective way to satisfy the increasing need for project management tools in the classroom arising from a general didactic trend towards group learning in cooperative education, design thinking, storytelling, student-organized teaching, and transdisciplinary learning in general.

In summary, Scrience and EduScrum effectively adapt the original Scrum to fit individual educational and scholarly needs. Although both frameworks are helpful tools to promote transdisciplinary research and learning by providing a clear and ready-to-implement collaboration and communication framework for interdisciplinary teams, small deviations from the described approaches matter and may alter their effectiveness. Thus, in the future both Scrience and EduScrum might benefit from further evidence-based consolidation in the form of published quality standards or – using the language of agile management (Schwaber and Sutherland 2020; Takeuchi and Nonaka 1986) – more definite "rules of the game".

References

Allan, Elizabeth. 2016. "I hate group work!": Addressing students' concerns about small-group learning. *InSight: A Journal of Scholarly Teaching* 11: 81–89.

Baham, Corey. 2019. Implementing Scrum wholesale in the classroom. *Journal of Information Systems Education* 30 (3): 141–59.

Beck, Kent, Mike Beedle, Arie van Bennekum, Alistair Cockburn, Ward Cunningham, Martin Fowler, James Grenning, Jim Highsmith, Andrew Hunt, Ron Jeffries, Jon Kern, Brian Marick, Robert C. Martin, Steve Mellor, Ken Schwaber, Jeff Sutherland, Dave Thomas. 2001. *Agile manifesto*. Available from https://agilemanifesto.org.

Cervone, H. Frank. 2011. Understanding agile project management methods using Scrum. *OCLC Systems & Services* 27 (1): 18–22.

Fernandes, Sandra, José Dinis-Carvalho, and Ana Teresa Ferreira-Oliveira. 2021. Improving the performance of student teams in project-based learning with Scrum. *Education Sciences* 11 (8): 2–16.

Fowler, Frederik M. 2019. *Navigating hybrid Scrum environments*. 2018. Berkeley: Apress.

Grelle, Darrin, and Eric Popp. 2021. Considering the interaction of individual differences and remote work contexts. *Industrial and Organizational Psychology* 14 (1–2): 244–47.

Henke, Jonn B., Samantha K. Jones, and Thomas A. O'Neill. 2022. Skills and abilities to thrive in remote work: What have we learned? *Frontiers in Psychology* 13: 893895.

Hicks, Michael, and Jeffrey S. Foster. 2010. SCORE: Agile research group management. *Communications of the ACM* 53 (10): 30–31.

Hidalgo, Enric Senabre. 2019. Adapting the Scrum framework for agile project management in science. *Heliyon* 5 (3): e01447.

Kudikyala, Udai Kumar, and Uma N. Dulhare. 2015. Using Scrum and Wikis to manage student major projects. *International Conference on MOOCs, Innovation and Technology in Education*, eds. Deepak Garg, Russ Meier, and Rob Reilly, 15–20. Available from https://ieeexplore.ieee.org/document/7375279.

Krawczyk-Bryłka, Beata. 2017. Comparative study of traditional and virtual teams. *TASK Quarterly* 21 (3): 233–45.

Krawczyk-Bryłka, Beata, and Henryk Krawczyk. 2019. The preferable ways of decision-making in IT teams. *Decision Making in Manufacturing and Services* 13 (1–2): 37–51.

Linos, Panagiotis K., Ryan Rybarczyk, and Nathan Partenheimer. 2020. Involving IT professionals in Scrum student teams. *Frontiers in Education Conference*. Available from https://doi.org/10.1109/FIE44824.2020.9274190.

Nesher Shoshan, Hadar, and Wilken Wehrt. 2022. Understanding "Zoom Fatigue": A mixed-method approach. *Applied Psychology* 71 (3): 827–52.

Neumann, Michael, and Lars Baumann. 2021. Agile methods in higher education: Adapting and using EduScrum with real world projects. *IEE Frontiers in Education Conference*. Available from https://doi.org/10.1109/FIE49875.2021.9637344.

Ochoa, Omar, Massood Towhidneiad, Timothy Wilson, James Pembridze, Erin Bowen, and Carlos Castro. 2021. *Adopting agility in academia through pilot projects*. Available from https://ieeexplore.ieee.org/document/9637458

Persson, Mia, Ivan Kruzela, Kristina Allder, Per Johansson, and Olof Johansson. 2011. On the use of Scrum in project driven higher education. *Proceedings of the 2011 International Conference on Frontiers in Education Computer Science & Computer Engineering*, ed. IEEE Computer Society, 59–64. Waikoloa, HI: IEEE. Available from http://worldcomp-proceedings.com/proc/p2011/FEC3416.pdf.

Pope-Ruark, Rebecca. 2012. We Scrum every day. *College Teaching* 60 (4): 164–69.

Pope-Ruark, Rebecca. 2017. *Agile faculty*. Chicago: University of Chicago Press.

Pope-Ruark, Rebecca, Michelle Eichel, Sarah Talbott, and Kasey Thornton. 2011. Let's Scrum: How Scrum methodology encourages students to view themselves as collaborators. *Teaching and Learning Together in Higher Education* 3 (3): 1–16.

Preußig, Jörg. 2015. *Agiles Projektmanagement*. Freiburg: Haufe.

Schwaber, Ken, and Jeff Sutherland. 2020. The Scrum guide. Available from https://scrumguides.org/docs/scrumguide/v2020/2020-Scrum-Guide-US.pdf.

Serrador, Pedro, and Jeffrey K. Pinto. 2015. Does agile work? A quantitative analysis of agile project success. *International Journal of Project Management* 33 (5): 1040–51.

Shalloway, Alan, Guy Beaver, and James Trott. 2010. *Lean-agile software development*. Upper Saddle River, NJ: Addison-Wesley.

Sloan, Dan. 2015. *Is the "growth mindset" an agile mindset?* Available from https://www.scrum.org/resources/blog/growth-mindset-agile-mindset.

Sutherland, Jeff, Anton Viktorov, Jack Blount, and Nikolai Puntikov. 2007. Distributed Scrum: Agile project management with outsourced development teams. In *Proceedings of the Annual Hawaii International Conference on System Sciences*, ed. IEEE Computer Society, 274a. Waikoloa, HI: IEEE.

Speiser, Dorothee, Maren Heibges, Laura Besch, Caren Hilger, Marie Keinert, Katharina Klein, Christine Olbrich, Julia Kußmaul, Caroline Neeb, Gudrun Rauwolf, Christine Schmid, Stephanie Stegen, Viola Westfal, Isabell Witzel, Benedikt Zang, Friederike Kendel, and Markus A. Feufel. 2023. iKNOW – A paradigmatic approach to support personalized counseling with digital health. *JMIR Formative Research 2023;7:e41179*. Available from https://www.doi.org/10.2196/41179.

Takeuchi, Hirotaka, and Ikujiro Nonaka. 1986. The new product development game. *Harvard Business Review* 64 (1): 137–46.

Thesing, Theo, Carsten Feldmann, and Martin Burchardt. 2021. Agile versus waterfall project management. *Procedia Computer Science* 181: 746–56.

Wintersteiger, Andreas. 2015. *Scrum*. Frankfurt am Main: entwickler.press.

Storytelling

Juliette Cortes Arevalo, Kathryn Adamson, Emanuele Fantini, Laura Verbrugge, and Roland Postma

Definition

All animals and plants communicate in one way or another. However, creating and telling stories to make sense of the world and one's experiences are uniquely human traits (Boyd 2018). The etymology of the word story relates to the Greek εἶδος (eîdos), which means the idea, form, or shape of things (Martin and Miller 1988). Through stories, one makes sense of past events and creates new worlds and possibilities for ourselves and others (Boyd et al. 2020). Storytelling is a subjective and engaging way of referring to an event or series of events through multisensory mediums such as audial, verbal, non-verbal, visual, and textual communication (Anderson 2010).

One can narrate, interact with, and exchange stories in various ways and from at least three perspectives: (1) the perspective of the *storyteller(s)* who, through their accounts, share knowledge, meaning, and emotions so that others can care, remember, retell, and share their story; (2) the perspective of the *audience* that may or may not be present at the time of the performance but perceives and reacts to the story(tellers); (3) the wider perspective of the *context* in which the story takes place is interpreted and influenced by the exchanged stories (Murray and Sools 2015).

Collaborative storytelling refers to group efforts to create, tell, and share stories, including but not limited to the story audience (Goldstein et al. 2015). Moreover, digital storytelling incorporates the technology of various multimedia modes such as graphics, audio, texts, videos, and animations to enhance the power of storytelling and ensure that the story is accessible through multiple dissemination channels (Bee Choo et al. 2020). Many different types of communication or texts can be analyzed as narratives or stories, including facial expressions, gestures, dance, and sound effects related to cultural, art-based, and role-playing forms of storytelling such as films, theater, photography, music, and games (Blackburn Miller 2020; Wang et al. 2017).

A *story* typically refers to a collection of events with a beginning, a middle, and an end, where something happens and changes the protagonist. *Narratives* are the means to describe such events, which as a whole, can also be referred to as a story. The narrative term also refers to the general argument about an issue and a text style (Moezzi et al. 2017). Narrative research analyzes the story's content, structure, and function (Murray and Sools 2015). In contrast, non-narrative styles refer to a collection of facts, typically told from an impersonal perspective, to build a position from referenced sources (Avraamidou and Osborne 2009). In both cases, stories and facts are not mutually exclusive categories. Stories are a form of data collection, analysis, and reporting that complement traditional sources and research methods (Rhodes and Brown 2005).

Any type of story includes the following six elements (Green et al. 2018): (1) characters and protagonist: usually humans, but possibly another animate actor, object, practice, or idea; (2) acts or events: typically performed by, or happening to, the protagonist through the story sequence; (3) setting: generic world, specific locality, or non-physical environment where the story takes place; (4) plot: sequence of events in which the protagonist pursues a goal that may include unexpected twists to help the audience experience the story's drama or emotion; (5) conflict: tension, problem, or urgent question that needs to be solved; and (6) outcome: the resolution of the protagonist's pursuit. In addition, storytelling's interactive,

Figure 1. Key questions to start with storytelling in research or education (Source: the authors).

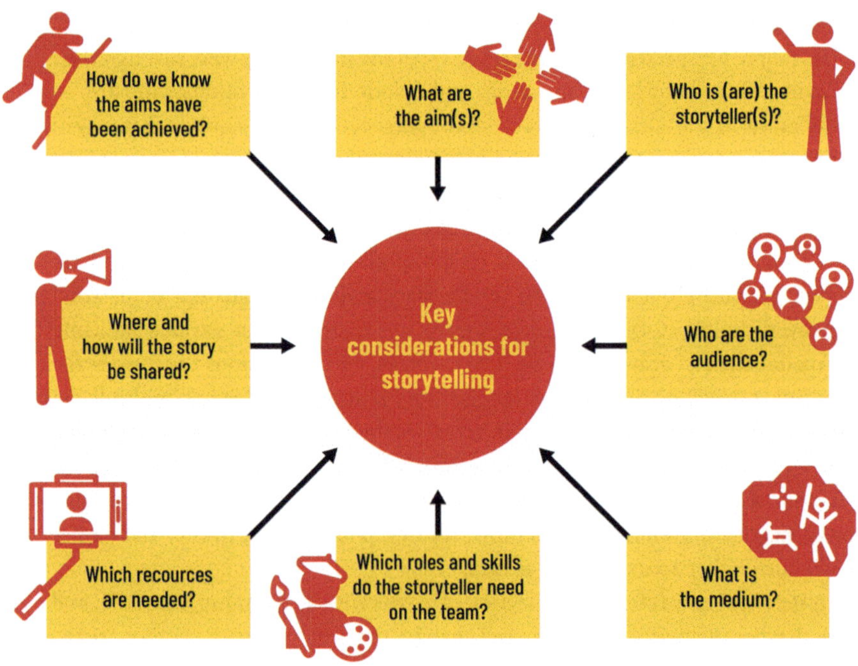

Table 1. Key example considerations based on experiences from education and research

Key questions	Examples	References
What are the aims?	Promote reflection, achieve empathy and identification, gather knowledge, change behavior, and challenge social constructs.	(Grimaldi et al. 2013; Rhodes and Brown 2005)
Who is (are) the storyteller(s)?	Perspective from which the story is told – the actual teller or intermediary who makes the performance.	(Moezzi et al. 2017; Murray and Sools 2015)
Who are the audience?	Individuals or groups for which the story is crafted include citizens, students, policymakers, academics, etc. Every audience group has its interest and knowledge levels concerning the story content.	(Reinermann et al. 2014; Stewart and Nield 2013)
What is the medium?	Storytelling mediums include discussion forums, digital media, books, podcasts, visual art, songs, movies, poetry, and theatre. Each medium has its characteristics and requirements regarding form, content, and length.	(Wang et al. 2017)
Which roles and skills does the storyteller need on the team?	Roles vary according to the influence of the team member on the storytelling, such as editor, advisor, designer, audience representative, etc. Relevant team member skills include imagination, improvisation, and creative and technical skills in writing, performing, or visual arts. Technical knowledge in the expertise domain of the story content. Knowledge about and relations with the intended audience.	(Blackburn Miller 2020; Goldstein et al. 2015)
Which resources are needed?	Multimedia equipment (cameras, microphones), editing software, online application, meeting space, art supplies, etc., according to the chosen medium.	(Wang and Zhan 2010)
Where and how will the story be shared?	Sometimes the storytelling medium and the channel of dissemination coincide, but not necessarily. Where and how the story will be shared limits the medium choices and story considerations.	(Blackburn Miller 2020)
How do we know the aims have been achieved?	Audience's feedback (expressions, responses, and actions). Effects such as enhanced knowledge, changed attitudes, preferences, and behavior are measured through audience questionnaires, surveys, observations, and tracking into short- or long-term studies.	(Wang et al. 2018)

persuasive, and living essence typically requires a strategic design to combine all elements into a "good story". Figure 1 includes key questions or considerations for effective storytelling, for which Table 1 gives some examples. To interact with the audience, the storyteller should have specific aims for which story elements are carefully chosen. Story creation is not linear, and choices may be revisited several times. Although often overlooked, the storytelling practice improves when measuring the extent to which the intended aims are achieved. Moreover, a responsible practice requires the critical and reflective recognition of the desired and undesired effects, either as a direct or indirect result of the storytelling.

Background

Stories have been part of everyday life for millennia via mainly oral and visual traditions. Through storytelling, individuals and communities have historically entertained, taught cultural norms, and built shared perspectives. Today's global communication happens via email, blogs, social networking, and video-sharing sites. Storytelling takes place in all cultures and a variety of forms. The storytelling research and practice include anthropology, philosophy, psychology, linguistics, history, library science, art, and media studies, among many others (Anderson 2010).

Modern storytelling has proliferated due to three important developments. After the invention of printing, printed stories became widely available. Subsequently, mass media such as the press, cinema, radio, and television have made stories accessible and marketable to many people. More recently, with computers and the internet, storytelling incorporated multiple media, including social media, so digital stories became easier to create and share, especially with more affordable technologies.

Storytelling has countless applications at the individual, community, and organizational levels and is used in the realms of education (Wang and Zhan 2010), research (Murray and Sools 2015), and innovation among many other applications (Sergeeva and Trifilova 2018). As a communication strategy, storytelling successfully leverages the intrinsic temporality, plurality, reflexivity, and subjectivity of stories for (Rhodes and Brown 2005): (1) sense-making, (2) communication, (3) identity and identification, (4) politics and power, and (5) learning and change. Stories are a means of recounting, interpreting, and making sense of events, individually or collectively, to position ourselves and others (McVee and Boyd 2015). Stories are effective by tapping into the emotions of personal experiences to mentally transport the audience into the storyteller's world (Morris et al. 2019). Thereby, stories help overcome the perceived abstract, often distant nature of knowledge, allowing people with different backgrounds to relate to each other's identity and understanding of the world (Bamberg and Georgakopoulou 2008).

The compelling characteristic of stories can also be used to grab attention and (political) support without much debate (Rhodes and Brown 2005). Concerning learning, story finding, telling, expanding, processing, and reconstructing encourage interactive learning about the actors, places, and events of an unfolding plot to reflect and decide on the role one wants to play. Importantly, imagination encourages creativity instead of merely tapping into preexisting stories to collaboratively (re)frame the past, present, and future (McDrury and Alterio 2003; van Hulst, 2012). In today's society, education and research increasingly require facilitating transformative and sustainability-oriented learning to counteract environmental problems and injustice (Lotz-Sisitka et al. 2015). Stories can elucidate desired futures (and fears) and help to analyze such scenarios to mobilize and empower transformation (Inayatullah 1998).

Transdisciplinary learning implies the inclusive, reflective, and active engagement and dialogue between multiple societal actors. These actors, as storytellers or audiences, are not limited to students, teachers, and researchers but include representatives of different types of knowledge. Stories can effectively promote dialogue and bridge across disciplines, societal demographics, and geographic scales and are, therefore, a powerful tool for transdisciplinary learning (Blackburn Miller 2020).

Debate and criticism

While storytelling can effectively communicate with, influence, and engage different audiences, ethical concerns are related to subjectivity, misuse, and responsibility. These concerns arise from moral principles that guide people in taking responsibility for their choices and actions based on how they may affect others (Rosenstand 2017).

First, stories are a subjective way to produce and shape reality. Especially when compared to scientific ways of working, storytelling uses literary and persuasive techniques rather than being limited to evidence-based claims and facts. Whether following a scientific method or not, stories and facts result from a simplification of the complexity of reality. This simplification could propagate stereotypes and preserve power imbalances by, for example, only considering the perspective of dominant or supportive groups (Solórzano and Yosso 2002). As a counterargument, storytelling can give voice to those who have previously been silenced and thus plays an important role in decolonization movements, among other movements to counteract power imbalances in research and education (Cunsolo Willox et al. 2013; Petheram et al. 2015; Zavala 2016).

Second, because the power of storytelling is so great, stories can easily be misused, for example to spread misinformation. Stories could mislead people when they believe the content without scrutiny. The spread of misinformation to mislead people is becoming more common in the current political and digital media environment, leading to distrust and misperceptions of scientific knowledge in society (Iyengar and Massey 2019). When used appropriately, fictive stories can assist in envisioning possible futures and establishing community engagement (Riedlinger et al. 2019). However, there are also words of caution as researchers, policymakers, and other audiences may fail to acknowledge the assumptions and simplifications underpinning their storytelling (Twyman et al. 2011). Misuse and misinformation require researchers to proactively develop strategies in their practice and communication to restore societal trust in science (Cardew 2020).

A critical stance is thus needed to engage in ethical storytelling as a storyteller or audience. When evaluating what makes a "good story", one should reflect on the intended purpose or outcome and consider how it may affect others. With this in mind, carefully plan and facilitate the storytelling process and consider the skills required of the storytelling team and the audience. While communication is indispensable in almost every profession, storytelling skills have received little attention in higher education. Developing such a skillset comes with its share of discomfort, as it takes time for researchers to "unlearn" trades of using jargon or focusing on methods (Green et al. 2018). Instead, more attention to critical thinking skills in education can aid the identification of effective characteristics of ethical storytelling and scientific practice (Glisson 2019).

Current forms of implementation in higher education

The following examples from around the globe are selected from a growing body of literature to illustrate the many forms of storytelling focusing on environmental applications. Many other relevant examples exist online or in community spaces but are not frequently documented.

In South Africa, Loots (2021) explores the decolonization of higher education dance curricula to transcend prescribed "western/northern" teaching models and incorporate locally situated dance styles. By combining verbal and physical dialogue, students can effectively express and explore personal narratives of race, gender, and health issues using culturally anchored dance styles. Similarly, visual and oral storytelling has been noted as a powerful, cost-effective tool for curriculum decolonization in the African higher education system by integrating global perspectives and local indigenous narratives to interrogate social injustice and marginalization and address existing social hierarchies (Mampane et al. 2018). Working across disciplines, visual storytelling was used to unite Australian indigenous communities and researchers to co-create scientific knowledge to inform policymaking. During workshops and interviews, photographs, videos, diagrams, and oral accounts were collated in order to better understand local perspectives on marine food dependence, climate change, and coastal adaptation (Petheram et al. 2015). Moreover, collaborative storytelling and theatrical practice helped researchers and local communities to co-explore and reimagine the effects on the lives of people in the USA and the United Kingdom of the projected results of climate change (Liguori et al. 2021; Shenk and Gutowski 2022) and to explore the understanding of air pollution in Kenya (West et al. 2021).

Adamson et al. (2021) use storytelling across generations to facilitate communication between researchers and schoolchildren from varied cultural and socioeconomic backgrounds in the United Kingdom. Follow-up evaluation showed that personal narratives and imaginative stories proved highly effective for (1) putting a "human face" to abstract or complex scientific ideas; (2) developing common ground between researchers and audiences via shared language and experience; and (3) using co-creation to remove perceived power imbalances. Two examples from the Netherlands use storytelling across geographical, disciplinary, and professional boundaries. The first teaches video storytelling to masters students worldwide (Fantini 2019), and the second develops StoryMaps collaboratively to communicate research outputs to practice (Cortes Arevalo et al. 2020; Kok et al. 2022). Both projects demonstrate that storytelling enhances transdisciplinary learning when students/researchers increase their understanding: (1) how to structure and illustrate a story using the most appropriate resources (style, characters, images, sounds, etc.); (2) storytelling as universal but not uniform practice. Clear consideration of the audience, message, and format ensures a balance between local requirements while drawing inspiration from global, mainstream communication platforms; (3) visual storytelling as an iterative process and collaborative craft. The storytelling team takes responsibility for their role (as editor, narrator, cameraperson, audio, sound engineer, etc.) while simultaneously negotiating their choices with team members and audiences with different backgrounds, understandings, and preferences.

While short-term storytelling effects, such as improving the engagement of the variety of actors relevant to transdisciplinary learning, are more frequently studied (Smeda et al. 2014), a key challenge for research and education is to measure how stories are directly or indirectly used by the audience and their long-term efficacy for behavioral change (Wang et al. 2018). One of the few long-term studies is Cordero et al. (2020), who surveyed pro-environmental decisions of university graduates in the USA after they completed a university sustainability course. The use of personal narratives embedded in the course highlights the relevance of environmental issues, meaning that students considerably reduced their carbon emissions long after course completion.

To conclude, by acting as storytellers, the researchers or students can acknowledge the subjectivity of their stories and reflect on the stereotypes and dominant narratives they can reproduce. They can identify how to better care for and communicate trustingly with audiences in multiple societal groups. As a universal but not uniform phenomenon, storytelling can facilitate communication across boundaries of knowledge, power, and identities. At the same time, storytelling remains situated in specific cultural contexts and traditions, making it relevant and suitable to address local specificities. More and better documentation of current and future storytelling applications within higher education and research is needed to study the effects of storytelling. Finally, responsibly exploring different storytelling methods and types help to value both the process and the output of transdisciplinary learning.

References

Adamson, Kathryn, Timothy Lane, Kris De Meyer, Matthew Carney, Leonora Oppenheim, Sina Panitz, Hannah Price, Emma Smith, and Gregory Watson. 2021. Enhancing Physical geography schools outreach: Insights from co-production and storytelling narratives. *Progress in Physical Geography: Earth and Environment* 45 (6): 907–30.

Anderson, Katie Elson. 2010. Chapter 28. Storytelling. In *21st century anthropology: A reference handbook*, ed. James Birx, 277–86. Thousand Oaks: Sage.

Avraamidou, Lucy, and Jonathan Osborne. 2009. The role of narrative in communicating science. *International Journal of Science Education* 31 (12): 1683–707.

Bamberg, Michael, and Alexandra Georgakopoulou. 2008. Small stories as a new perspective in narrative and identity analysis. *Text & Talk* 28 (3): 377–96.

Bee Choo, Yee, Tina Abdullah, and Abdullah Mohd Nawi. 2020. Digital storytelling vs. oral storytelling: An analysis of the art of telling stories now and then. *Universal Journal of Educational Research* 8 (May): 46–50.

Blackburn Miller, Jennifer. 2020. Transformative learning and the arts: A literature review. *Journal of Transformative Education* 18 (4): 338–55.

Boyd, Brian. 2018. The evolution of stories: From mimesis to language, from fact to fiction. *WIREs Cognitive Science* 9 (1): e1444.

Boyd, Ryan L., Kate G. Blackburn, and James W. Pennebaker. 2020. The narrative arc: Revealing core narrative structures through text analysis. *Science Advances* 6 (32): eaba2196.

Cardew, Gail. 2020. People will not trust unkind science. *Nature* 578 (7793): 9.

Cordero, Eugene, Diana Centeno, and Anne Marie Todd. 2020. The role of climate change education on individual lifetime carbon emissions. *PLOS ONE* 15 (2): e0206266.

Cortes Arevalo, Juliette, Laura Verbrugge, Anneke Sools, Marcela Brugnach, Rik Wolterink, Pepijn van Denderen, Jasper Candel, and Suzanne Hulscher. 2020.

Storylines for practice: A visual storytelling approach to strengthen the science–practice interface. *Sustainability Science* 15: 1013–32.

Cunsolo Willox, Ashlee, Sherilee Harper, and Victoria Edge. 2013. Storytelling in a digital age: Digital storytelling as an emerging narrative method for preserving and promoting indigenous oral wisdom. *Qualitative Research* 13 (2): 127–47.

Fantini, Emanuele. 2019. Just like a movie: Teaching visual storytelling on water. *J-READING Journal of Research and Didactics in Geography* 2: 63-70.

Glisson, Lane. 2019. Breaking the spin cycle: Teaching complexity in the age of fake news. *Portal: Libraries and the Academy* 19 (3): 461–84.

Goldstein, Bruce Evan, Anne Taufen Wessells, Raul Lejano, and William Butler. 2015. Narrating resilience: Transforming urban systems through collaborative storytelling. *Urban Studies* 52 (7): 1285–303.

Green, Stephanie, Kirsten Grorud-Colvert, and Heather Mannix. 2018. Uniting science and stories: Perspectives on the value of storytelling for communicating science. *FACETS*, 3(1): 164-173..

Grimaldi, Silvia, Steven Fokkinga, and Ioana Ocnarescu. 2013. Narratives in design: A study of the types, applications, and functions of narratives in design practice. In *Proceedings of the 6th International Conference on Designing Pleasurable Products and Interfaces* – DPPI '13, 201. Newcastle upon Tyne: ACM.

Inayatullah, Sohail. 1998. Causal layered analysis: Poststructuralism as method. *Futures* 30 (8): 815–29.

Iyengar, Shanto, and Douglas Massey. 2019. Scientific communication in a post-truth society. *Proceedings of the National Academy of Sciences* 116 (16): 7656–61.

Kok, Matthijs, Juliette Cortes Arevalo, and Martijn Vos. 2022. *Towards improved flood defences*. Delft: TU Delft OPEN Books.

Liguori, Antonia, Lindsey McEwen, James Blake, and Michael Wilson. 2021. Towards "creative participatory science": Exploring future scenarios through specialist drought science and community storytelling. *Frontiers in Environmental Science* 8 (February): 589856.

Loots, Lliane. 2021. Decolonising dance pedagogy? Ruminations on contemporary dance training and teaching in South Africa set against the specters of colonisation and apartheid. *Theatre, Dance and Performance Training* 12 (2): 184–97.

Lotz-Sisitka, Heila, Arjen Wals, David Kronlid, and Dylan McGarry. 2015. Transformative, transgressive social learning: Rethinking higher education pedagogy in times of systemic global dysfunction. *Current Opinion in Environmental Sustainability, Sustainability Science*, 16 (October): 73–80.

Mampane, Ruth, Margaret Omidire, and Folake Ruth Aluko. 2018. Decolonising higher education in Africa: Arriving at a glocal solution. *South African Journal of Education* 38 (4):1-9.

Martin, Kathleen, and Etta Miller. 1988. Storytelling and science. *Language Arts* 65 (3): 255–59.

McDrury, Janice, and Maxine Alterio. 2003. Chapter 7. Expanding stories through reflection. In *Learning through Storytelling Higher Education: Using Reflection & Experience to Improve Learning*, 1st edition. London: Routledge, 105–120.

McVee, Mary, and Fenice Boyd. 2015. *Exploring diversity through multimodality, narrative, and dialogue: A framework for teacher reflection*. New York: Routledge.

Moezzi, Mithra, Kathryn Janda, and Sea Rotmann. 2017. Using stories, narratives, and storytelling in energy and climate change research. *Energy Research & Social Science* 31 (September): 1–10.

Morris, Brandi, Polymeros Chrysochou, Jacob Dalgaard Christensen, Jacob Orquin, Jorge Barraza, Paul Zak, and Panagiotis Mitkidis. 2019. Stories vs. facts: Triggering emotion and action-taking on climate change. *Climatic Change* 154 (1): 19–36.

Murray, Michael, and Anneke Sools. 2015. Chapter 9. Narrative research. In *Qualitative research in clinical and health psychology*, eds. Poul. Rohleder and Antonia Lyons, 133–54. London: Palgrave Macmillan.

Petheram, Lisa, Natasha Stacey, and Ann Fleming. 2015. Future sea changes: Indigenous women's preferences for adaptation to climate change on South Goulburn Island, Northern Territory (Australia). *Climate and Development* 7 (4): 339–52.

Reinermann, Julia-Lena, Sarah Lubjuhn, Martine Bouman, and Arvind Singhal. 2014. Entertainment-education: Storytelling for the greater, greener good. *International Journal of Sustainable Development* 17 (2): 176–91.

Rhodes, Carl, and Andrew Brown. 2005. Narrative, organizations, and research. *International Journal of Management Reviews* 7 (3): 167–88.

Riedlinger, Michelle, Luisa Massarani, Marina Joubert, Ayelet Baram-Tsabari, Marta Entradas, and Jennifer Metcalfe. 2019. Telling stories in science communication: Case studies of scholar-practitioner collaboration. *Journal of Science Communication* 18 (5): N01.

Rosenstand, Nina. 2017. Chapter 1: Thinking about values. In *The moral of the story: An introduction to ethics*, 30. 8th edition. New York: McGraw Hill.

Sergeeva, Natalya, and Anna Trifilova. 2018. The role of storytelling in the innovation process. *Creativity and Innovation Management* 27 (4): 489–98.

Shenk, Linda, and William Gutowski Jr. 2022. Mind the gaps! Climate scientists should heed lessons in collaborative storytelling from William Shakespeare. *WIREs Climate Change* n/a (n/a): e783.

Smeda, Najat, Eva Dakich, and Nalin Sharda. 2014. The effectiveness of digital storytelling in the classrooms: A comprehensive study. *Smart Learning Environments* 1 (1): 6.

Solórzano, Daniel, and Tara Yosso. 2002. Critical race methodology: Counter-storytelling as an analytical framework for education research. *Qualitative Inquiry* 8 (1): 23–44.

Stewart, Iain, and Ted Nield. 2013. Earth stories: Context and narrative in the communication of popular geoscience. *Proceedings of the Geologists' Association, Geoconservation for Science and Society* 124 (4): 699–712.

Twyman, Chasca, Evan Fraser, Lindsay Stringer, Claire Quinn, Andrew Dougill, Todd Crane, and Susannah Sallu. 2011. Climate science, development practice, and policy interactions in dryland agroecological systems. *Ecology and Society* 16 (3): 14.

Van Hulst, Merlijn. 2012. Storytelling, a model of and a model for planning. *Planning Theory* 11 (3): 299–318.

Wang, Qingchun, Sara Coemans, Richard Siegesmund, and Karin Hannes. 2017. Arts-based methods in socially engaged research practice: A classification framework. *Art/Research International: A Transdisciplinary Journal* 2 (2): 5–39.

Wang, Shuyan, and Hong Zhan. 2010. Enhancing teaching and learning with digital storytelling. *International Journal of Information and Communication Technology Education (IJICTE)* 6 (2): 76–87.

Wang, Susie, Adam Corner, Daniel Chapman, and Ezra Markowitz. 2018. Public engagement with climate imagery in a changing digital landscape. *Wiley Interdisciplinary Reviews: Climate Change* 9 (2): e509.

West, Sarah, Cressida Bowyer, William Apondo, Patrick Büker, Steve Cinderby, Cindy Gray, Matthew Hahn, et al. 2021. Using a co-created transdisciplinary approach to explore the complexity of air pollution in informal settlements. *Humanities and Social Sciences Communications* 8 (1): 1–13.

Zavala, Miguel. 2016. Decolonial methodologies in education. In *Encyclopedia of educational philosophy and theory*, ed. Michael A. Peters, 1–6. Singapore: Springer.

Student-Organized Teaching

Judith Bönisch, Frank Becker, Laurenz Blömer, Sanjeet Raj Pandey, Baiba Prūse, and Johannes Vollbehr

Definition

As student-organized teaching has until now been a poorly researched and documented field, we propose a working definition on the subject based on our experience as practitioners: Student-organized teaching can be defined as a form of learning and teaching at higher education institutions in which students actively and self-determinedly design learning processes for themselves and others. Thus, they are learners and teachers at the same time and shape their learning spaces and the learning environment of their universities. Student-organized teaching is opposite to the hierarchical approach of the knowing teacher and the unknowing student. Student-organized teaching follows the idea that knowledge can be self-acquired, jointly developed in a group, gained in social interaction, and oriented to a concrete object. On this basis, it demands individual power to shape the joint learning process.

The concept of student-organized teaching can be connected to a social constructivist approach of teaching and learning (Dudley-Marling 2012; Singer-Brodowski 2016, 109–12) and to student autonomy (for a discussion on autonomy see Holmes 2021). The extent to which this is permitted by study regulations and universities varies greatly and changes over time, and in the context of societal change as a whole.

The formats of student-organized teaching can be systematized according to various dimensions, e.g. degrees of freedom (can students decide about contents, methods, project goals, and scope of their work or just some of those?), disciplinary affiliation (discipline-based or inter- or transdisciplinary?), or integration into curricula (compulsory or elective?). Student-organized teaching can include different methods and learning concepts, so overlaps with project teaching, problem- and inquiry-based learning (Jonassen and Hung 2012; Pedaste et al. 2015), or service learning (Dolgon et al. 2017) are common. Tutors in student-organized teaching are usually other students.

As student-organized teaching is a heuristic and essentially self-designed form of teaching, the degrees of freedom represent a continuum: On the one end courses exist in which students have complete freedom to choose their own content and methods and receive sufficient material and nonmaterial support from the university. The professional or didactic support of the students is primarily aimed at enabling them to act independently in and with a project group and to solve their self-defined problem, considering their individual knowledge, previous experience, and competencies. External control of the process, for example by professors, is not intended. Student-organized teaching on this side of the continuum is committed to self-determined (Blaschke 2012; Kenyon and Hase 2013) and self-organized learning (Harri-Augstein and Thomas 1991; Low and Jin 2012). On the other side, the degrees of freedom are rather low because, for example, content and methods are closely prescribed.

We consider the formats with the greater degrees of freedom to be more promising with regard to the context of transdisciplinary learning as reflection processes, e.g. on hierarchical structures, different needs and perspectives or communication patterns are necessary to a far greater extent in order to be able to manage these forms of student-organized teaching. Reflection processes like those are of fundamental importance for the acquisition of transdisciplinary competences (Pearce et al. 2018; Wiek et al. 2011).

Background

In our experience, many student-organized courses and projects do not or not explicitly refer to any historic context or history as they last for a few years only (constrained by the capacities of the initiators, the manageable time frame of the initiators' education program, or the limited funding provided by universities). Student-organized teaching often happens without theorizing on their own chosen approach and with little documentation. Some projects, however, do explicitly refer to a historic background (e.g. Bönisch and Energieseminar 2021, 9). An appreciative stocktaking on a larger scale is certainly necessary here.

The student movement of the 1960s and the following politicization of the student body had a strong impact on self-organization processes and student organization in general (Altbach 2007, 329). The movement sought to open universities, to combine theory with practice, and to place knowledge and knowledge production in social contexts. From this starting point, numerous student initiatives, associations, and self-governance structures were established and tutoring structures were introduced or expanded (Bönisch 2021, 9; Della Porta et al. 2021, 19). At the same time, an ongoing discussion continued on whether replacing regular teaching staff by students could foster increasing economization of higher education institutions (Heyner 2014, 44; Topping 1996, 321). In these processes, university didactics and teaching staff argued that opportunities of innovation and reform could get lost when student tutors just repeat the content of lectures (Heyner 2014, 44 and 50).

However, the demand for change in higher education institutions both to meet their social responsibility and to qualify students for the challenges of their future life can be discussed as a field of learning for these challenges and changes. The link between student-organized teaching and transdisciplinarity is established from a transdisciplinary competence dimension. Which competences are necessary to tackle current problems? Several attempts try to systematize "transdisciplinary competences", such as the Inner Development Goals (Jordan 2021) or the Green Comp European competence framework (Bianchi et al. 2022). Key competences such as communicating, dealing with ambiguities, and self-reflexivity can be better acquired in a self-organized setting than in a traditional unidirectional course (see Hawtrey 2007). In self-organized teaching projects, students are confronted with negotiation processes on various levels, which they otherwise face rarely in their studies: students negotiate with students over the course contents, students work with academic teaching staff to implement the course, and students cooperate with professors in the final assessment of the performance achieved in these courses. In essence, student-organized teaching is a contribution to learning for a transdisciplinary practice: *My counterpart is not me. My values, goals and actions do not a priori coincide with those of my non-university partners.*

Debate and criticism

Research on student-organized teaching at higher education institutions is rare, even though self-organized learning has been broadly discussed since the 1980s (Singer-Brodowski 2016, 112). This failure is due to the often-marginalized position of student-organized courses at higher education institutions and the lack of related scientific organs such as peer-reviewed journals or scientific societies.

Student-organized teaching is not beneficial per se. One should ask in what ways the different forms of self-organized teaching are useful and how they reflect the social conditions of research. What intentions are being pursued with a course? And are these intentions transparent to the students? According to our perception, at least in the Western world more and more models appear that claim to satisfy student-organized processes in one form or another. Whether specific degree programs provide sufficient freedom to integrate this demanding teaching–learning format is questionable – to put it bluntly: Are students "allowed" to make "mistakes" within the framework of their studies? Or is the study program determined by "credit points"? And do students have the courage, the necessary skills and reflection techniques, to enter this failure-based learning process in the performance-driven surroundings?

The question of failure results in the question of quality measurement and evaluation. For some education institutions, it is lack of a professor to guarantee the quality of a course, while others complain about too little technical depth in student-led courses. We suggest rethinking quality and quality measurement in student-organized courses: Can the quality of the learning process be measured in other ways than passing an exam in the end of the course? Should facts and figures really play such a central role in higher education? What are the success criteria?

In addition, universities rarely provide "blank spaces", i.e. open spaces for teaching and learning that are not pre-structured, e.g. by narrow boundaries or the influence of teaching experts. We observe that "blank spaces" of self-organized teaching have recently been discovered as "didactically fillable voids", whereby didactics comes across as a supposedly neutral tool. However, student-led self-organized teaching projects are usually the result of a debate between students and the university. Containing these spaces and projects didactically establishes a relation of power that impedes the development of students' self-organization. We consider this to be non-emancipatory, as it does not correspond to the humanitarian world view universities are committed to. This development has monetary consequences as well: In education systems that do not appreciate open and student-led spaces, budgets are cut or projects no longer receive funding. Spaces for student-organized teaching are often in a precarious situation.

Apart from that, self-organized teaching empirically works with methods and skills, e.g. in moderation, presentation, or decision-making. The challenge for academic staff in supervising student-led courses is to develop a support structure to ensure that students receive information and support when needed without governing the self-organized structure of the students' work.

There are also limits of self-orchestration. How is the framework defined that can be filled by student self-organized teaching? What is the role of student tutors? Do tutors decide on the grading of participants? Do tutors decide on content and outcome or is this a group decision?

Students' self-organized teaching projects can turn out to be a benefit for universities: They can easily be used as an incubator for innovation in teaching and successful projects can be incorporated into regular teaching through appropriate processes.

Current forms of implementation in higher education

As typologies of student-organized teaching, documentation of student-organized teaching is difficult to find and hardly a topic for peer-reviewed journals. As an example, the project lab *Sustainable Handprints* at TU Berlin is a four-semester student-organized platform for fellow students to work in interdisciplinary groups on interest-driven projects regarding education for sustainable development in interaction with relevant stakeholders. Bachelor and master students from different universities engage in the course and work towards common solutions on how to implement sustainability in practical and playful ways. Socially and personally relevant topics are the subject of interest-driven research in the project lab. With skill- and knowledge-sharing methods all participants gain insights into various science disciplines and diverse perspectives on how to approach the complexity of sustainability. The project lab opens the space for knowledge exchange between students and relevant actors, such as experts of organizations, researchers, project lab alumni, teachers, and pupils (Project Lab Sustainable Handprints 2023).

Another example is based on an interdisciplinary course constructed around the concept of citizen science and local ecological knowledge at the University of Iceland. The way the students were involved in the learning process was different from traditional academic courses, as they were directly linked to the local communities as well as to the coauthors of the peer-reviewed studies. The course was available for international and Icelandic students at various study levels, including PhDs. In order to keep the interaction open among the tutor and the students, several ice-breaking methods were integrated into the teaching, including a common coffee break (informal) at the end of each class.

Degree of guidance in "Sustainable Handprints": In the beginning, a high degree of guidance was given by the tutors. In the later stages the room was opened for self-guided group work. With the autonomy and support provided through the tutors as well as the project lab community, groups created a strong drive to realize their project ideas. The tutors acted as coaches who guided the project groups during their interest-driven projects. To keep the balance between blank space and formally predefined guidelines was a great challenge for the tutors.

Degree of guidance in "citizen science and local ecological knowledge": The short time allocated for the course (one week) set some restrictions in terms of how much guidance was needed to be set by the tutor. The format of the final report and the layout for the note and question page was already pre-set for the students to directly dive into the subject. The success of the course was very much dependent on students' curiosity to question and summarize the discussions. Each lecture began with an online call with the experts (both academic and community level). In order to gain long-lasting benefit from the notes, the students were invited to co-organize a blog post which was later published on the eu.citizen.science web page (Gupta et al. 2022).

Science communication and network in "Sustainable Handprints": Starting with a creativity phase in which easily accessible information sources were consulted, a basic knowledge foundation of the sustainability topics being researched was created. Design Thinking (e.g. Brown 2008) elements, like creating personas, helped the students to empathize with different actor perspectives. This turned out to be a useful foundation to enable transdisciplinary knowledge transfer.

Science communication and network in "citizen science and local ecological knowledge": The tutor opens up the network by connecting students with the authors of the articles the students were reading and community members. One way to achieve this was through collaboration on a blog post. The course involved an assessment which integrated the needs of researchers from the university as well as student interests. Besides providing the potential topics for the students, the researchers also joined the reporting session at the end of the course.

Evaluation in "Sustainable Handprints": With accompanying reflections, students were bound to self-reflect their group process and constantly iterate their learning objectives. Applying feedback methods regularly made receptive adaptations possible and ensured constant improvement of the common learning process. At the end of every project lab semester, evaluation criteria were developed with the whole group and students were given the responsibility to grade each other's project presentations. This shift in perspective and power relations between teacher and learner created an empowering experience for the students.

Evaluation in "citizen science and local ecological knowledge": The students were given the opportunity to evaluate the course by anonymous feedback as well as discuss the teaching methods during the informal coffee break. As acknowledged by the students in the evaluation notes, this type of practice in terms of question-making helped them understand the method and see a difference in communicating between academics and non-academic knowledge-keepers.

In summary, in both initiatives credits in the European Credit Transfer and Accumulation System were offered to the students; however, the outcomes of the courses went way beyond academic confines. Evaluation as well as the equal importance of informal and formal communication was reflected in both cases.

Other examples of student-organized teaching activities can be found in various regions of the world, e.g. student-led seminars in universities in Europe (CEMUS 2023; Duke University School of Law 2023; Utrecht University 2023), or the United States (Loyola University 2023; University of California 2023), or student-organized symposiums (Boston University 2023). These examples vary in terms of disciplinary affiliation – with more examples that are rooted in a certain discipline and allow students to define content freely but set close boundaries in terms of the general framework. The benefits of student-organized teaching, the potential for solutions to the pressing problems of our time, and the potential emancipatory effects would justify putting more effort into this research.

References

Altbach, Philip G. 2007. Student politics. Activism and culture. In *International handbook of higher education*, eds. James J. F. Forest and Philip G. Altbach, 329–45. Dordrecht: Springer.

Bianchi, Guia, Ulrike Psiotis, and Marcelino Cabrera. 2022. *GreenComp. The European sustainability competence framework*. Publications Office of the European Union. Available from https://publications.jrc.ec.europa.eu/repository/handle/JRC128040.

Blaschke, Lisa Marie. 2012. Heutagogy and lifelong learning. A review of heutagogical practice and self-determined learning. *International Review of Research in Open and Distributed Learning* 13 (1): 56–71.

Bönisch, Judith, and Energieseminar. 2021. *Anders denken, gemeinsam handeln. Ein Konzept für Selbstorganisierte Lehre*. Hamburg: Tredition.

Boston University, ed. 2023. Bioinformatics student-organized symposium. Available from https://www.bu.edu/bioinformatics/special-events/student-organized-symposium.

Brown, Tim. 2008. Design thinking. *Harvard Business Review* 86 (6): 84–92.

CEMUS [Centre for Environment and Development Studies], ed. 2023. *Student-led education*. Available from http://www.cemus.uu.se/about.

Della Porta, Donatella, Lorenzo Cini, and César Guzmán. 2021. *Contesting higher education: Student movements against neoliberal universities*. Bristol: Bristol University Press.

Dolgon, Corey, Tania D. Mitchell, and Timothy K. Eatmann. 2017. *The Cambridge handbook of service learning and community engagement*. Cambridge: Cambridge University Press.

Dudley-Marling, Curt. 2012. Social construction of learning. In *Encyclopedia of the sciences of learning*, ed. Norbert M. Seel, 3095–98. Boston: Springer.

Duke University School of Law, ed. 2023. *Ad hoc seminars*. Available from https://law.duke.edu/study/adhoc.

Gupta, Parnika, Hlynur Steinsson, Theresa Henke, Walter L. Brent van der Hell, Moira Aileen Brennan, Diana Sól Editudóttir, Kayla Maureen Þorbjörnsson, Lina Andrea Johansson, Marina Ermina, Aron Alexander Þorvarðarson, Guðrún Heiður Ísaksdóttir, Bjargey Anna Guðbrandsdóttir, and Baiba Prūse. 2022. *Reflections from the first ever course on Citizen science and Local ecological knowledge at the University of Iceland*. Available from https://eu-citizen.science/blog/2022/04/20/reflections-first-ever-course-citizen-science-and-local-ecological-knowledge-university-iceland.

Harri-Augstein, Sheila, and Laurie Thomas. 1991. *Learning conversations: The self-organised learning way to personal and organizational growth*. London: Routledge.

Hawtrey, Kim. 2007. Using experiential learning techniques. *Journal of Economic Education* 38 (2): 143–52.

Heyner, Marco. 2014. Zur Aktualität historischer tutorieller Konzepte. Universität Hamburg. *Universitätskolleg-Schriften* 5: 43–57.

Holmes, Andrew G. D. 2021. Can we actually assess learner autonomy? The problematic nature of assessing student autonomy. *Shanlax International Journal of Education* 9 (3): 8–15.

Jonassen, David, and Woei Hung. 2012. Problem-based learning. In *Encyclopedia of the sciences of learning*, ed. Norbert M. Seel, 2687–90. Boston: Springer.

Jordan, Thomas. 2021. *The inner development goals: Background, method and the IDG framework*. Available from https://static1.squarespace.com/static/600d80b3387b98582a60354a/t/627bb821f4978468f9f311ba/1652275238451/220511_IDG_Report_Full.pdf.

Kenyon, Chris, and Stewart Hase. 2013. Heutagogy fundamentals. In *Self-determined learning: Heutagogy in action*, eds. Stewart Hase and Chris Kenyon, 7–17. London: Bloomsbury.

Loyola University, ed. 2023. *Student-led seminars*. Available from https://www.loyola.edu/department/center-humanities/grants-funded-opportunities/student/student-led-seminars.

Low, Renae, and Putai Jin. 2012. Self-organized learning. In *Encyclopedia of the sciences of learning*, ed. Norbert M. Seel, 3008–10. Boston: Springer.

Pearce, BinBin, Carolina Adler, Lisette Senn, Pius Krütli, Michael Stauffacher, and Christian Pohl. 2018. Making the link between transdisciplinary learning and research. In *Transdisciplinary theory, practice and education: The art of collaborative research and collective learning*, eds. Dena Fam, Lina Neuhauser, and Paul Gibbs, 167–84. Cham: Springer.

Pedaste, Margus, Mario Mäeots, Leo A. Siiman, Ton de Jong, Siswa A. N. van Riesen, Ellen T. Kamp, Constantinos C. Manoli, Zacharias C. Zacharia, and Eleftheria Tsourlidaki. 2015. Phases of inquiry-based learning: Definitions and the inquiry cycle. *Educational Research Review* 14: 47–61.

Project Lab Sustainable Handprints, ed. 2023. *Sustainable Handprints*. Available from https://naha.app.

Singer-Brodowski, Mandy. 2016. *Studierende als GestalterInnen einer Hochschulbildung für nachhaltige Entwicklung. Selbstorganisierte und problembasierte Nachhaltigkeitskurse und ihr Beitrag zur überfachlichen Kompetenzentwicklung Studierender.* Berlin: Berliner Wissenschafts-Verlag.

Topping, Keith. 1996. The effectiveness of peer tutoring in further and higher education. A typology and review of literature. *Higher Education* 32: 321–45.

University of California, ed. 2023. *Student-organized seminars*. Available from https://neuroscience.ucdavis.edu/student-organized-seminars.

Utrecht University, ed. 2023. *Student-led honours seminar.* Available from https://students.uu.nl/en/university-college-utrecht/academics/curriculum-enrichment-options/student-led-honours-seminar.

Wiek, Arnim, Lauren Withycombe, and Charles L. Redman. 2011. Key competencies in sustainability. A reference framework for academic program development. *Sustainability Science* 6: 203–18.

Transdisciplinarity

Ulli Vilsmaier, Juliana Merçon, and Esther Meyer

Definition

The term transdisciplinarity has been complementing the landscape of research approaches since the 1970s. It joins a steadily growing list of terms that refer to the concept of discipline – such as multidisciplinarity, pluridisciplinarity, cross-disciplinarity, interdisciplinarity, supra-disciplinarity, antidisciplinarity, meta-disciplinarity, and post-disciplinarity. Discipline refers to a field of research defined by content and institutions (Hacking 2010). It derives from the Old Latin *disciplina*, where it means "instruction, tuition, teaching", and in a metonymic sense also "learning, knowledge, science, discipline" (Lewis and Short 2020). The term discipline, however, has another, metaphorical meaning that goes back to Christian origins. It means to educate, to discipline, and to punish. A *discipulus*, a *discipula*, is a disciple, a pupil. And an undisciplined person is considered one who does not fit into existing orders and follow established rules. This context of meaning occupies an important position in the discussion of social orders (Foucault 1970; Horkheimer and Adorno 1972). It is of particular importance for the topic of transdisciplinary higher education, especially since the research areas designated by the concept of discipline are followed by educational organizations. They serve the diffusion and reproduction of disciplinary organized expertise in social orders (Nowotny 1999). The prefix *trans* is also taken from Latin, where, in connection with verbs of movement, it refers to going "beyond", and, in connection with verbs of rest, it means lying "beyond" or "on the farther side of" (Lewis and Short 2020, 1097).

Etymologically, the term transdisciplinarity can thus be read in two ways: (1) as a positioning term, denoting lying across or beyond the disciplines; (2) as a movement term, describing a movement out of the discipline. These different etymological readings of the term are reflected in the various discourse streams on transdisciplinarity. Transdisciplinary research is conceptualized as complementary, existing alongside disciplinary, multidisciplinary, and interdisciplinary forms of research and as one that not only refers back to to the basic building block of modern scientific organization, but does so to change disciplines or – even more comprehensively – the disciplining of science. Verbs such as transcend, transgress, and transform are used to characterize transdisciplinarity (Klein 2014). What they have in common is that they all carry an element of movement, although different paths and goals are addressed. With regard to transdisciplinarity in higher education, this etymological distinction is relevant in several respects. It raises questions of how educational organizations must be situated and structured within existing knowledge orders to be transdisciplinary and how disciplinary knowledge bases and research practices matter in teaching and learning transdisciplinarily.

Background

Taking the concept of discipline as the starting point for the discussion on transdisciplinarity ties in with numerous works from philosophy and science studies (Bernstein 2015; Osborne 2015). However, these were not the fields in which the term was first used, but in the context of an OECD conference that addressed issues of education and innovation (Apostel et al. 1972). The emergence of terms, however, is not a singular moment; rather, it reflects conditions that enable certain forms of thought and practices to appear. By 1970, an intense examination of the claims, practices, and organizational forms of science had taken place for a long time. And precursors can be identified in both theorizing and research practice that are written in one way or another in transdisciplinary terms (Osborne 2015; Streck et al., forthcoming). Conceptual elaborations of transdisciplinarity, however, have been a long time coming. In what follows, we elaborate on main discourse streams that shaped the conceptual evolvement of transdisciplinarity.

Mittelstraß (1987) first framed transdisciplinarity as a principle of research in the late 1980s. It was intended to serve the overcoming of cognitive boundaries through constrictions in the organizational system of science, to become effective where subject or disciplinary perspectives are too narrow to solve problems. Transdisciplinarity is grounded in the critique of the internal organization of science and research, but remains oriented to the idea of occidental reason and scientific rationality. In stark contrast, a Charter of Transdisciplinarity was adopted in 1994, which is grounded in open rationality and based on a dialogue of epistemic cultures between sciences, philosophies, arts, literature, poetry, and religions (Nicolescu 2002). What is considered transdisciplinary is "the semantic and practical unification of the meanings that *traverse* and *lie beyond* different disciplines" (Nicolescu 2002, 149) and based on in a specific vision, attitude, ethics, and open mindedness. The inclusive logic that underlies this discourse shakes central axioms of modern science and is oriented toward their transformation.

In the 1990s, the concept of transdisciplinarity is positioned in the formulation of a Mode 2 knowledge production. Michael Gibbons and colleagues (1994) distinguish between a classical, occidental complex of ideas, methods, values, and norms, referred to as Mode 1, which emerged from the search for universally valid explanatory principles, and Mode 2, which is produced in the context of concrete application. Transdisciplinarity is understood as research that is "grounded in a shared axiomatics and permeation of different disciplinary methods of knowledge" and oriented towards the production of socially robust knowledge (Nowotny 1999, 106, own translation). Mode 2 is seen as a response to downsides of Mode 1 knowledge production – the concealment of historical contingency, strategic essentializations through posits such as objectivity, universality, and purity of method, the appropriation of the concept of research for a social institution (Gibbons et al. 1994) and the associated reinforcement of the "sense[s] of superiority of the Western world" (Nowotny 1999, 77, own translation). The authors thus bring into the field those critiques of the constitution of occidental-modern science that have been developed, among others, in post- and decolonial studies as well as in feminist and gender studies.

The relationship of science to other areas of society is at the center of a discourse that is currently unfolding, predominantly in sustainability-related research fields. In the face of highly complex and pressing problems, the question of how to do research becomes a question of sustainability in itself. Transdisciplinarity is conceptualized as society-oriented research that complements the spectrum of research forms. It is realized in cooperation between scientists and practitioners. Transdisciplinary research should bridge the growing gap between science and the public, promote social learning and negotiation processes, consider scientific and life-world problems as well as abstract and case-specific knowledge in participatory processes, and make knowledge efficiently accessible for decision-making (Hirsch Hadorn et al. 2008; Scholz 2011). This is framed in the so-called ISOE model of the transdisciplinary research process, elaborated by the Institute for Social-Ecological Research (ISOE) (Jahn et al. 2012). It describes the integration of life-world and science-centered approaches to problems. Methodological approaches to this are introduced in Bergmann et al. (2012) and principles and practices by Lang et al. (2012). Transdisciplinarity is understood as a reflexive, integrative principle oriented towards scientific methods, in which a clear science-centeredness is brought to bear.

The search for adequate answers to change, acceleration, spread, and aggravation of problems proves to be a shared motif in the development of transdisciplinarity. While diagnoses of its justification are similar, very different therapeutic proposals can be identified. On the one hand, one opts for flexibilization and expansion of the internal organization of science and research, while holding on to scientific rationality. Here, transdisciplinarity is about overcoming the drawing of boundaries *within* an institutionalized body of scientific knowledge. On the other hand, science's claim to legitimacy as the highest knowledge system is fundamentally questioned and an expansion of participants in knowledge production and related decision-making processes is demanded. In consequence, boundaries of the scientific system itself are tackled. In research approaches at the science–society interface we can observe two fundamentally different ways of dealing with epistemic-political questions regarding the value and legitimization of different knowledges: An additive understanding of transdisciplinarity is that scientific knowledge production is embedded in larger social research constellations, but scientific rationality remains unaffected. An entangled understanding of transdisciplinarity, however, is grounded in an open relationship between epistemic cultures that does not grant primacy to any specific form of knowledge generation, which raises significant epistemological, methodological, and ethical–political questions, and opens up a space between institutions and knowledge cultures (Merçon 2022; Vilsmaier et al. 2017).

However, the increasing fanning out of transdisciplinarity discourses by no means results solely from the disintegration of established orders and problems to be tackled. Technological developments have opened up possibilities for participation in knowledge production that have led to far-reaching social shifts. Forms and actors involved in the production of knowledge have multiplied almost exponentially and mechanisms of justification and legitimation have also changed as a result. The concept of knowledge society and debates on the democratization of knowledge mark these shifts. For the probing of the discourses on transdisciplinarity, the reference to socio-technical transformations is significant insofar as it helps to broaden the view. The complexes of questions that evolve around transdisciplinarity are by no means to be negotiated in purely academic terms. Rather, they represent a task for society as a whole.

Debate and criticism

Since the 2000s, discourses of transdisciplinarity have proliferated across a broad spectrum of research fields. In addition to diversifying in sustainability sciences, the subject has become established in multiple fields, such as technology impact research, urban, regional, agricultural, and landscape research, medical research and epidemiology, architecture and design, gender and justice research, as well as in the arts and at the interface between science and art.

A conceptualization of transdisciplinarity that understands the cooperation of scientists with non-scientific actors as definitionally constitutive became widespread. Schmidt (2021) observes a domination of "instrumental or strategic viewpoints" in discourses of transdisciplinarity and a loss of the critical momentum that has been a "cornerstone" when discourses emerged. However, more recent works increasingly take up the foreshortening and shadowing of essential epistemological, methodological, and ethical–political questions. Work on power relations, social and epistemic control, and social justice in transdisciplinary research processes is helping to illuminate these blind spots (e.g. Fritz and Meinherz 2020; Herberg and Vilsmaier 2020; Kareem et al. 2022). This is also true for conceptual and analytical approaches, such as research on methods (e.g. Defila and Di Giulio 2019; Pereira et al. 2021); quality criteria, impact, and evaluation (e.g. Lux et al. 2019); and the normative dimension of transdisciplinary research (e.g. Popa et al. 2015; Rosendahl et al. 2015).

Working on interfaces to related or neighboring research traditions also dynamizes the discourse. These include (participatory) action research, intervention research, integration and implementation science, science of team science, citizen science, and artistic research, amongst others. Critical engagements with the heavily Europeanized concept of transdisciplinarity by researchers from Africa, Asia, Oceania, and Latin America also bring to bear limits to the transferability of more techno-scientistic approaches to transdisciplinary research (e.g. Van Breda and Swilling 2018) and point to the danger that the concept itself could develop hegemonic power by displacing (at least discursively) non-Western research traditions (De Santolo 2018). Culturality, difference, multilingualism – related to regional, epistemic, and institutional origins – as well as work on post- and decoloniality (De Santolo 2018; Merçon 2022) make a contribution in considering sometimes rather abstract and static assignments, positing in a more differentiated way, and unleashing the socio-political and onto-epistemological potential of this form of research.

Interventions from the humanities in those discourses that have developed strongly out of transdisciplinary research practice offer particular potential for this. They strengthen the linkage of the conceptual unfolding of transdisciplinarity back to larger historical discourse contexts (Osborne 2015). That which is inherently transdisciplinary is worked up in the thought of Michel Foucault, Jaques Derrida, Michel Serres, Gilles Deleuze, Felix Guattari, and the educational approach of Paulo Freire (Serna 2016; Vilsmaier et al. 2020), among others. A central criticism is the extensive, theoretical underexposure of the concept of the problem in transdisciplinary research (Meyer 2020). The constitutive grounding of transdisciplinary research in lifeworld problems rests on a drawing of boundaries that the research form purports to overcome. These paradoxes point to the modern legacy of transdisciplinarity. It is the task at hand to clarify them. However, they in no way diminish the importance of testing transdisciplinary forms of research and teaching and of exploring new institutional configurations.

Current forms of implementation in higher education

The discourse of transdisciplinarity took its origin in calls for reforms of the educational system (Apostel et al. 1972). In this respect, too, the appearance of the term can be located in the context of larger social upheavals. With the 1968 movement, reform pedagogical approaches experienced a strong upswing, and the learner as a person gained importance, as did the experiential and dialogical moment of learning. Since then, learning in formal and informal environments and also as a constitutive component of research has been processed in the transdisciplinarity discourse: as cooperative, mutual, situated, case-based, recursive, circular, and transformative.

With regard to implementations in higher education, individual-, group-, and process-centered approaches can be identified. The focus on the individual as a "transdisciplinary subject" requires not only the education of the intellect, according to Nicolescu (2002), but also of the emotions and the body. Transdisciplinary education – far from being limited to university education – has to be practiced as an attitude. Forming a transdisciplinary orientation requires learning environments that enable engagement with values, norms, beliefs, conceptual skills, and knowledge (Stokols 2014). We find transdisciplinary learning formats in all sorts of thematic fields and methodological approaches, ranging from dialogue centered *Empathetic–Reflective–Dialogical Restorying* in human rights education in South Africa (Jarvis 2018) to methodologically complex integrative formats such as the *Transdisciplinary Case Study Approach* (Scholz and Tietje 2002; Stauffacher et al. 2008), Living Labs (Fam et al. 2018), and the *Intercultural Education Approach* implemented in Mexico (Merçon and Alatorre Frenk 2019). Conceived as student-based research, these can be classified as inquiry-based learning formats (Mieg 2019), sharing the focus on societal problems with problem-based learning and the process-oriented organizational form with project-based learning approaches. They organize team research between students, university teachers, and actors from other sectors and enable students to conduct a transdisciplinary research process and to practice working in heterogeneous groups. However, the implementation of transdisciplinary case studies is dependent on curricular freedom.

This addresses a neuralgic point in the establishment of transdisciplinary forms of higher education. If study programs are highly interdisciplinary and application-oriented, for example in the sustainability sciences, there are greater opportunities and legitimacy for integrating extensive transdisciplinary courses into the curricula. In disciplinarily narrower fields of study, conflicts of objectives with disciplinary bodies of knowledge and teaching of methods can arise. In addition, possibilities for creating transdisciplinary learning spaces between study programs are often limited by administrative–legal hurdles. The question of institutionalizing transdisciplinarity in higher education is primarily framed as a debate on the right timing. Two diametrically opposed positions confront each other: While some consider a solid disciplinary education indispensable to qualify for (inter-) and transdisciplinary research, others see the earliest possible point in time as elementary in order to avoid disciplinary constrictions.

There will be no simple answer to the questions raised in this chapter, and, above all, no single answer. However, discussions of these issues will always concern knowledge orders and social regimes, values, and power relations and, ultimately, worldviews and conceptions of human nature. And these must be held in high esteem. Transdisciplinary research, teaching, and learning do not yet enjoy widespread approval. So far, they are marginal phenomena and are often perceived as competition to existing institutional orders and orientations – and an attack on values that have been established over long periods of time in the academic world and the social fabric at large. Efforts to implement transdisciplinarity in higher education, research, and societal transformation are confronted with persistent structures, while at the same time by rapid socio-technical change and its ecological, cultural, political, and economic consequences. Careful introspection of transdisciplinary research, teaching, and learning practices, as well as work on theoretical and methodological consolidations of transdisciplinarity, will help not only to celebrate it as a reinvention, but also to bear consequences – including the institutional, identity, and power shifts it entails. Undoubtedly, this is an intergenerational endeavor that requires one thing above all: epistemic curiosity and a breaking out of a "circle of certainty" (Freire 1996, 21) to critically change what exists and creatively engage with what is to come.

References

Apostel, Léo, Guy Berger, Asa Briggs, and Guy Michaud, eds. 1972. *Interdisciplinarity: Problems of teaching and research in universities*. Paris: OECD.

Bergmann, Matthias, Thomas Jahn, Tobias Knobloch, Wolfgang Krohn, Christian Pohl, and Engelbert Schramm. 2012. *Methods for transdisciplinary research: A primer for practise*. Frankfurt am Main: Campus.

Bernstein, Jay H. 2015. Transdisciplinarity: A review of its origins, development, and current issues. *Journal of Research Practice 11* (1): R1.

De Santolo (Garrwa and Barunggam), Jason. 2018. Shielding indigenous worlds from extraction and the transformative potential of decolonizing collaborative research. In *Transdisciplinary theory, practice and education: The art of collaborative research and collective learning*, eds. Dena Fam, Linda Neuhauser, and Paul Gibbs, 203–19. Cham: Springer.

Defila, Rico, and Antonietta Di Giulio, eds. 2019. *Transdisziplinär und transformativ forschen, vol. 2. Eine Methodensammlung*. Wiesbaden: Springer.

Fam, Dena, Linda Neuhauser, and Paul Gibbs, eds. 2018. *Transdisciplinary theory, practice and education: The art of collaborative research and collective learning*. Cham: Springer.

Foucault, Michel. [1966] 1970. *The order of things*. New York: Pantheon.

Freire, Paulo. 1996. *Pedagogy of the oppressed*. London: Penguin.

Fritz, Livia, and Franziska Meinherz. 2020. Tracing power in transdisciplinary sustainability research: An exploration. *GAIA – Ecological Perspectives for Science and Society* 29: 41–51.

Gibbons, Michael, Camille Limoges, Helga Nowotny, Simon Schwartzmann, Peter Scott, and Martin Trow. 1994. *The new production of knowledge: The dynamics of science and research in contemporary societies*. Los Angeles: Sage.

Hacking, Ian. 2010. Verteidigung der Disziplin. In *Interdisziplinarität. Theorie, Praxis, Problem*, eds. Michael Jungert, Elsa Romfeld, Thomas Sukopp, and Uwe Voigt, 193–206. Darmstadt: WBG.

Herberg, Jeremias, and Ulli Vilsmaier. 2020. Social and epistemic control in collaborative research – Reconfiguring the interplay of politics and methodology. *Social Epistemology* 34: 309–18.

Hirsch Hadorn, Gertrude, Holger Hoffmann-Riem, Susette Biber-Klemm, Walter Grossenbacher-Mansuy, Dominique Joye, Christian Pohl, Urs Wiesmann, and Elisabeth Zemp, eds. 2008. *Handbook of transdisciplinary research*. Berlin: Springer.

Horkheimer, Max, and Theodor W. Adorno. 1972. *Dialectic of enlightenment*. New York: Herder and Herder.

Jahn, Thomas, Matthias Bergmann, and Florian Keil. 2012. Transdisciplinarity: Between mainstreaming and marginalization. *Ecological Economics* 79: 1–10.

Jarvis, Janet. 2018. Restorying for transdisciplinarity: A proposed teaching–learning strategy in a context of human rights education. *Journal for Transdisciplinary Research in Southern Africa* 14: 1–9.

Kareem, Buyana, Alice McClure, Jaqueline Walubwa, Kweku Koranteng, Paul Isolo Mukwaya, and Anna Taylor. 2022. Power dynamics in transdisciplinary research for sustainable urban transitions. *Environmental Science & Policy* 131: 135–42.

Klein, Julie Thompson. 2014. Discourses of transdisciplinarity: Looking Back to the future. *Futures* 63: 68–74.

Klein, Julie Thompson, Walter Grossenbacher-Mansuy, Rudolf Häberli, Alain Bill, Roland W. Scholz, and Myrtha Welti, eds. 2001. *Transdisciplinarity: Joint problem solving among science, technology, and society. An effective way for managing complexity*. Basel: Birkhäuser.

Lang, Daniel J., Arnim Wiek, Matthias Bergmann, Michael Stauffacher, Pim Martens, Peter Moll, Mark Swilling, and Christopher J. Thomas. 2012. Transdisciplinary research in sustainability science: Practice, principles, and challenges. *Sustainability Science* 7: 25–43.

Lewis, Charlton T., and Charles Short. 2020. *A Latin dictionary: Founded on Andrews' edition of Freund's Latin dictionary*. Chapel-en-le-Frith: Nigel Gourlay.

Lux, Alexandra, Martina Schäfer, Matthias Bergmann, Thomas Jahn, Oskar Marg, Emilia Nagy, Anna-Christin Ransiek, and Lena Theiler. 2019. Societal effects of transdisciplinary sustainability research – How can they be strengthened during the research process? *Environmental Science & Policy* 101: 183–91.

Merçon, Juliana. 2022. Investigación transdisciplinaria e investigación-acción participativa en clave decolonial. *Utopía y Praxis Latinoamericana* 27 (98): 1–19.

Merçon, Juliana, and Gerardo Alatorre Frenk. 2019. Educación, interculturalidad y sustentabilidad. Una experiencia de posgrado en México. *Inter-ethnica* 22 (1): 142–61.

Meyer, Esther. 2020. The problematic of transdisciplinary sustainability sciences. In *Thinking the problematic*, eds. Oliver Leistert and Isabell Strickl, 69–92. Bielefeld: transcript.

Mieg, Harald A., ed. 2019. *Inquiry-based learning – undergraduate research. The German multidisciplinary experience*. Cham: Springer.

Mittelstraß, Jürgen. 1987. Die Stunde der Interdisziplinarität? In *Interdisziplinarität. Praxis – Herausforderungen – Ideologie*, ed. Jürgen Kocka, 152–58. Frankfurt am Main: Suhrkamp.

Nicolescu, Basarab. 2002. *Manifesto of transdisciplinarity*. Albany: State University of New York Press.

Nowotny, Helga. 1999. *Es ist so. Es könnte auch anders sein*. Frankfurt am Main: Suhrkamp.

Osborne, Peter. 2015. Problematizing disciplinarity, transdisciplinary problematics. *Theory, Culture & Society* 32: 3–35.

Pereira, Laura, Per Olsson, Lakshmi Charli-Joseph, Olive Zgambo, Nathan Oxley, Patrick Van Zwanenberg, J. Mario Siqueiros-García, and Adrian Ely. 2021. Transdisciplinary methods and T-Labs as transformative spaces for innovation in social-ecological systems. In *Transformative pathways to sustainability*, ed. Adrian Eli, 53–64. London: Routledge.

Popa, Florin, Mathieu Guillermin, and Tom Dedeurwaerdere. 2015. A pragmatist approach to transdisciplinarity in sustainability research: From complex systems theory to reflexive science. *Futures* 65: 45–56.

Rosendahl, Judith, Metheus A. Zanella, Stephan Rist, and Jes Weigelt. 2015. Scientists' situated knowledge: Strong objectivity in transdisciplinarity. *Futures* 65: 17–27.

Schmidt, Jan C. 2021. *Philosophy of interdisciplinarity: Studies in science, society and sustainability*. London: Routledge

Scholz, Roland. W., and Olaf Tietje. 2002. *Embedded case study methods: Integrating quantitative and qualitative knowledge*. Thousand Oaks: Sage.

Serna, Edgar M. 2016. La Transdisciplinariedad en el Pensamiento de Paulo Freire. *Revista de Humanidades* 33: 213–43.

Stauffacher, Michael, Thomas Flüeler, Pius Krütli, and Roland W. Scholz. 2008. Analytic and dynamic approach to collaboration: A transdisciplinary case study on sustainable landscape development in a Swiss Prealpine region. *Systemic Practice and Action Research* 21: 409–22.

Streck, Danilo, Streck, Danilo Romeu, Sandro de Castro Pitano, and Nilda Stecanela (forthcoming): Pronunciar el mundo, juntos: la conectividad, principios de investigación en Paulo Freire. In *Aprendiendo de Paulo Freire para la investigación transdisciplinaria: principios, métodos, y experiencias*, eds. Ulli Vilsmaier, Juliana Meçon, Loni Hensler, and Gerald Faschingeder. Ciudad de México: CopIt-arXives.

Stokols, Daniel. 2014. Training the next generation of transdisciplinarians. In *Enhancing communication and collaboration in interdisciplinary research*, eds. Michael O'Rourke, Stephen Crowley, Sanford D. Eigenbrode and J. D. Wulfhorst, 56–81. London: Sage.

Van Breda, John, and Mark Swilling. 2018. The guiding logics and principles for designing emergent transdisciplinary research processes: Learning experiences and reflections from a transdisciplinary urban case study in Enkanini informal settlement, South Africa. *Sustainability Science* 14: 823–41.

Vilsmaier, Ulli. 2021. Transdisziplinarität. In *Handbuch Transdisziplinäre Didaktik*, eds. Tobias Schmohl and Thorsten Philipp, 333–46. Bielefeld: transcript.

Vilsmaier, Ulli, Vera Brandner, and Moritz Engbers. 2017. Research in-between: The constitutive role of cultural differences in transdisciplinarity. *Transdisciplinary Journal of Engineering & Science* 8: 169–79.

Vilsmaier, Ulli, Gerald Faschingeder, and Juliana Merçon. 2020. Learning from Paulo Freire for inter- and transdisciplinary research. *Journal für Entwicklungspolitik XXXVI* (3): 4–18.

Transformative Learning

Sadaf Taimur and Katie Ross

Definition

Throughout childhood – based on our families, language, history, culture, schooling, and all life experiences – an overarching, complex, and mostly unconscious set of beliefs of how the world works are developed. Akin to our own personal philosophy, this entirely unique "worldview" developed over the first two decades of life, is comprised of "frames of reference" (or hidden-meaning structures of assumptions) which we consider to be common-sense, unquestionable truths about what is good, right, true, and valuable. We view and comprehend our experiences through these frames of reference and, consequently, frames of reference shape our feelings, perceptions (views), expectations, cognition, and subsequently guide our actions (Mezirow 1991).

Sometimes, however, during life, we experience moments and processes that highlight our unconscious beliefs and instigate significant and lasting changes in our worldview, or more specifically, the "frames of reference" that comprise our worldview (Cranton 1994; Mezirow 1991, 1995, 1996, 2000, 2007). After watching his wife experience such a dramatic shift in her frames of reference, John Mezirow, Professor of Education at Columbia University, coined the term *transformative learning* to describe this process of deep learning. Mezirow (2003, 58) defined transformative learning as "learning that transforms problematic frames of reference – sets of fixed assumptions and expectations (habits of mind, meaning perspectives, mindsets) – to make them more inclusive, discriminating, open, reflective, and emotionally able to change".

Transformative learning is not a simple process. Humans have the propensity to reject perspectives which are not aligned with their own frames of reference and consider new perspectives illogical or as aberrations (Kaplan et al. 2016). Hence, a disorienting dilemma, which challenges a person's worldview, is often the catalyst for transformative learning. Disorientation happens when someone experiences something not yet contained within their "frames of reference", and hence affects them in deep and profound ways. However, reflection, discourse, and other transrational meaning-making processes can support people through this disorientation to shift previous perspectives and assumptions (Cranton 2016; Feller 2015).

These processes – discourse, reflection, action – describe a type of learning where people make their own meaning of an experience and then use this interpretation to guide their actions or decision-making. According to John Mezirow (2003), critical reflection via discourse allows individuals to inquire into their existing frames of reference, leading to new or revised interpretations of experiences that guide our understanding and action. In essence, in transformative learning, "meaning" converts into three orders of learning: learning about the world (*what*), learning about their own worldview (*why*), and insights of how then to act in the world (*how*) (Cranton 2016, 28; Ross 2020). Therefore, transformative learning is not only about adding to the existing knowledge base, but it requires being aware of one's own and others' assumptions or perspectives and subsequently evaluating their relevance (critically) via reflection (Mezirow 2000), which may lead to expanding their worldview (Taimur and Onuki 2020, 2022).

Background

While John Mezirow's work is foundational to the theory and facilitation of transformative learning, there is a growing ecology of transformative learning theories building from other foundational scholars (Stuckey et al. 2013). Other contributors to transformative learning theory recognized in adult education literature include Paulo Freire, Carl Jung, Laurent Daloz, John Dirkx, and Patricia Cranton, and are briefly summarized here: *Paulo Freire's* transformative learning is focused on individual and social liberation. In his social-emancipatory transformative learning, Freire (1970) argues that conscientization, or raising awareness about systemic forms of oppression, is key and leads to social liberation. Carl Jung's (1921) concept of transformative learning is grounded in individualization. As a type of psychoanalytical transformative learning, an individual becomes aware of their own processes of formation, differentiation, and different selves operating within the psyche for the development of their individual personality (Boyd and Myers 1988).

According to Laurent Daloz (1986), transformative learning is a process that occurs between the cognitive developmental phases when the changing world requires learners to have new meaning structures to make meaning (Dirkx 1998). *John Dirkx* explores how transformative learning occurs through subjective reframing or self-reflection (rather than Mezirow's focus on objective reframing or critical reflection), using soul and subconscious mind work, to support an evolution in frames of reference (Dirkx 2008; Dirkx et al. 2006). Patricia Cranton supported individual transformations while taking the social context of the individuals into account, but the focus of her work was how individuals transform in light of their own personality (Taylor and Tisdell 2020). Collectively, these theories illustrate several reasons why transformative learning is an integral part of transdisciplinary learning processes.

Firstly, transformative learning and transdisciplinary learning share an intention for transformative change. The word *transformation* appeared in 15th century Latin and French, specifically in reference to Christianity and ideas of liberation and conversion (Lange 2015). *Trans* means "beyond or across" and *formare* means "to form"; thus transformation is understood as "undergoing a change in form". The shared assumption of *transformation* across both learning theories is that change, specifically radical change, is beneficial for societal improvement, as compared to continuity or custom. Specifically, processes of transdisciplinary learning can actively seek transformative outcomes in three ways, including a change in situation, change in stocks and flows of knowledge, and transformative learning for all involved (Mitchell et al. 2015). The outcomes of transformative learning across both learning theories are similar, e.g. reflection and reconstruction of perspectives, values, and norms (Mitchell et al. 2015; Young and Karme 2015), giving more importance to social justice and environmental resources (Moyer et al. 2016), transformation of worldview and perspective (Feriver et al. 2016; Papenfuss and Merritt 2019; Ross and Mitchell 2018), and experiencing self-awareness (Taimur et al. 2022). Both learning theories seek transformative learning via continuous learning between internal interpretation, i.e. *why*, and external action, i.e. *how* (Müller et al. 2005; Ross and Mitchell 2018).

Secondly, transformative learning theory helps explain to educators and students alike why transdisciplinary learning is so often challenging for students. Students have often unconsciously learned through formal education that learning happens in a school, is discipline-based, and usually a single right answer is to be provided. In contrast, the first experience of a transdisciplinary course challenges many of these "common sense" learning "frames of references", e.g. beliefs about what learning is, who it is for, and how it is done. Many students can experience emotional responses to the challenges of their beliefs about what "learning" is. Similarly challenging, in transdisciplinary learning, participants engage with other actors in discourse and reflection to shift their perspectives and establish a shared, emergent understanding. Engagement in critical discourse and reflection to shift point of view is an emotionally disturbing process, where learners may feel uncomfortable, surprised, tormented, embarrassed, and emotional. Transformative learning theory guides educators and students on ways to honor, process, and use these "disorienting dilemmas" inherent in a transdisciplinary process as part of the meaning-making and learning in the experience, towards more inclusive, open, and reflective frames of reference.

Thirdly and more specifically, transdisciplinary learning requires critical evaluation of diverse perspectives, which Mezirow's processes of discourse and critical reflection can effectively support. For example, transdisciplinary learning to collaboratively address complex situations can involve a three-step (creative, descriptive, and normative) learning cycle. As a first creative step, each participant comes to the table with their own perspectives, knowledge, experiences, purpose, and interpretations (represented by pentagons in Figure 1) guided by their worldview (represented by circles within the pentagons in Figure 1). In the second descriptive step, participants share their internal perspectives in the form of proposed actions. Transformative learning processes allow individuals to be aware of their own worldviews and perspectives and be more open to others' perspectives, and hence can support step two. In the third normative step, these actions are discussed between the participants, leading to the convergence of viewpoints to create new integrated knowledge, concepts, and ideas. In this third step, transformative learning processes engage participants in critical discourse and help participants reflect on all proposed actions to converge towards a new integrated intellectual framework to establish a shared understanding between all the actors.

Similarly, many other transdisciplinary learning processes, when collaboratively undertaken in shared inquiry and dialogue, can lead to the conditions for transformative learning, including those summarized by Ross and Mitchell (2018):

Figure 1. Three-step transdisciplinary learning cycle and transformative learning (adapted from Müller et al. 2005, 202).

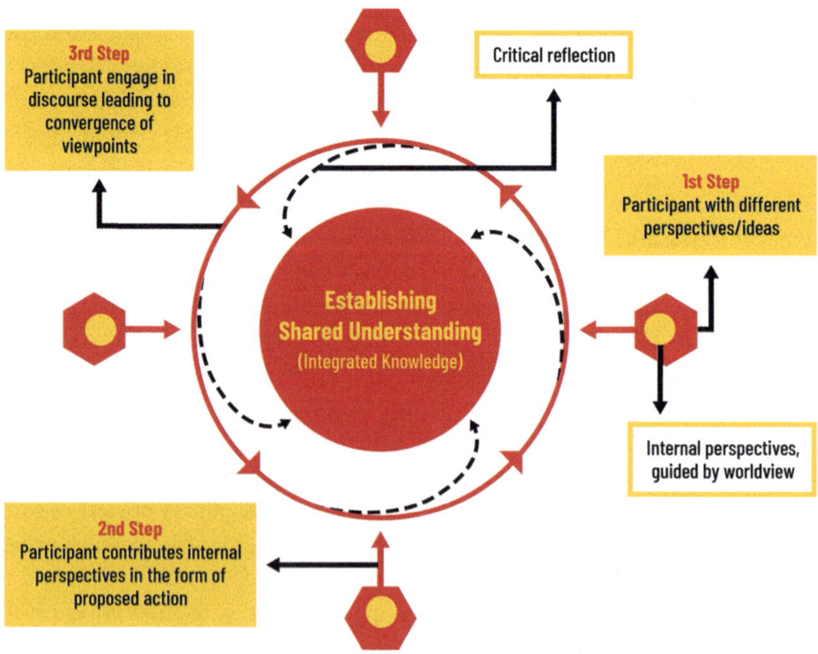

Checkland's Soft Systems Methodology and its notion of purpose (Checkland and Poulter 2010); Meadows' System Thinking and its notion of intervention points (Meadows 1999); Inayatullah's Causal Layered Analysis and its notion of myths and metaphors (Inayatullah 2008); Snowden's Cynefin model and its notion of complexity (Snowden and Boone 2007); Kooiman's meta-governance and its notions of values (Kooiman and Jentoft 2009).

Debate and criticism

From the 1990s onwards, Mezirow's conceptualization of transformative learning has changed the way we understand and design adult learning. However, Mezirow's transformative learning theory was not received without criticism. Scholars argued his theory focused too exclusively on individual transformations and the rational process of learning while avoiding the social and emotional sides of learning (Cranton 2016; Mälkki 2010). In response, Mezirow was receptive, but largely retained his original line of thought. The work of Daloz (1986), Dirkx (1998, 2002), and Cranton (1994, 2016), however, removed discourse as a mandatory condition for transformative learning.

The transformative learning field continues the dialogue of how to engage with the emotional side of learning, such as the role of empathizing. Empathizing is the ability to subjectively share and experience others' feelings or psychological states (Taylor 2007; Willis 2012), or, in simple words, putting oneself in another's shoes. While Mezirow's transformative learning theory does not pay much attention to empathizing, Mezirow has roughly referred to empathizing by using other terms as facets critical for transformative learning to occur. For example, having an open mind, bracketing or letting go of prejudgments, seeking common ground, and listening empathetically (Mezirow 2003).

Other scholars have sought to theorize the role of empathizing much more explicitly in transformative learning, for example in dealing with emotions in a group setting when going through the critical reflection phase of the transformative learning process and subsequently creating a safe and trustworthy space for critical discourse and reflections. Empathizing helps learners to be more open, and to identify and understand others' perspectives, decreasing the likelihood of prejudgment and increasing the opportunity to establish shared understanding. Research has shown that perspectival transformation increases the ability to empathize with others (Gravett 2004) – which is particularly essential for transdisciplinary learning as learners have to be empathetic when considering stakeholders' perspectives and when engaging in discourse and reflection to establish a shared understanding.

The ethical dimensions of transformative learning are also debated. Unfortunately, in formal settings (in universities), educators are often instructive, telling their students what needs to be done rather than nudging learners on a journey of critical reflection to instigate transformative learning. According to both Mezirow's work (1991) and subsequent educators like Moore (2005), an educator cannot decide on the specific outcome of transformative learning because predetermination of outcomes by an educator may lead to coercion, indoctrination, or brainwashing, more than transformation. Pluralism of thought should be encouraged and discussed instead of concealed. The question needing consideration is: can transformative learning be implemented in authoritarian regimes or places with radical policies? In authoritarian regimes, the perspectives not aligned to the perspectives of the regime are concealed forcefully; therefore, the outcome of learning is predetermined. If transformative learning is implemented in such settings, this may lead to brainwashing and oppression through manipulation. The learners may be able to think autonomously but not critically – they will only be able to think in one direction as diverse perspectives are not provided to them. This is not aligned to the basic ethical dimensions of transformative learning, i.e. pluralism of thought, autonomous thinking, critical discourse, and reflection; therefore, the outcome cannot be normative.

Current forms of implementation in higher education

In both formal and non-formal settings, transformative learning and transdisciplinarity are usually integrated around action-oriented projects, also termed problem-based learning (Biberhofer and Rammel 2017; Nielsen 2020; Taimur and Onuki 2022; Wynn and Okie 2017). For example, the *Sustainability Challenge* course fostered transformative learning while promoting transdisciplinarity to drive sustainable urban development by exposing learners to interdisciplinary teamwork, interacting with diverse perspectives from diverse actors, involving creative and collaborative problem-solving (problem-based learning). This course was conducted under the coordination of the Regional Centre of Expertise on Education for Sustainable Development (RCE), located at Vienna University of Economics and Business. Since 2010, the course has been offered as a collaborative project between four Viennese universities, which encourages cooperation between learners, university partners, and practitioners to establish a shared understanding of urban development and create solution concepts to respond to these challenges (Biberhofer and Rammel 2017).

In another example, problem-based learning was implemented in the secondary-level social studies course by preservice teachers at Kennesaw State University, in the United States. This course (a) engaged stakeholders to expose learners to multiple truths (ontologies) and introduce the problem from multiple perspectives; (b) recognized the conflicting and competing positions; (c) generated solutions via deliberation on potential outcomes; and (d) guided reflection on types of thinking used by students (Wynn and Okie 2017). Teachers regarded problem-based learning as a transformative pedagogy as it allowed the teachers to create an environment for open discourse encouraging learners to think differently by considering different perspectives and see their relationship with the teachers differently (Wynn and Okie 2017).

Taimur and Onuki (2022) used design thinking, comprised of five stages (adapted from Plattner 2010), as a pedagogy to implement transformative learning in a semester-long university course in Japan and Germany. Both courses aimed to deal with sustainability challenges in a specific context (Kashiwa-no-ha, Japan, and Hude-Oldenburg, Germany). Throughout the implementation of design thinking for transformative learning, learners worked in diverse teams and consulted with the relevant stakeholders to identify the problem, ideate and prototype solutions, and present the problem with the corresponding solution. In this case, design thinking promoted consulting transdisciplinarity by implementing transformative learning via design thinking.

In conclusion, educators and students in higher education can co-facilitate ethical and supportive transformative learning within transdisciplinary learning. To support the undetermined nature of outcomes in transformative and transdisciplinary learning, educators must take the role of facilitators instead of being instructors. Before facilitating the transformative learning process, educators can reflect on: (1) Is it ethical for me to present my own perspective, which may influence the learners? (2) Is it ethical to decide which of the learners' beliefs should be questioned? (3) Is it ethical to facilitate transformative learning when the results may include hopeless or dangerous actions? (Taimur and Onuki 2020, 244). Therefore, educators must also ensure that a trusting, comfortable, and safe space is created before exposing learners to the transformative learning experience. Educators and participants should also discuss the uncomfortable nature of the transformative learning process and the role of empathy and compassion when engaging with others, which makes learners more mindful of their own behavior in the process.

References

Biberhofer, Petra, and Christian Rammel. 2017. Transdisciplinary learning and teaching as answers to urban sustainability challenges. *International Journal of Sustainability in Higher Education* 18 (1): 63–83.

Boyd, Robert D., and J. Gordon Myers. 1988. Transformative education. *International Journal of Lifelong Education* 7 (4): 261–84.

Checkland, Peter, and John Poulter. 2010. Soft systems methodology. In *Systems approaches to managing change: A practical guide*, eds. Martin Reynolds and Sue Holwell, 191–242. London: Springer.
Cranton, Patricia. 1994. *Understanding and promoting transformative learning: A guide for educators of adults*. San Francisco: Jossey-Bass.
Cranton, Patricia. 2016. *Understanding and promoting transformative learning: A guide to theory and practice*. 3rd edition. Sterling, VA: Stylus.
Daloz, Laurent. 1986. *Effective teaching and mentoring: Realizing the transformational power of adult learning experiences*. San Francisco: Jossey-Bass.
Dirkx, John M. 1998. Transformative learning theory in the practice of adult education: An overview. *PAACE Journal of Lifelong Learning* 7: 1–14.
Dirkx, John M. 2008. The meaning and role of emotions in adult learning. *New Directions for Adult and Continuing Education* 120: 7–18.
Dirkx, John. M., John Mezirow, and Patricia Cranton. 2006. Musings and reflections on the meaning, context, and process of transformative learning: A dialogue between John M. Dirkx and Jack Mezirow. *Journal of Transformative Education* 4 (2): 123–39.
Feller, Amanda. E. 2015. Where experience meets transformation. In *Putting the local in global education: Models for transformative learning through domestic off-campus programs*, ed. Neil W. Sobania, 52–72. Sterling, VA: Stylus.
Freire, Paulo. 1970. *Pedagogy of the oppressed*. New York: Seabury.
Gravett, Sarah. 2004. Action research and transformative learning in teaching development. *Educational Action Research* 12 (2): 259–72.
Inayatullah, Sohail. 2008. Six pillars: Futures thinking for transforming. *Foresight* 10 (1): 4–21.
Jung, Carl G. 1971. *Psychological types*. Princeton, NJ: Princeton University Press.
Kaplan, Jonas, T., Sarah I. Gimbel, and Sam Harris. Neural correlates of maintaining one's political beliefs in the face of counterevidence. *Scientific Reports* 6: n39589.
Kooiman, Jan, and Svein Jentoft. 2009. Meta-governance: Values, norms, and principles, and the making of hard choices. *Public Administration* 87 (4): 818–36.
Lange, Elizabeth. 2015. The ecology of transformative learning: Transdisciplinary provocations. Journal of Transformative Learning 3 (1): 28–34.
Mälkki, Kaisu. 2010. Building on Mezirow's theory of transformative learning: Theorizing the challenges to reflection. *Journal of Transformative Education* 8 (1): 42–62.
Meadows, Donella. 1999. *Leverage points: Places to intervene in a system*. Vermont: The Sustainability Institute.
Mezirow, John. 1991. *Transformative dimensions of adult learning*. San Francisco: Jossey-Bass.
Mezirow, John. 1995. Transformation theory of adult learning. In *In defense of the lifeworld*, ed. Michael. R. Welton, 39–70. New York: State University of New York Press.

Mezirow, John. 1996. Contemporary paradigms of learning. *Adult Education Quarterly* 46 (3): 158–72.
Mezirow, John. 2000. *Learning as transformation: Critical perspectives on a theory in progress*. San Francisco: Jossey-Bass.
Mezirow, John. 2003. Transformative learning as discourse. *Journal of Transformative Education* 1 (1): 58–63.
Mezirow, John. 2007. Update on transformative learning. In *Radical learning for liberation* 2, eds. Ted Fleming, Brid Connolly, David McCormack, and Anne Rya, 19–24. Maynooth: MACE National University.
Mitchell, Cynthia M., Dana Cordell, and Dena Fam. 2015. Beginning at the end: The outcome spaces framework to guide purposive transdisciplinary research. *Futures* 65: 86–96.
Moore, Janet. 2005. Is higher education ready for transformative learning? A question explored in the study of sustainability. *Journal of Transformative Education* 3 (1): 76–91.
Moyer, Joanne M., A. John Sinclair, and Lisa Quinn. 2016. Transitioning to a more sustainable society: Unpacking the role of the learning–action nexus. *International Journal of Lifelong Education* 35 (3): 313–29.
Müller, Daniel B., Sybrand P. Tjallingii, and Kees J. Canters. 2005. A transdisciplinary learning approach to foster convergence of design, science and deliberation in urban and regional planning. *Systems Research and Behavioral Science: The Official Journal of the International Federation for Systems Research* 22 (3): 193–208.
Nielsen, Niels M. 2020. Problem-oriented project learning as a first year experience: A transformative pedagogy for entry level PPL. *Education Sciences* 10 (1): 6.
Papenfuss, Jason, and Merritt, Eileen. 2019. Pedagogical laboratories: A case study of transformative sustainability education in an ecovillage context. *Sustainability* 11 (14): 3880.
Plattner, Hasso. 2010. *Bootcamp bootleg*. Institute of Design at Stanford. Available from https://dschool.stanford.edu/resources/design-thinking-bootleg.
Polk, Merritt. 2015. Transdisciplinary co-production: Designing and testing a transdisciplinary research framework for societal problem solving. *Futures* 65: 110–22.
Ross, Katie E. 2020. *Transforming the ways we create change: Experiencing and cultivating transformative sustainability learning*. Available from https://opus.lib.uts.edu.au/handle/10453/149105.
Ross, Katie E., and Cynthia Mitchell, 2018. Transforming transdisciplinarity: An expansion of strong transdisciplinarity and its centrality in enabling effective collaboration. In *The art of collaborative research and collective learning: Transdisciplinary theory, practice and education*, eds. Dena Fam, Linda Nuehauser, and Paul Gibbs. Cham: Springer.
Snowden, David., and Mary Boone. 2007. A leader's framework for decision making. *Harvard Business Review* 85 (11): 69–76.

Stuckey, Heather. L., Edward W. Taylor, and Patricia Cranton. 2013. Developing a survey of transformative learning outcomes and processes based on theoretical principles. *Journal of Transformative Education* 11 (4): 211–28.

Taimur, Sadaf, and Motoharu Onuki. 2020. Effective implementation of sustainability education in higher education settings via transformative learning approach: Literature review and framework proposal. In *Paradigm shifts in 21st century teaching and learning*, ed. Senol Orakci, 230–51. Pennsylvania: IGI Global.

Taimur, Sadaf, and Motoharu Onuki. 2022. Design thinking as digital transformative pedagogy in higher sustainability education: Cases from Japan and Germany. *International Journal of Educational Research* 114: 101994.

Taimur, Sadaf, Motoharu Onuki, and Huma Mursaleen. 2022. Exploring the transformative potential of design thinking pedagogy in hybrid setting: A case study of field exercise course, Japan. *Asia Pacific Education Review* 23 (4): 571–93.

Taylor, Edward W. 2007. An update of transformative learning theory: A critical review of the empirical research (1999–2005). *International Journal of Lifelong Education* 26 (2): 173–91.

Taylor, Edward W., and Elizabeth Tisdell, ed. *Patricia Cranton and transformative learning theory: An integrated perspective.* Harrisburg: Penn State University. Available from https://journals.sagepub.com/pb-assets/cmscontent/AEQ/CrantonDraft7.1.17TaylorTisdell.pdf.

Willis, Peter. 2012. An existential approach to transformative learning. In *The handbook of transformative learning: Theory, research, and practice*, eds. Edward. W. Taylor and Patricia Cranton, 212–28. San Francisco: Jossey-Bass.

Wynn Sr, Charles T., and William Okie. 2017. Problem-based learning and the training of secondary social studies teachers: A case study of candidate perceptions during their field experience. *International Journal for the Scholarship of Teaching and Learning* 11 (2): n16.

Young, Suzanne, and Tina Karme. 2015. Service learning in an indigenous not-for-profit organization. *Education and Training* 57 (7): 774–90.

Authors

Kathryn Adamson is a researcher in Physical Geography at Manchester Metropolitan University. Her research focuses on the impacts of climate change on glaciers and rivers. Through her interests in science communication, she works with a range of media outlets, and has designed and led a number of public engagement projects, exhibitions, and workshops for geoscientists. Kathryn listens to a lot of punk music.

Gameli Adzaho lives in Accra and serves as the country technical lead for the Research and Innovation Systems for Africa Fund, a United Kingdom government initiative aimed at strengthening local research and innovation ecosystems. He is passionate about public participation in science and innovation and has played key roles in advancing this goal with community-oriented organizations. Previously, Gameli was a Research Fellow of the Weizenbaum Institute for the Networked Society, Berlin.

Gerardo Alatorre Frenk is a professor and researcher at the Universidad Veracruzana in Xalapa. As an agronomist and anthropologist, he has been involved since the 1980s in multi-actor and transdisciplinary action research aimed at sustainable management of water, watersheds, and forests. Gerardo participates in citizen organizations such as Forest Stewards Network of Xalapa and *Agua para Tod@s Agua para la Vida*. He loves to improvise on his piano.

Marie Alavi focuses on transdisciplinary learning of future skills and attitudes in the *Beyond Disciplinary Boundaries* project. Her involvement in the *Horizon2020 Path2Integrity* project and evaluations on teaching research integrity fostered her enthusiasm for reflective, values-based education promoting ethically sound responses to modern challenges. Marie is not sure whether it was Plato or encountering children's ways of understanding that shaped her method of exploring cognitive potential.

Alhassan Yakubu Alhassan is research fellow in the Department of Sociology and Social Work and the Centre for Digital Transformation at the University of Agder, Kristiansand. He holds a masters in Sociology from the University of Saskatchewan in Canada and a masters in Global Development and Planning from the University of Agder. Alhassan is also a guest lecturer at the Kwame Nkrumah University of Science and Technology in Ghana.

Lucy Allen is a transdisciplinary practitioner who helps students, educators, and industry partners tackle complex challenges by developing their creativity, innovation, and boundary-crossing capacities. Her research focuses on the future of education and the role of the arts in its transformation. Lucy has previously taught in the performing arts space and co-founded a start-up in the chronic illness support sphere. She is also on a quest to befriend uncertainty by performing improvisational theatre.

Andrea Altepost is a sociologist and mathematician with professional experience in human resources development. She is fascinated by how people interact with each other and how people and technology interact. Therefore, her research focus is on the empirical analysis of human–technology interaction in the context of work and the participatory design of work systems. For conference trips, Andrea always packs a puzzle book to relax by solving crossword and syllable puzzles, Sudoku, and the like in the evening.

Julia Backhaus is a trained multi-disciplinarian and has become a convinced transdisciplinarian. For this reason, she joined the Living Labs Incubator, within the Excellence Strategy of RWTH Aachen University, in 2021, where she studies and supports the (co-)production of knowledge, socio-technical innovation, and societal transformation in and through living labs. To her, life feels like a living lab most of the time.

Philip Barth is a science enthusiast. Formally educated in physics and biology, he now works in university didactics, where he splits his duties between faculty development and curriculum development. By night, he is a podcaster and co-produces the podcast *Kritisches Denken*. The contribution on *Critical Thinking* in this volume came about in the aftermath of an interview he did with coauthor Jonas Pfister.

Michael Beck is a sociologist who is interested in educational inequalities, social justice, and research methodology. Having spent a lot of time researching inequalities in the Swiss educational system, with brief detours into political participation research, he is today the head of the Institute Education and Society at St. Gallen University of Teacher Education. Michael enjoys playing in a rock band with his old band mates, although they are economists.

Frank Becker works at TU Berlin's Science Shop *kubus*. As a trained electrician, trade union secretary, management consultant, co-founder of a repair cafe and a community garden, he has learned to combine working cultures. Frank's areas comprise cooperation networks of sustainability, regional economy, reuse, and social innovations. He is addicted to cooking due to the combination of a meditative attitude, healthy food, and sustainable living. He takes recipes as suggestions, not as prescriptions.

Laurenz Blömer studied industrial economics at TU Berlin and works as head of product development in the sustainability education startup *PROSUMIO*. In his masters studies, he focused on sustainability topics und experienced the importance of learning beyond the lectures to apply knowledge in practice. In 2020, Laurenz started the project lab *Sustainable Handprints* together with his colleagues at the TU Berlin. He loves to travel through Europe by train as a digital nomad.

Judith Bönisch works as a coordinator for student-organized teaching projects at TU Berlin. Her background in social and cultural anthropology and intercultural communication helps her to mediate between the formal structure of the university and students' ideas, and thus create unique learning opportunities together. For Judith, student self-organization is a very powerful emancipatory tool. In her spare time, she enjoys watching the rabbits she shares the living room with.

Stefan Böschen is a chemical engineer and sociologist. He is an interdisciplinarian and enjoys working in interdisciplinary environments. In such contexts, he feels right at home. RWTH Aachen University offers an exciting field of activity for this. However, two grandchildren are teaching him how fresh life is beyond the university.

Bernd Böttger was born in 1966 in Karlsruhe. He studied chemistry at TU Karlsruhe and finished his experimental laboratory work on Electrochemistry in liquid ammonia in 1994. Afterwards he went to Córdoba for a two years to work on the electrochemistry of thin films. Since 1998 he has been working with *ACCESS* in Aachen on virtual microstructure experiments and thermodynamics and spends much time with writing articles on seemingly unrelated topics.

Bonny Brandenburger is a research associate in the research group Education for the Digital World at the Weizenbaum Institute for the Networked Society, Berlin, and the University of Potsdam. In her dissertation, she explores the question of how open teaching and learning processes can be designed in the context of university-integrated makerspaces. She further conducts research on democracy education, open hardware, critical making, and citizen science. Bonny likes to dance to Latin American rhythms and enjoys long walks in nature.

Michael Brennan is a professor at the Ulster Business School, Ulster University Belfast, Northern Ireland. He teaches creativity and innovation, and his research has a focus on sustainable development through transdisciplinary problem-solving. Michael has a strong record in researching business models and socio-technical systems – supported by United Kingdom research council funding. He previously worked as an archaeologist with a particular interest in prehistoric landscapes and technologies.

Katja Brundiers works as a clinical associate professor in the School of Sustainability at Arizona State University. Her teaching focuses on professional skills and competencies in sustainability collaborations. She conducts transdisciplinary research on sustainable disaster mitigation in partnership with city partners and nonprofit organizations. Katja's mode of transportation is her bicycle, which is a source of daily joy and a reminder of the sustainability challenges facing us.

Jonas Bylund is affiliated to the Department of Urban Planning and Environment at the KTH Royal Institute of Technology Stockholm as researcher. He was part of the JPI Urban Europe Management Board from 2013 to 2022. Jonas's main responsibility was science-policy communication and developing urban research and innovation funding calls. He is trained in human geography and social anthropology, with a specific research focus on the knowledge practices in planning and environmental sciences.

Alessio Cavicchi is a full professor of agribusiness, rural development, and branding at the Department of Agriculture, Food and Environment of the University of Pisa. His professional experience includes coordinating several *Erasmus+* and other funded projects, and working as a researcher in many national and international research projects. Alessio is also a lover of good food and wine, especially if good means that it has been produced by people who care about the environment and the welfare of their workers, and who are respectful of the landscape.

Alexander Chmelka discovered during his studies in actor-network theory the conditions under which he could be proud to be called a sociologist. He is currently writing a making-of for the term *engaged learning* and teaches at Otto-von-Guericke-University in Magdeburg in the social sciences undergraduate program. As a holder of the 3rd Kyu in Shotokan Karate, Alexander agrees with Pierre Bourdieu's statement that sociology is a martial art.

Frederik Cloppenburg is Head of Nonwovens in the Institute of Textile Technology at RWTH Aachen University. He leads the institute's research activities in the field of nonwovens and paper products and production and, as chief digital officer, is also responsible for digitization and all the institute's digitalization research projects. Frederik is a passionate sailor and long-distance runner.

Wendy Jo Coones is on the academic and research staff of the Center for Image Science at the University for Continuing Education Krems. She coordinates the Erasmus Mundus joint master in Media Arts Cultures, lectures and publishes in the fields of museology, digital arts, and cultural heritage, and prepares international students for emerging futures in the cultural and creative sectors. Wendy's creative diversion is writing speculative fiction short stories with museum narratives.

Juliette Cortes Arevalo is a researcher and lecturer at the Technology, Policy and Management faculty of the TU Delft. Her work focuses on the collaborative modeling and management of climate-related hazards in Europe and abroad. While doing so, storytelling has become essential to facilitate knowledge exchange between research, government, and local communities. Juliette believes everybody has a story to tell, and the sum of everyone's efforts can make a difference.

Barbara Doran is a lecturer, artist, and creativity connector. She specializes in expanding how we understand and harness creative intelligence in the art of living. With over 25 years of experience, Barbara has worked in the realms of collective well-being, including public health, urban and regional planning, healthy psychology, and the arts. Barbara sees creativity as a powerful pathway to personally and collectively identifying unseen opportunities that help us respond to the complex challenges before us.

Marc Dusseldorp is an interim professor of philosophy, technology, and sustainability at Darmstadt University of Applied Sciences and a senior research scientist at a sustainability-oriented battery research start-up. He has been engaged in the (transdisciplinary) theory, practice, and teaching of sustainable development and technology assessment for many years. Beyond academia, Marc has been teaching yoga for 18 years – an approach to personal sustainability that he also promotes in university teaching.

Frances Fahy is a geographer and energy social scientist at the University of Galway on the picturesque west coast of Ireland, where she leads several research teams on sustainable consumption and energy citizenship research. A Fulbright scholar, Frances previously served as Head of Geography at the University of Galway. Frances has coordinated more than 20 funded research projects and is the current lead of the Horizon 2020 project, *EnergyPROSPECTS*. A retired harpist, Frances enjoys transdisciplinary research, and she co-created her son Ben.

Emanuele Fantini is senior lecturer and researcher in Water Politics and Communication at IHE Delft Institute for Water Education with extensive experience in working with journalists, photographers, and science communicators. He is the editor of the water governance blog *FLOWs* and hosts the podcasts *The Sources of the Nile, Si dice acqua*, and *The House of Water*. Committed to crossover projects outside academia, Emanuele has 12 years of experience as a certified bedtime storyteller.

Markus Feufel has a background in Organizational and Human Factors Psychology and heads the Division of Ergonomics at TU Berlin. He introduced Scrum to the division and is interested in working out a *Scrum for Science – Scrience* guide to help research groups to adapt the framework for managing their transdisciplinary projects more effectively. Markus thinks about how agile methods may help to improve public administration.

Steven M. Finn is driven by the challenge of successfully and equitably feeding a global population of 10 billion by 2050 within planetary boundaries, anchored by a core focus on leading educational efforts and partnerships across sectors to scale food waste reduction to transform food systems and accelerate progress toward Target 12.3 and the Sustainable Development Goals. In his downtime, he engages in many projects to make old things new while also re-reading many literature classics.

Alessandro Fusco is Future Food Japan Director at the Future Food Institute. After 10 years of corporate experience as senior designer and innovation manager, he founded a start-up that has transformed the way we experience wine through storytelling. Alessandro has taught at several universities, among them Politecnico di Milano. He currently leads educational and innovation initiatives in Japan to regenerate the food ecosystem. Alessandro is an improv theatre actor, constantly applying the philosophy of '*yes, and*'.

Julia Gäckle is a landscape architect and research associate in the Institute of Landscape Architecture at RWTH Aachen University. In her research she deals with the topics of local identification and appropriation of urban space through temporary interventions. Julia also works as a project leader at Treibhaus Landschaftsarchitektur in Hamburg.

Gabriele Gramelsberger is a philosopher of science and technology at the RWTH Aachen University, interested in the digitization of science. She is co-head of the Käte Hamburger Kolleg *Cultures of Research – an International Center for Advanced Studies*. Gabriele's recent book is devoted to the philosophy of the digital. She likes to watch biologists observing software observing fish.

Thomas Gries has a background in mechanical engineering and economics and is the director of the Institut für Textiltechnik of RWTH Aachen University, a leading institution in digitalization and biotransformation of the textile sector. In addition to his work as a scientist and science manager, he holds several management positions in leading international technological companies of various size (start-ups, mid-size, large scale) as well as nonprofit organizations. Thomas enjoys spinning wheels, both at work and in his free time.

Mary Griffith is a perpetual foreigner and approaches communication in unique ways. She is a lecturer at the Universidad de Málaga in Spain, researching bilingualism. Most recently she has been exploring how humans process language to develop machine learning tools using artificial intelligence. Her 20-year-old self could never have imagined being on a junior year abroad for the last 30 years. Perhaps she did it all backwards, studying Spanish in the United States and English in Spain, going back to university with four kids under ten. But, indeed, higher education is a lifelong process.

Heidi Grobbauer is the managing director of *KommEnt – Society for Communication, Development, and Dialogic Education* in Vienna and a member of the Austrian Strategy Group for Global Education. Her work focuses on the implementation of continuing education programs for teachers and other education professionals, such as the Master's Program Global Citizenship Education at the Alps-Adriatic University in Klagenfurt. All this leaves too little time for literature (adult and children's) and jazz.

Jutta Gutberlet was born in Germany and raised in São Paulo. She started out as a biologist and converged towards geography, providing her with a diverse skill set to connect the social, economic, and political with the environmental dimension. Jutta is passionate about participatory and action-oriented research approaches pursuing the goals of sustainable development, resilient communities, and global environmental health. She loves hiking, camping in wilderness, and mountains are the best.

Christoph Heckwolf is a sociologist at the Chair of Sociology of Technology and Organization at RWTH Aachen University. With an emphasis on transdisciplinary contexts, his current work focuses on sustainability, the bioeconomy, and the textile industry. His work is grounded in network theory, social network analysis, and qualitative social science, which he integrates into transformative research, e.g. real-world experiments and labs.

Maren Heibges, an anthropologist of science and technology, is the deputy head of the Division of Ergonomics at TU Berlin. She conceived of *Scrum for Science – Scrience* during her management of a transdisciplinary research project in her unit. Maren also works as a trainer for both agile science facilitation and mixed and qualitative methods. Maren is the co-owner of two exceptionally large rabbits, which reside in her back-garden and have been rescued from a balcony.

Loni Hensler practices the art of facilitating participatory processes of collective learning and action towards a shared territory management. Her focus is on transformative learning, diverse values about nature, conflict transformation, and construction of horizontalities from a collaborative action-research approach. Loni studied economy, environmental and sustainable sciences, and alternative pedagogies. Currently, she contributes to care of nature and people in the solidarity economy network *La Gira*, the NGO Mexican Center for Environmental Law, and the Forest Stewards Network of Xalapa. Loni loves to practice karate.

Vera Husfeldt is a scientist with long experience in education research. For eight years, until 2019, she was head of the Quality Development Department at the Swiss Conference of Cantonal Ministers of Education and was responsible for education monitoring and educational standards. Since 2020, she has been a professor at the University of Applied Sciences of the Grisons in Chur, where she heads the research area of education informatics. As a contrast, Vera is studying theology to appear a little wiser.

Melanie Jaeger-Erben is an environmental sociologist working at Brandenburg University of Technology. Her research focuses on the interface of technological, social, and environmental change, particularly on challenges and possibilities for sustainable consumption and production. In her research, Melanie experiments with novel methodological and conceptual approaches. She regularly visits the Discworld.

Isa Jahnke is professor of learning technologies and founding vice-president for academic and international affairs at the newly founded University of Technology Nürnberg. She studied social sciences and was doctoral researcher in informatics at TU Dortmund. Isa was professor of Interactive Media and Learning at Umeå University and director of the Information Experience Lab at the University of Missouri-Columbia. She likes to run and did a marathon, some years ago.

Chhavi Jatwani is a design innovation project leader contributing to a sustainability transition in the food sector. She leverages design thinking, systems thinking, and creative problem solving for complex food system challenges. These days

Chhavi is busy with developing a new planet-centered innovation methodology and mentoring an outstanding team of associate designers changing the world and how we eat.

Thies Johannsen teaches and works in the span between humanities, arts, and social sciences on the one end and science, technology, engineering, and mathematics on the other. His research focuses on competences for innovation along with knowledge and technology transfer. After working on various research projects and in research administration, he has developed and coordinated cooperative education programs with practical partners from business and public administrations. Thies is a passionate plant guardian of green souvenirs from his travels.

Stefan John is a researcher with the Living Labs Incubator located at the Human Technology Center within RWTH Aachen University. For the Living Labs Incubator project, he is networking, researching, and supporting living labs. In his academic work, Stefan is analyzing (power) structure in living labs and researches about modes and understandings of experimentation in these (knowledge) infrastructures in the contemporary knowledge society.

Marina Joubert is a science communication researcher, lecturer, and trainer at Stellenbosch University. Her research interests focus on scientists' roles in public communication of science, science representations in mass media, and online interfaces between science and society. Marina is the deputy editor of the *Journal of Science Communication* published by SISSA Medialab. She is passionate about oceans and enjoys snorkeling in South Africa's kelp forests.

Katharina Jungnickel is an industrial engineer and has developed an instrument to measure agility for teams. She is also a certified Scrum Master and has given various workshops on agile management to small and medium-sized enterprises. Currently, Katharina is a research fellow in a transdisciplinary research project at the Division of Ergonomics at TU Berlin, supporting the adaption of *Scrum for Science – Scrience*. In her spare time, she cuts and paints wood in her garden.

László Z. Karvalics is a passionate omnivore and manufacturer of multidisciplinary information culture research, as a historian and linguist, teaching in different Hungarian universities. An advocate of lifelong research and the initiation-like integration of young generations into local and global problem-solving projects, László is also an expert on the cultural history of toothpicks and a determined collector of old toothpicks and toothpick holders.

Konstantin S. Kiprijanov works as project coordinator and public engagement researcher for the project *Public Engagement with Planetary Health* at the Natural History Museum in Berlin. He is also engaged in teaching and training activities, and acts as a strategy consultant to the Museum's directorate. Konstantin holds a PhD in history and philosophy of science from the University of Leeds. His favorite pastime activities are cooking and reading pretentious books in countryside pubs.

Julie Thompson Klein passed away during the production of this book. She was professor of humanities emerita at Wayne State University, international research affiliate at the Transdisciplinarity Lab at ETH Zurich, and, later, visiting foreign professor at Shimane University in Japan, Fulbright professor at Tribhuvan University in Nepal, Foundation Visitor at the University of Auckland, and distinguished scholar in residence at the University of Victoria. Julie was the author and editor of numerous books, articles, and chapters on inter- and transdisciplinary research, and education. She was president of the Association for Interdisciplinary Studies and a recipient of the Kenneth Boulding Award for outstanding scholarship on interdisciplinarity. Her openness to share her experience with uncountable scholars from all career stages around the world was outstanding. Julie's death is a tremendous personal and intellectual loss.

Giedre Kligyte's expertise is in transdisciplinary education. In her research, she investigates how university education can be transformed to advance more sustainable and socially just futures. In particular, Giedre explores how different perspectives and relationships across boundaries can be creatively leveraged to create third spaces: spaces where difference, experimentation, and co-creation are embraced to stimulate mutual learning, new ways of thinking, and creativity. Giedre has worked in universities across three continents and five national higher education systems.

Tibor Koltay, after having worked as an information professional, began teaching Library and Information Science at various universities in Hungary, first part-time, then for 15 years full-time. He has always enjoyed writing professional papers and books, as much as he loves listening to good classical music and blues.

Kerstin Kremer is a researcher in science education at Justus Liebig University Giessen. She is involved in biology teacher education and research about various challenges of science education in the 21st century, like the reflection of the interplay between science and society and the role of data literacy and sustainability literacy for future citizens. When traveling, Kerstin loves to visit art exhibitions.

Ines Langemeyer conducts research in the fields of science education, the philosophy of education, educational psychology, and vocational education at the Karlsruhe Institute of Technology. She focuses on teaching and learning relations research, exploring the many roles digital and other modern technologies play in mediating developing knowledge. Ines is interested in daily learning practices, and epistemic cultures in higher and other adult education. In her research, she draws on historical and critical psychological theorizing.

Martine Legris has always been a keen and curious observer of the world around her, even as a child. As an adult, she was drawn to history and sociology. Martine studied both disciplines at university and began to practice participatory action research. Unable to remain in a single discipline, she began to work in interdisciplinary teams, using both social and engineering sciences. Martin now works at the University of Lille and is part of the its science shop. She loves dancing, a fun way to recharge.

Bem Le Hunte is an internationally published novelist and Director of Teaching and Learning in the TD School at the University of Technology Sydney. She draws on creative, cultural, and spiritual wisdom for her work as the Course Director of the Bachelor of Creative Intelligence and Innovation – a transdisciplinary degree that combines with 25 other degrees. She describes herself as a liminal human – half Indian, half English, and Australian by choice, and advocates for human unity in all her work.

Marlene Mader works as a lecturer and research assistant in the Transdisciplinarity Lab at ETH Zurich. She is responsible for a stakeholder- and project-oriented sustainability lecture for first-year bachelor students in environmental sciences. Marlene has a background in Environmental Systems Sciences and is engaged in higher education for sustainable development since 2009. Her family runs a small apiary in a mountain village.

Sonia Massari has more than 20 years of experience as an educator, researcher, consultant, and designer in the fields of human–food interaction design, sustainability education, design thinking, and creative methods for innovative agri-food systems. She is a research fellow in the Department of Agriculture, Food and Environment at the University of Pisa, co-founder of FORK Food Design, and the director of the Food Future Institute's academy. Sonia loves exploring the world through the sincere and creative eyes of her two amazing children.

Kaisa Matschoss works as a university researcher at the University of Helsinki in the Centre for Consumer Society Research. Her research interests lie in tensions in the everyday life of citizens in energy transition, (living lab approaches to) experimentation and innovation in the energy sector, including sustainable household practices, as well as policies for sustainable consumption. Kaisa likes to hug trees and talk to flowers, butterflies, and bees.

Jacqueline Melvold is a passionate scientist, educator, and transdisciplinary practitioner who aims to create more desirable and sustainable futures via the transformation of education that can empower young people. Her research focuses on the future of education, where she explores innovative and creative approaches to create a space for mutual learning between educators, academics, students, and professional partners. In her spare time, Jacqueline can be found kayaking in the National Parks in Australia.

Jimlea Nadezhda Mendoza is from the Philippines, daughter of a fisherman and public-school teacher. She works at Ca' Foscari University of Venice, is passionate about exploring the local ecological knowledge of fishers and about contributing to building bridges between academia and local communities through storytelling, conveying the acquired oral histories about the freshwater resources to a wider audience. Jimlea loves to eat fish all day long.

Juliana Merçon was born and raised in Brazil but has lived for the past 12 years in Mexico. She studied psychology, philosophy, and education, before exploring the field of agroecology. Collaborating towards social and environmental justice is at the heart of her (academic) life. For Juliana, transdisciplinarity and participatory action research are not only epistemological approaches as they also entail ways of being and deeply connecting with people, places, and time. Her eight-year-old daughter Michelle is her main source of fun, learning, and inspiration.

Esther Meyer was born in Fürth, and has a background in philosophy, economics, and sustainability sciences. She is committed to supporting transdisciplinary change agents and those who move at the institutional boundaries of education by teaching, knowledge systematization, consultation, creating networks, and methodological accompaniment. As a lifelong learner, Esther completed training in organizational development and consultancy. As a passionate dancer, she has recently started going to football stadiums to watch fan choreographies.

Joost Meyer discovered his scientific focus through his work at the forest lab *Waldlabor* in Cologne. With a team of young scientists, he founded the *Willowprint* project, where he and his colleagues seek for sustainable biogenic materials for 3D printing. In addition to his work for science and teaching, Joost uses sculptural means to explore the meaning of the natural environment in human perception. He enjoys the Scandinavian summer in Denmark's North Jytland, swimming in the rain.

Stefanie Meyer has been leading the Case Study Office at Leuphana University in Lüneburg since 2019. She is responsible for the coordination and further development of the structural and content-related conception of case studies in the areas of transdisciplinary research and teaching in bachelor and master programs at the School of Sustainability.

Ewald Mittelstädt has been a founder since his youth. Later he studied business management, slavistics, political science, and public law in Münster as well as Economics in Krakow. At TU Dortmund, he worked in teacher education and did research on chaos theory. Following postdoctoral positions in Freiburg and Kiel, Ewald now teaches and conducts research at South Westphalia University of Applied Sciences in Iserlohn. Ewald is interested in contemporary and street art and feels dedicated to the maxim: Life is as colorful as you dare to paint it.

Manon Mostert-van der Sar is director of Stadslab Rotterdam, the fab lab of Rotterdam University of Applied Sciences, and author of *Hey Teacher, Find Your Inner Designer.* Manon is a maker and designer of education. She trains teachers, coaches instructors, designs cross-institute modules, and developed a didactic compass, but still loves to teach students. At home Manon is a real maker, rebuilding her own farmhouse. She is not afraid of bricklaying, sawing, milling, and all kinds of carpentry.

Daniel Münderlein is a trained landscape architect. His research focuses on the perception and further development of urban landscapes and the integration of productive open-space systems. Special attention is paid to the interrelationships between landscape, health, and well-being as well as the development of innovative methods for visual communication in spatial planning. Daniel loves taking pictures of urban landscapes and plays the electric guitar louder than the neighbors would like.

Olena Mykolenko works as a lecturer and researcher in the field of entrepreneurship at Karasin University in Kharkiv. She is currently doing her post-doctoral research on humane entrepreneurship with a special focus on leadership and sustainability. Olena is also involved in a European research project on sustainability thinking and entrepreneurship. She likes theater, and – despite challenging circumstances – she regularly visits it in her city together with her lovely daughter.

Oliver Parodi, philosopher, cultural scientist, and civil engineer, is head of the Karlsruhe Transformation Center for Sustainability and Cultural Change at the KIT Institute for Technology Assessment and Systems Analysis. As an expert in transformative sustainability research, he was the initiator and has been director of the real-world lab *Quartier Zukunft – Labor Stadt* since 2012. A matter close to his heart is personal sustainability in research and teaching – but in private, he is still far away from achieving it.

BinBin J. Pearce is an assistant professor for policy analysis and design at TU Delft's Faculty of Technology, Policy and Management. Her transdisciplinary research interests include public participation processes and policy design for the energy transition, collaborative decision making for sustainable development, joint problem-framing processes, and developing curriculum based on integrated systems and design thinking.

Daniela Peukert has a background in product design and sustainability sciences and holds the professorship of design for sustainability transformation at Tomorrow University of Applied Sciences in Berlin. In her research, she explores prototyping practices and material metaphors to foster integration and knowledge co-production within transdisciplinary processes and works towards a more-than-human-centered-design approach. Daniela enjoys being outdoors on boards and bikes.

Jonas Pfister came to philosophy through conversations during adolescence, stayed with it because of a guy who talked about the philosophy of Paul Grice – a topic which eventually became one of his main areas of research. He taught for several years at the universities of Berne, Lucerne, and others, and is now teaching at the University of Innsbruck. He is the author of introductions to philosophy, to philosophical methodology, and to critical thinking – all in German. He hopes to write more.

Thorsten Philipp studied Romance languages, art history, and political science in Munich, Aix-en-Provence, Vienna, and Brescia. As an advisor to the presidium at TU Berlin, his mission is to promote transdisciplinary learning. A lecturer with a non-disciplinary agenda at various universities, such as the University of Freiburg, Thorsten explores pop music as a sounding board for sustainability communication. Background music by Dat Adam, Monolink, PNL, Tommy Guerrero, and Nils Petter Molvær accompanied him while working on this handbook.

Roland Postma is a geographer, storyteller, and professional video-maker. He has been researcher, director, and editor-in-chief of a number of leading programs for Dutch National Television for children and series about science and nature. He also gives on-location and online training courses to students, and public and private organizations. Roland's emphasis is on telling stories with content that matter: Stories that captivate the viewers in such a way that they are willing to listen to the message.

Monique Potts is a learner, thinker, teacher, mother, and healer whose research focuses on understanding resilience and experiential learning for young people in the context of uncertain futures and climate disruption. Her background is in public media, education, and digital media. Living on Gadigal land of the Eora Nation in the inner west of Sydney, Monique has a long and checkered work background, including dressing up as a koala to collect money for wildlife, being a clown, and fire-eating.

Susanne Pratt is a transdisciplinary artist, educator, and researcher working at the intersection of foresight, environmental change, and transdisciplinary learning. Her recent research focuses on participatory arts-led means of reimagining practices of care, multi-species relations, and regenerative futures. She grew up on a kiwifruit orchard in Aotearoa and currently lives and works in Australia on unceded Gadigal Land.

Baiba Prūse is research associate at Vrije Universiteit Amsterdam, where her work involves discussions regarding public engagement in research, food systems, and local ecological knowledge aspects. She is also a coach for the course on interdisciplinary community service learning and is eagerly following the student knowledge exchange on the course topics. Collaboration is the motto of Baiba's approach to life and this process is directly reflected in her work. Part of her contribution to this book received funding from the European Research Council under the EU Horizon 2020 research and innovation program.

Sanjeet Raj Pandey is a skilled full-stack software developer and graduate in computer science. As a research associate in the service-centric networking at TU Berlin, he teaches in the field of web-technology and his research is focused on Web3 for privacy in digital systems. Sanjeet is the head of software engineering in the sustainability education startup *PROSUMIO* and a core member of the *Sustainable Handprint* project lab at TU Berlin. Sanjeet loves to use fast carsharing with his friends.

Eliana Rodrigues is a dedicated researcher in the areas of Ethnobotany and Ethnopharmacology at the Federal University of São Paulo. She coordinates the activities of the Center for Ethnobotanical and Ethnopharmacological Studies. Eliana has dedicated her studies to the area of traditional use and management of medical plants, from the perspective of participatory ethnobotanism. She loves making mosaics on the walls of her house.

Katie Ross is curious about ways to facilitate change towards more beautiful, equitable futures. As a transdisciplinary writer and facilitator at *Soils for Life*, and an adjunct fellow at the University of Technology Sydney, she integrates transformative learning processes into collaborative transdisciplinary action learning. One of the best perks in Katie's role is meeting incredible farmers and experiencing their delicious food, grown in living soils, brimming with the nutrients we need.

Sara Roversi is an author, entrepreneur, public speaker, and activist. She has been a serial entrepreneur since 2003, formulating multiple for-profit and not-for-profit ventures. In 2010, Sara shifted her focus onto sustainable food systems to increase awareness and understanding through education, research projects, and innovation challenges. She loves connecting and working with people and organizations on food systems transformation and radical change.

Alexander Ruser is professor of Sociology and Director of the Centre for Digital Transformation at the University of Agder in Kristiansand. His research focuses on the social role of expertise and the impact of scientific knowledge on environmental discourses and climate politics. Alex owns too many guitars and always keeps an emergency half-acoustic in his office.

Olivia Rütti-Joy is a senior researcher at the Université de Fribourg and the research assistant of the rector at the St. Gallen University of Teacher Education. Building on her background in Media Studies, English Literature, and English language teaching, she now pursues research in foreign language teaching and learning, applied linguistics, language testing, educational assessment, and teacher education. If she's not behind her screen, Olivia is most likely somewhere in the mountains.

Lyda Patricia Sabogal-Paz is an associate professor at São Carlos School of Engineering, University of São Paulo. Her research is centered on selection, design, and assessment of drinking water treatment for small and medium-sized communities. She has a strong record on water research with a focus on technology assessment and economic evaluation. Brazilian and United Kingdom research councils have been financing her research. Lyda enjoys traveling and discovering new cultures.

Jennifer Schluer is an assistant professor for Teaching English to Speakers of Other Languages and Advanced Academic English at Chemnitz University of Technology. As an applied linguist and teacher educator, she specializes in digital teaching and digital feedback methods as well as in language awareness, multilingualism, and culture learning. Jennifer can be a cat or cow and a dog or dolphin, but has never turned into a camel or crow on her yoga mat so far.

Marco Schmitt is a sociologist with a strong background in network research and theory working at the Chair of Sociology of Technology and Organization at RWTH Aachen University. He has worked in a variety of transformation and innovation projects involving public administration, industry, politics, research, and civil interest groups. For Marco, transdisciplinary thinking is the most striking success factor in enabling change, but so hard to achieve.

Stephan Schmidt-Wulffen is senior professor for Art Theory at the Free University Bozen-Bolzano. As the head of the Academy of Fine Arts Vienna – among others – for many years he developed university programs in art and design. His research focuses on artistic practice and its history since the early 20th century.

Tobias Schmohl is passionate about the transformative potential of learning. He believes that by fostering critical thinking we can empower individuals to make positive impact on the world around them. Tobias has spent years exploring the intersection of higher education and education research to better understand how to create meaningful learning experiences. His children's love for inquiry-based projects, exploring their passions, inspired him to pursue progressive and avant-garde teaching and learning methods.

Erika Solimeo is an environmental legal specialist. She focused her earlier career on global water policy and the human right to water, which evolved toward an analysis of marine conservation and ocean ecosystems. At the Future Food Institute in Rome, she directly followed topics such as nutrition, climate change, earth regeneration, Mediterranean diet, water safety, and security. Erika has always been known in the family as Don Chiscotte for having a soft spot for the hard causes.

Alexander Sonntag works as a research assistant at the Living Labs Incubator of the Human Technology Center at RWTH Aachen University. He is currently participating in the project *TRANSFER* in which he is observing mechanisms of knowledge transfer in Living Labs. Alexander Sonntag studies Governance of Technology and Innovation at RWTH Aachen University.

Anja Steglich is a landscape architect and consultant for urban transformation and transfer. She has many years of experience in transformative research and transdisciplinary teaching. Anja coordinates the development of TU Berlin's *Stadt-Manufaktur*, a platform and center for Living Labs in the Executive Board of the TU of Berlin. Her work focuses on urban transformation, water-sensitive design strategies, and productive infrastructure development. Anja develops stage designs and learns from dance.

Sadaf Taimur is a researcher exploring how to create effective learning processes and environments for sustainability that lead to individual and organizational transformation. She has researched transformative learning approaches to improve pedagogies in Higher Education for Sustainability at the University of Tokyo. In addition, Sadaf is interested in reading non-fiction and exploring different perspectives and cultures.

Wing-Shing Tang's research interests focus on experimenting mainstream urban thinkers with socio-historical processes and patterns of urban development in Hong Kong, mainland China, and India, and, concomitantly, formulating new ontology, epistemology, and methodology for this cause. Teaching high-school students on the spatial dialectics of injustice had been experimented with in Hong Kong by taking them around urban communities on local tours.

Ewald Terhart is professor Emeritus in the Institute of Educational Science at the University of Münster. His academic fields are research on teaching and general didactics, as well as teacher professionalism and teacher education. He serves on the editorial boards of several academic journals in these research areas. Due to his expertise in teacher education, he has been a member or chair of several commissions on the evaluation and reform of teacher education in Germany and other European countries.

Axel Timpe is a landscape architect and research and teaching associate at RWTH Aachen University's Institute of Landscape Architecture. He is coordinating *H2020 IA proGIreg*, the transdisciplinary green infrastructure research project *CoProGrün*, and had a coordinating role in *COST Action Urban Agriculture Europe*. Axel's research focus is on green infrastructure, nature-based solutions, and urban agriculture and their co-production with local stakeholders. In addition to design studio teaching, Axel developed the edX MOOC *Nature-Based Urban Regeneration*.

Peter Troxler is a research professor at Rotterdam University of Applied Sciences. He studies how digital technologies enable new ways of designing and manufacturing and challenge and influence incumbent practices in industry and education. Peter has worked internationally in the energy industry, as a design consultant, and in higher education. He enjoys photographing weird inner-city moments and sharing them with the world.

Valentin Unger is a senior researcher at St. Gallen University of Teacher Education. His research interests stem from his dual role as a teacher and researcher. Valentin's research focuses on the empirical study of educational learning processes, especially in language subjects, research on the teaching profession, and research on the impact of crises on education. Besides his very cool research job, he has an even cooler hobby: classical music.

Karen van den Berg is professor of Art Theory and Curating at Zeppelin University in Friedrichshafen, and head of the university's arts program. Research and fellowships have taken her to international universities and research institutions, including the Chinati Foundation in Marfa, IKKM Bauhaus University in Weimar, and the Department of Comparative Literature at Stanford University. Karen's research focuses on art, politics, and activism; artistic episteme and studio practice. She is a real mother animal.

Laura Verbrugge is an environmental scientist aiming to improve collaborations across disciplines and especially between the natural and social sciences. She developed educational material about invasive alien species for Dutch (vocational) schools and coordinated communication and monitoring campaigns between government and stakeholder organizations. Currently, Laura works as a grant writer at Aalto University.

Matteo Vignoli is building the future through research, education, and business all centered on the application of design thinking experimenting with the world-renowned *Reggio Approach* in higher education. He is associate professor at Università di Bologna, teaches in several master programs and business schools, and works on innovation projects with global organizations. According to Matteo, involving the client and future user during co-design sessions is the secret sauce of the world's most visionary companies.

Ulli Vilsmaier grew up in the Austrian Alps and has a background in Geography. She conducts and studies inter- and transdisciplinary research and higher education. Her work is based on a responsive logic and emphasizes complementarities between different ways of knowing, acting, and being. Ulli accompanies teams of researchers and educators in transforming research and higher education towards inter- and transdisciplinarity. While her professional life allows her to travel and learn with colleagues from different world regions, she loves to spend her spare time mountaineering in the Alps.

Maximilian Voigt works for the Open Knowledge Foundation Germany on the topics of Open Education and Open Hardware. He is focused on the potential of open technologies and infrastructures for the circular society. Maximilian is a board member of the Association of Open Workshops – the German umbrella organization of makerspaces – and studied engineering, journalism, and the philosophy of technology. When he is not writing, he sings in a choir, repairs things, or sails the Baltic Sea.

Johannes Vollbehr is an experienced carpenter and purpose entrepreneur. At TU Berlin, he studies vocational education, being a scholar in the professional talents program of the German Federal Ministry of Education and Research. Johannes works as head of business development for the sustainability education startup *PROSUMIO*. In 2020, he initiated and facilitated the project lab *Sustainable Handprints* together with his colleagues at TU Berlin. Johannes loves to ride his bike on Berlin's roads, being part of the critical mass.

Hildrun Walter is a researcher at the University of Graz with her research interests in science communication at the interface to science education. She is particularly dedicated to evaluating communication programs in order to assess their potential to evoke emotions and foster scientific interest. She is concerned with the reflection of open science practices and participatory approaches in research. Like a herb lady, she loves to spice up the vegetables from the garden in colorful dishes.

Christine Wamsler is professor at Lund University Centre for Sustainability Studies and Director of the Contemplative Sustainable Futures Program. She is an internationally renowned expert in sustainable development and associated (material and cognitive) transformation processes, with 25 years of experience, both in theory and practice. Christine has led many international projects and published more than 200 papers and book chapters on these issues. She draws inspiration from nature and her two children, who help her daily take better care of spiders, flowers, and the world at large.

Hendrik Weiner researches urban transformation and co-design processes and develops concepts and projects for the collaborative development of urban spaces. He organizes transdisciplinary working and teaching settings to connect practice and theory, students with citizens, universities with cities. He teaches at the Brandenburg University of Technology Cottbus-Senftenberg and at the Otto-von-Guericke-University Magdeburg and runs the office *raumdialog*. Hendrik enjoys bicycle tours with his two children.

Annika Weiser is coordinating the sustainability-focused responsibility module for all first-year students at Leuphana University Lüneburg. With a background in transdisciplinary sustainability research, she has long been working – and teaching – in various transdisciplinary research projects such as *Lüneburg 2030+*. She invests large parts of her spare time in trying to grow, with varying success, the best pimientos de padrón north of the Alps.

Christina West, a geographer, philosopher, and artist, founded the *Energy Academy* in DELTA, an experimental real-world lab of the energy transition, for the

University of Applied Sciences Darmstadt. She is Chairwoman of Urban Innovation-Rethinking City. At Heidelberg University, she headed two real-world labs, was interim professor for the Didactics of Geography at Koblenz-Landau University, and visiting scholar at EUAP Hong Kong Baptist University. Her favorite mode is exploring the world while dancing.

Ulrike Weyland is a professor in the Institute of Educational Science at the University of Münster. Her research focuses on teacher professionalism and teacher education, competence modeling and measurement, career orientation, and digitization, with a special focus on vocational education in the health care professions and their educational personnel. She serves on the editorial boards of academic journals in these research areas. In addition, she has served on various commissions as an expert on issues related to the design of internships in teacher education.

Mary Whalen was the managing director of EPIZ – Center for Global Education in Berlin until 2019. Presently, her work focuses on international dialog forums on the implementation of Global Education, fostering a critical discussion of Global Citizenship Education in European civil society, and biographical approaches to feminist civic education. She divides her time between the magical Wet Tropics in Queensland and the pulsating diversity of the city of Berlin.

Claudia Wiepcke studied economics, Slavic studies, and politics at the University of Münster, where she also received her doctorate. She worked as a research assistant at TU Dortmund and has been professor of Business and Economic Education since 2008. Claudia researches and teaches at the University of Education in Karlsruhe. She is interested in art and creates content about art and museums.

Johannes Wildt studied Psychology, Pedagogy and Sociology. Since 1970 he has worked as a professional educationalist and academic developer at different universities, mainly at TU Dortmund, heading its Center for Teaching and Learning. Johannes put his main emphasis on innovative learning concepts like project-oriented studies and transformative learning, participatory curriculum development, learner-centered further education, and consulting. He likes to dance at a lot of weddings.

Tyson Yunkaporta is an academic, an arts critic, and a researcher who is a member of the Apalech Clan in far north Queensland. His work focuses on applying Indigenous methods of inquiry to resolve complex issues and explore global crises. He carves traditional tools and weapons and also works as a senior lecturer in Indigenous Knowledges at Deakin University in Melbourne.

Eike Zimpelmann's research is mainly focused on the fields of vocational teacher education and the subjective perception teachers develop regarding their academic disciplines and their work. Further, he is interested in vocational didactics, education policy, and education for sustainable development. Eike is passionate about understanding how people construct their reality. Probably, that's why he likes spending his time with roleplaying games where he can try out behaving like other people.

Reviewers

Søren Smedegaard Bengtsen, Aarhus
Mattias Bingerud, Gothenburg
Eglė Butkevičienė, Kaunas
Raphaela Casata, Passau
Hugo Caviola, Bern
Ana M. Corbacho, Montevideo
Irina Dannenberg, Oldenburg
Christian Decker, Hamburg
Uwe Elsholz, Hamburg
Mikael Enelund, Gothenburg
Dieter Euler, St. Gallen
Julianna Faludi, Budapest
Andra-Ioana Horcea-Milcu, Kassel
Tobias Jenert, Paderborn
Ásthildur Jónsdóttir, Helsinki
Jonas Kellermeyer, Berlin
Florian Keppeler, Aarhus
Renate G. Klaassen, Delft
Thomas M. Klotz, Munich
Maximilian Lantelme, Munich
Oddrun Maaø, Trondheim
Claudia Mendes, Hamburg
Magnus Merkle, Ås
Ingmar Mundt, Berlin
Linda Neuhauser, Berkeley
Iris-Niki Nikolopoulos, Tübingen
Philipp Ulrich Abele, Munich
Susanne Pratt, Sydney
Gerrit Rößler, Berlin
Susanna Sancassani, Milan
Gamel Sankarl, Accra
Marco Scalvini, London
Bjørn Sortland, Trondheim
Sofia Sousa, Porto
Steve Taylor, London
Douglas Tetteh Ayitey, Accra
Katharina Thies, Lemgo
Rune Tranås, Trondheim
Ignacio Yusim, Valencia

Promoted by

Berlin University Alliance

Senatsverwaltung
für Wissenschaft, Gesundheit,
Pflege und Gleichstellung
BERLIN

Technische
Universität
Berlin

Bundesministerium
für Bildung
und Forschung

Freie Universität Berlin

CHARITÉ
UNIVERSITÄTSMEDIZIN BERLIN

HUMBOLDT-UNIVERSITÄT ZU BERLIN

ENHANCE
European Universities of Technology Alliance

STIFTERVERBAND

TH OWL
TECHNISCHE HOCHSCHULE
OSTWESTFALEN-LIPPE
UNIVERSITY OF
APPLIED SCIENCES
AND ARTS

randstad stiftung
impulse für unsere
arbeitskultur

Fördert • Verbessert • Unterstützt
Hochschulgesellschaft
Ostwestfalen-Lippe e.V.

Pädagogik

Tobias Schmohl, Thorsten Philipp (Hg.)
Handbuch Transdisziplinäre Didaktik

2021, 472 S., kart., 7 Farbabbildungen
39,00 € (DE), 978-3-8376-5565-0
E-Book: kostenlos erhältlich als Open-Access-Publikation
PDF: ISBN 978-3-8394-5565-4
ISBN 978-3-7328-5565-0

Andreas Germershausen, Wilfried Kruse
Ausbildung statt Ausgrenzung
Wie interkulturelle Öffnung und Diversity-Orientierung in Berlins Öffentlichem Dienst und in Landesbetrieben gelingen können

2021, 222 S., kart., 8 Farbabbildungen
25,00 € (DE), 978-3-8376-5567-4
E-Book: kostenlos erhältlich als Open-Access-Publikation
PDF: ISBN 978-3-8394-5567-8

Andreas de Bruin
Mindfulness and Meditation at University
10 Years of the Munich Model

2021, 216 p., pb.
25,00 € (DE), 978-3-8376-5696-1
E-Book: available as free open access publication
PDF: ISBN 978-3-8394-5696-5

Leseproben, weitere Informationen und Bestellmöglichkeiten finden Sie unter www.transcript-verlag.de